Error Control and Adaptivity
in Scientific Computing

NATO Science Series

A Series presenting the results of activities sponsored by the NATO Science Committee. The Series is published by IOS Press and Kluwer Academic Publishers, in conjunction with the NATO Scientific Affairs Division.

A. Life Sciences	IOS Press
B. Physics	Kluwer Academic Publishers
C. Mathematical and Physical Sciences	Kluwer Academic Publishers
D. Behavioural and Social Sciences	Kluwer Academic Publishers
E. Applied Sciences	Kluwer Academic Publishers
F. Computer and Systems Sciences	IOS Press
1. Disarmament Technologies	Kluwer Academic Publishers
2. Environmental Security	Kluwer Academic Publishers
3. High Technology	Kluwer Academic Publishers
4. Science and Technology Policy	IOS Press
5. Computer Networking	IOS Press

NATO-PCO-DATA BASE

The NATO Science Series continues the series of books published formerly in the NATO ASI Series. An electronic index to the NATO ASI Series provides full bibliographical references (with keywords and/or abstracts) to more than 50000 contributions from internatonal scientists published in all sections of the NATO ASI Series.
Access to the NATO-PCO-DATA BASE is possible via CD-ROM "NATO-PCO-DATA BASE" with user-friendly retrieval software in English, French and German (WTV GmbH and DATAWARE Technologies Inc. 1989).

The CD-ROM of the NATO ASI Series can be ordered from: PCO, Overijse, Belgium

Error Control and Adaptivity in Scientific Computing

edited by

Haydar Bulgak

Research Center of Applied Mathematics,
Selcuk University,
Konya, Turkey

and

Christoph Zenger

Institut für Informatik,
Technische Universität München,
Germany

Kluwer Academic Publishers

Dordrecht / Boston / London

Published in cooperation with NATO Scientific Affairs Division

Proceedings of the NATO Advanced Study Institute on
Error Control and Adaptivity in Scientific Computing
Antalya, Turkey
August 9–21, 1998

A C.I.P. Catalogue record for this book is available from the Library of Congress.

ISBN 0-7923-5808-2 (HB)
ISBN 0-7923-5809-0 (PB)

Published by Kluwer Academic Publishers,
P.O. Box 17, 3300 AA Dordrecht, The Netherlands.

Sold and distributed in North, Central and South America
by Kluwer Academic Publishers,
101 Philip Drive, Norwell, MA 02061, U.S.A.

In all other countries, sold and distributed
by Kluwer Academic Publishers,
P.O. Box 322, 3300 AH Dordrecht, The Netherlands.

Printed on acid-free paper

TABLE OF CONTENTS

PREFACE

The numerical simulation of processes in science and engineering has become one of the main sources of our understanding of those processes. The growth of computer speed and storage capacity by a factor of ten every five years allows the simulation of more and more complicated phenomena. We expect that this development continues in the near future.

Modern simulation tools are not only used by experts. In this situation, reliability becomes an important issue. Here, reliability means that it is not sufficient for a simulation package to print out some numbers claiming that these numbers are the desired results, but we need in addition an estimate for the error. These errors may have several sources: Errors in the model, errors in the discretization, rounding errors, and others.

Unfortunately, this is not the state of the art for today's simulation packages, and a lot has to be done to improve the situation. This was the goal of this NATO Advanced Study Institute on "Error control and adaptivity in scientific computing", which brought together experts from several areas of numerical analysis. The title reflects the close connection of error estimation and error control. Only if the error can be estimated, it is possible to look for possibilities to reduce the error. In continuous models (e.g. partial differential equations), this is done by adaptively refining the discretization. This technique can reduce the amount of work by a big factor, because the grid is only refined at places where the approximation of the solution needs a high resolution.

The contributions of the conference covered many aspects of the subject. Error estimates and error control in numerical linear algebra algorithms (which are closely related to the concept of condition numbers), interval arithmetic, and adaptivity for continuous models were the main topics. This combination of areas of research which usually are not closely connected turned out to be very fruitful.

With people from many western and eastern countries discussing, learning, and working together the Summer School was a successful event both from a professional and a social point of view.

It is our privilege to thank the NATO Scientific Committee for their advice in the preparation of the conference and for providing generously the financial funds. We also thank all those people who helped to make the Summer School a big success. These are the participants, the lecturers, our hosts in Antalya region, the staff from Konya, and especially Dr. Victor Ganzha and Klaus Wimmer who took care successfully of all questions of organization. Finally, we thank Mrs. Wil Bruins from Kluwer Academic Publishers, who was a very pleasant partner to work with when preparing these lecture notes.

Konya and Munich, March 1999

Haydar Bulgak and *Christoph Zenger*

LIST OF CONTRIBUTORS

G. Alefeld
Institut für Angewandte Mathematik
Universität Karlsruhe
76128 Karlsruhe, Germany
goetz.alefeld@math.uni-karlsruhe.de

A.M. Blokhin
Sobolev Institute of Mathematics
Koptyuga 4
Novosibirsk 630090, Russia
Blokhin@math.nsc.ru

C. Brezinski
Laboratoire d'Analyse Numérique et d'Optimisation
Université des Sciences et Technologies de Lille
59655-Villeneuve d'Ascq cedex, France
Claude.Brezinski@univ-lille1.fr

H. Bulgak
Research Center of Applied Mathematics
Selcuk University
Konya, Turkey
bulgk@karatay1.cc.selcuk.edu.tr

H.J. Bungartz
Institut für Informatik
Technische Universität München
D-80290 München
bungartz@in.tum.de

P.M. van Dooren
Department of Mathematical Engineering
Université catholique de Louvain
Belgium
Vandooren@anma.ucl.ac.be

K.A. Gallivan
Computer Science Department
Florida State University
Florida, USA

E. Grimme
Intel Corporation
Santa Clara
California, USA

B. Karasözen
Department of Mathematics
Middle East Technical University
06531 Ankara, Turkey
bulent@rorqual.cc.metu.edu.tr

A. Quarteroni
Département de Mathématiques
Ecole Polytechnique Fédérale de Lausanne
1015 Lausanne, Switzerland

R. Rannacher
Institut für Angewandte Mathematik
Universität Heidelberg
INF 293
D-69120 Heidelberg, Germany
rannacher@iwr.uni-heidelberg.de

S.M. Rump
Institut für Informatik
Technical University Hamburg-Harburg
Eißendorfer Str. 38
D-21071 Hamburg, Germany
rump@tu-harburg.de

S. Steinberg
Department of Mathematics and Statistics
University of New Mexico
Albuquerque NM 87131-1141 USA
stanly@math.unm.edu

A. Valli
Department of Mathematics
University of Trento
38050 Povo (Trento), Italy

O.B. Widlund
Courant Institute of Mathematical Sciences
251 Mercer Street
New York, NY 10012, USA
widlund@cs.nyu.edu

Ch. Zenger
Institut für Informatik
Technische Universität München
D-80290 München
zenger@in.tum.de

INDEX

xii

INTERVAL ARITHMETIC TOOLS FOR RANGE APPROXIMATION AND INCLUSION OF ZEROS

G. ALEFELD
Institut für Angewandte Mathematik
Universität Karlsruhe
76128 Karlsruhe
e-mail: goetz.alefeld@math.uni-karlsruhe.de

1. Introduction

In this paper we start in section 2 with an introduction to the basic facts of interval arithmetic: We introduce the arithmetic operations, explain how the range of a given function can be included and discuss the problem of overestimation of the range. Finally we demonstrate how range inclusion (of the first derivative of a given function) can be used to compute zeros by a so-called enclosure method.

An enclosure method usually starts with an interval vector which contains a solution and improves this inclusion iteratively. The question which has to be discussed is under what conditions is the sequence of including interval vectors convergent to the solution. This will be discussed in section 3 for so-called Newton-like enclosure methods. An interesting feature of inclusion methods is that they can also be used to prove that there exists no solution in an interval vector. It will be shown that this proof needs only few steps if the test vector has already a small enough diameter. In the last section we demonstrate how for a given nonlinear system a test vector can be constructed which will very likely contain a solution.

A very important point is, of course, the fact that all these ideas can be performed in a safe way (especially with respect to rounding errors) on a computer. We can not go into any details in this paper and refer instead to the survey paper [14] by U. Kulisch and W. Miranker.

1

H. Bulgak and C. Zenger (eds.), Error Control and Adaptivity in Scientific Computing, 1–21.
© 1999 *Kluwer Academic Publishers. Printed in the Netherlands.*

2. On Computing the Range of Real Functions by Interval Arithmetic Tools

Let $[a] = [a_1, a_2]$, $b = [b_1, b_2]$ be real intervals and $*$ one of the basic operations 'addition', 'subtraction', 'multiplication' and 'division', respectively, that is $* \in \{+, -, \times, /\}$. Then we define the corresponding operations for intervals $[a]$ and $[b]$ by

$$[a] * [b] = \{a * b \mid a \in [a], b \in [b]\}, \tag{1}$$

where we assume $0 \notin [b]$ in case of division.

It is easy to prove that the set $I(\mathbb{R})$ of real intervals is closed with respect to these operations. What is even more important is the fact that $[a] * [b]$ can be represented by using only the bounds of $[a]$ and $[b]$.

The following rules hold:

$$[a] + [b] = [a_1 + b_1, a_2 + b_2],$$

$$[a] - [b] = [a_1 - b_2, a_2 - b_1],$$

$$[a] \times [b] = [\min\{a_1 b_1, a_1 b_2, a_2 b_1, a_2 b_2\}, \max\{a_1 b_1, a_1 b_2, a_2 b_1, a_2 b_2\}].$$

If we define

$$\frac{1}{[b]} = \left\{\frac{1}{b} \mid b \in [b]\right\} \qquad \text{if } 0 \notin [b],$$

then

$$[a]/[b] = [a] \times \frac{1}{[b]}.$$

From (1) it follows immediately that the introduced operations for intervals are *inclusion monotone* in the following sense:

$$[a] \subseteq [c], [b] \subseteq [d] \Rightarrow [a] * [b] \subseteq [c] * [d]. \tag{2}$$

If we have given a rational function (that is a polynomial or a quotient of two polynomials), and a fixed interval $[x]$, and if we take an $x \in [x]$ then, applying inclusion monotonicity repeatedly, we obtain

$$x \in [x] \Rightarrow f(x) \in f([x]). \tag{3}$$

Here $f([x])$ denotes the interval which one obtains by replacing the real variable x by the interval $[x]$ and evaluating this expression following the rules of interval arithmetic. From (3) it follows that

$$R(f; [x]) \subseteq f([x]) \tag{4}$$

where $R(f; [x])$ denotes the range of f over $[x]$. $f([x])$ is usually called (an) *interval arithmetic evaluation* of f over $[x]$.

(4) is the fundamental property on which nearly all applications of interval arithmetic are based. It is important to stress what (4) really is delivering: Without

any further assumptions is it possible to compute lower and upper bounds for the range over an interval by using only the bounds of the given interval.
The concept of interval arithmetic evaluation can by generalized to more general functions without principal difficulties.

EXAMPLE 1.

Consider the rational function

$$f(x) = \frac{x}{1-x} \quad , \qquad x \neq 1$$

and the interval $[x] = [2,3]$. It is easy to see that

$$R(f;[x]) = \left[-2, -\frac{3}{2}\right],$$
$$f([x]) = [-3,-1],$$

which confirms (4).
For $x \neq 0$ we can rewrite $f(x)$ as

$$f(x) = \frac{1}{\frac{1}{x}-1}, \quad x \neq 0, x \neq 1$$

and replacing x by the interval $[2,3]$ we get

$$\frac{1}{\frac{1}{[2,3]}-1} = \left[-2, -\frac{3}{2}\right] = R(f;[x]).$$

□

From this example it is clear that the quality of the interval arithmetic evaluation as an enclosure of the range of f over an interval $[x]$ is strongly dependent on how the expression for $f(x)$ is written. In order to measure this quality we introduce the so-called Hausdorff *distance* between intervals:
Let $[a] = [a_1, a_2]$, $[b] = [b_1, b_2]$, then

$$q([a],[b]) = \max\left\{|a_1 - b_1|, |a_2 - b_2|\right\}.$$

Furthermore we use

$$d[a] = a_2 - a_1$$

and call $d[a]$ *diameter* of $[a]$.

How large is the overestimation of $R(f;[x])$ by $f([x])$?
This question is answered by the following

4

THEOREM 1. (MOORE [17])

Let there be given a continuous function $f : D \subset \mathbb{R}$ and assume that the interval arithmetic evaluation exists for all $[x] \subseteq [x]^0 \subseteq D$. Then

$$q(R(f;[x]), f([x])) \le \gamma \cdot d[x], \quad \gamma \ge 0,$$
$$d\, f([x]) \le \delta \cdot d[x], \quad \delta \ge 0.$$

□

We do not discuss here which functions are allowed in order that the interval arithmetic evaluation exists. The theorem states that if it exists then the Hausdorff distance between $R(f;[x])$ and $f([x])$ goes linearly to zero with the diameter $d[x]$. Similarly the diameter of the interval arithmetic evaluation goes linearly to zero if $d[x]$ is approaching zero.

On the other hand we have seen in the second part of Example 1 that $f([x])$ may be dependent on the expression which is used for computing $f([x])$. Therefore the following question is natural:

Is it possible to rearrange the variables of the given function expression in such a manner that the interval arithmetic evaluation gives higher than linear order of convergence to the range of values?

We consider first the simple example

EXAMPLE 2.

Let $f(x) = x - x^2$, $x \in [0,1] = [x]^0$.
It is easy to see that for $0 \le r \le \frac{1}{2}$ and $[x] = [\frac{1}{2} - r, \frac{1}{2} + r]$ we have

$$R(f;[x]) = \left[\frac{1}{4} - r^2, \frac{1}{4}\right]$$

and

$$f([x]) = \left[\frac{1}{4} - 2r - r^2, \frac{1}{4} + 2r - r^2\right].$$

From this it follows

$$q(R(f;[x]), (f([x]))) \le \gamma \cdot d[x] \text{ with } \gamma = 1,$$

and

$$d\,f([x]) \le \delta \cdot d[x] \text{ with } \delta = 2$$

in agreement with Theorem 1.
If we rewrite $f(x)$ as

$$x - x^2 = \frac{1}{4} - \left(x - \frac{1}{2}\right)\left(x - \frac{1}{2}\right)$$

and plug in the interval $[x] = [\frac{1}{2} - r, \frac{1}{2} + r]$ on the right hand side then we get the interval $[\frac{1}{4} - r^2, \frac{1}{4} + r^2]$ which, of course, includes $R(f; [x])$ again, and

$$q\left(R(f; [x]), \left[\frac{1}{4} - r^2, \frac{1}{4} + r^2\right]\right) = r^2 = \frac{1}{4}(d[x])^2 .$$

Hence the distance between $R(f; [x])$ and the enclosure interval $[\frac{1}{4} - r^2, \frac{1}{4} + r^2]$ goes quadratically to zero with the diameter of $[x]$. □

The preceding example is an illustration for the following general result.

THEOREM 2 (*The centered form*)

Let the function $f : \mathbb{R} \to \mathbb{R}$ *be represented in the 'centered form'*

$$f(x) = f(z) + (x - z) \cdot h(x) \tag{5}$$

for some $z \in [x] \subseteq D$. *If we define*

$$f([x]) := f(z) + ([x] - z) \cdot h([x])$$

then

a) $R(f; [x]) \subseteq f([x])$

and

b) $q(R(f; [x]), f([x])) \leq \kappa \cdot (d[x])^2, \quad \kappa \geq 0 .$ □

b) is called '*quadratic approximation property*' of the centered form. For rational functions it is not difficult to find a centered form. See [21], for example.
After having introduced the centered form it is natural to ask if there are forms which deliver higher than quadratic order of approximation of the range. Unfortunatly this is not the case as has been shown recently by P. Hertling [11]. See also [18].
Nevertheless in special cases one can use so-called generalized centered forms to get higher order approximations of the range. See [8], e.g. . Another interesting idea which uses a so-called 'remainder form of f' was introduced by Cornelius and Lohner [10].
In passing we note that the principal results presented up to this point also hold for functions of several variables.

As a simple example for the demonstration how the ideas of interval arithmetic can be applied we consider the following problem:
Let there be given a differentiable function $f : D \subset \mathbb{R} \to \mathbb{R}$ and an interval $[x]^0 \subseteq D$ for which the interval arithmetic evaluation of the derivative exists and does not contain zero: $0 \neq f'([x]^0)$. We want to check whether there exists a zero x^* in $[x]^0$, and if it exists we want to compute it by producing a sequence

of intervals containing x^* with the property that the lower and upper bounds are converging to x^*. (Of course, checking the existence is easy in this case by evaluating the function at the endpoints of $[x]^0$. However, the idea following works also for systems of equations. This will be shown in the next section).

For $[x] \subseteq [x]^0$ we introduce the so-called *Interval-Newton-Operator*

$$N[x] = m[x] - \frac{f(m[x])}{f'([x])}, \quad m[x] \in [x], \tag{6}$$

and consider the following iteration method

$$[x]^{k+1} = N[x]^k \cap [x]^k, \quad k = 0, 1, 2, \ldots, \tag{7}$$

which is called *Interval-Newton-Method*.
The properties of the method are described in the following result.

THEOREM 3.

Under the above assumptions the following hold for (7) :

a) $x^* \in [x]^0, f(x^*) = 0 \Longrightarrow \left\{ [x]^k \right\}_{k=0}^{\infty}$ is well defined, $x^* \in [x]^k$, $\lim\limits_{k \to \infty} [x]^k = x^*$.

If $d\, f'([x]) \leq c \cdot d[x], [x] \subseteq [x]^0$, then $d[x]^{k+1} \leq \gamma \left(d[x]^k \right)^2$.

b) $N[x]^{k_0} \cap [x]^{k_0} = \emptyset$ (= empty set) for some $k_0 \geq 0$ iff $f(x) \neq 0, x \in [x]^0$. □

Hence, in case a), the diameters are converging quadratically to zero. On the other hand, if the method (7) breaks down because of empty intersection after a finite number of steps then it is *proved* that there exists no zero of f in $[x]^0$. From a practical point of view it would be interesting to have qualitiative knowledge about the size of k_0 in this case. This will be discussed in the next section in a more general setting.

3. Enclosing Solutions of Nonlinear Systems by Newton-like Methods

At the end of the last section we introduced the so-called Interval-Newton-Method for a single equation. In order that we can introduce this and similar methods for systems of simultaneous equations we have to discuss some basic facts about interval matrices and linear equations with intervals as coefficients. For a more general discussion of this subject we refer to [6], especially chapter 10.
An interval matrix is an array with intervals as elements. Operations between interval matrices are defined in the usual manner.

If $[A] = ([a_{ij}])$ and $[B] = ([b_{ij}])$ are interval matrices and if $c = (c_i)$ is a real vector then

$$[A]([B]c) \subseteq ([A][B])c. \tag{8}$$

This was proved in [22], p. 15. If, however, c is equal to one of the unit vectors e^i then

$$[A]([B]e^i) = ([A][B])e^i. \tag{9}$$

Assume now that we have given an n by n interval matrix $[A] = ([a_{ij}])$ which contains no singular matrix and an interval vector $[b] = ([b_i])$. By applying formally the formulas of the Gaussian algorithm we compute an interval vector $[x] = ([x_i])$ for which the relation

$$\{x = A^{-1}b \mid A \in [A], b \in [b]\} \subseteq [x]$$

holds. See [6], section 15, for example. Here we assumed that no division by an interval which contains zero occurs in the elimination process. Some sufficient conditions for this are contained in [6]. See also [15]. It is an open question to find necessary and sufficient conditions for the feasibility of the Gaussian elimination process in the case of an interval matrix.

Subsequently we denote by

$$IGA([A], [b])$$

the result of the Gaussian algorithm applied to the interval matrix $[A]$ and the right hand side $[b]$, whereas

$$IGA([A])$$

is the interval matrix whose i-th column is obtained as $IGA([A], e^i)$ where e^i denotes the i-th unit vector again. In other words: $IGA([A])$ is an enclosure for the inverses of all matrices $A \in [A]$.

Now, let there be given a mapping

$$f : [x] \subset D \subset \mathbb{R}^n \to \mathbb{R}^n \tag{10}$$

and assume that the partial derivatives of f exist in D and are continuous. If $y \in [x]$ is fixed then

$$f(x) - f(y) = J(x) \cdot (x - y), \quad x \in D, \tag{11}$$

where

$$J(x) = \int_0^1 f'(y + t(x - y))dt, \quad x \in [x]. \tag{12}$$

Note that J is a continuous mapping of x for fixed y. Since $t \in [0, 1]$ we have $y + t(x - y) \in [x]$ and therefore

$$J(x) \in f'([x]) \tag{13}$$

where $f'([x])$ denotes the interval arithmetic evaluation of the Jacobian of f.

In analogy to (6) we introduce the Interval-Newton-Operator $N[x]$. Suppose that $m[x] \in [x]$ is a real vector. Then

$$N[x] = m[x] - IGA(f'([x]), f(m[x])) . \tag{14}$$

The Interval-Newton-Method is defined by

$$[x]^{k+1} = N[x]^k \cap [x]^k , \quad k = 0, 1, 2, \dots . \tag{15}$$

Analogously to Theorem 3 we have the following result.

THEOREM 4.

Let there be given an interval vector $[x]^0$ and a continuously differentiable mapping $f : [x]^0 \subset D \subseteq \mathbb{R}^n \to \mathbb{R}^n$ and assume that the interval arithmetic evaluation $f'([x]^0)$ of the Jacobian exists. Assume that $IGA(f'([x]))$ exists (which is identical to assuming that the Gaussian algorithm is feasible for $f'([x]^0)$). Assume that $\rho(A) < 1$ (ρ denotes the spectral radius of the matrix A) where

$$A = |I - IGA(f'([x]^0)) \cdot f'([x]^0)| . \tag{16}$$

(I denotes the identity. The absolute value of an interval matrix, say $[H] = ([h_{ij}])$, is defined elementwise by $\|[H]\| = (|h_{ij}|)$. For a single interval $[h] = [h_1, h_2]$ we define $\|[h]\| = \max\{|h_1|, |h_2|\}$).

a) *If f has a (necessarily unique) zero x^* in $[x]^0$ then the sequence $\{[x]^k\}_{k=0}^{\infty}$ defined by (15) is well defined, $x^* \in [x]^k$ and $\lim_{k \to \infty} [x]^k = x^*$.*
   Moreover, if

$$d f'([x])_{ij} \leq c \cdot \|d[x]\|, \quad c \geq 0, \quad 1 \leq i, j \leq n, \tag{17}$$

   for $[x] \subseteq [x]^0$ (where $d[x]$ is a real vector which is obtained by forming componentwise the diameter) then

$$\|d[x]^{k+1}\| \leq \gamma \|d[x]^k\|^2, \quad \gamma \geq 0 . \tag{18}$$

b) *$N[x]^{k_0} \cap [x]^{k_0} = \emptyset$ (= empty set) for some $k_0 \geq 0$ iff $f(x) \neq 0, x \in [x]^0$.*

□

Note that in contrast to the onedimensional case we need the condition (16). Because of continuity reasons this condition always holds if the diameter $d[x]^0$ of the given interval vector ('starting interval') is componentwise small enough (and if $f'([x]^0)$ contains no singular matrix) since because of Theorem 1 we have $A = 0$ in the limit case $d[x]^0 = 0$. H. Schwandt [22] has discussed a simple example in the case $\rho(A) \geq 1$ which shows that for a certain interval vector (15) is feasible, $x^* \in [x]^k$, but $\lim_{k \to \infty} [x]^k \neq x^*$.

In case a) of the preceding theorem we have by (18) quadratic convergence of the diameters of the enclosing intervals to the zero vector. This is the same favorable behaviour as it is well known for the usual Newton-Method. If there is no solution x^* of $f(x) = 0$ in $[x]^0$ this can be detected by applying (18) until the intersection becomes empty for some k_0. From a practical point of view it is important that k_0 is not big in general. Under natural conditions it can really be proved that k_0 is small if the diameter of $[x]^0$ is small:

Note that a given interval vector $[x] = ([x_i])$ can be represented by two real vectors x^1 and x^2 which have as its components the lower and upper bounds of $[x_i]$, respectively. Similarly we also write $[x] = ([x_i]) = [x^1, x^2]$ and $N[x] = [n^1, n^2]$ for the Interval-Newton-Operator (14). Now it is easy to prove that

$$N[x] \cap [x] = \emptyset$$

iff for at least one component i_0 either

$$\left(n^2 - x^1\right)_{i_0} < 0 \tag{19}$$

or

$$\left(x^2 - n^1\right)_{i_0} < 0. \tag{20}$$

Furthermore it can be shown that

$$x^2 - n^1 \leq \mathcal{O}\left(\|d[x]\|^2\right) + A^2 f(x^2) \tag{21}$$

and

$$n^2 - x^1 \leq \mathcal{O}\left(\|d[x]\|^2\right) - A^1 f(x^1) \tag{22}$$

provided (17) holds. Here A^1 and A^2 are two real matrices contained in $IGA(f'([x]^0))$ and $\mathcal{O}\left(\|d[x]\|^2\right)$ denotes a real vector whose components all have the order $\mathcal{O}\left(\|d[x]\|^2\right)$. Furthermore if $f(x) \neq 0$, $x \in [x]$, then for sufficiently small diameter $d[x]$ there is at least one $i_0 \in \{1, 2, \ldots, n\}$ such that

$$\left(A^1 \cdot f(x^1)\right)_{i_0} \neq 0 \tag{23}$$

and

$$\text{sign}\left(A^1 \cdot f(x^1)\right)_{i_0} = \text{sign}\left(A^2 \cdot f(x^2)\right)_{i_0}. \tag{24}$$

Assume now that $\text{sign}\left(A^1 \cdot f(x^1)\right)_{i_0} = 1$. Then for sufficiently small diameter $d[x]$ we have $\left(n^2 - x^1\right)_{i_0} < 0$ by (22) and by (19) the intersection becomes empty. If $\text{sign}\left(A^1 \cdot f(x^1)\right)_{i_0} = -1$ then by (21) we obtain $x^2 - n^1 < 0$ for sufficiently small $d[x]$ and by (20) the intersection becomes again empty.

If $N[x]^{k_0} \cap [x]^{k_0} = \emptyset$ for some k_0 then the Interval-Newton-Method breaks down and we speak of *divergence* of this method. Because of the terms $\mathcal{O}\left(\|d[x]\|^2\right)$ in (21) and (22) we can say that in the case $f(x) \neq 0$, $x \in [x]^0$, the Interval-Newton-Method is *quadratically divergent*.

We demonstrate this behaviour by a simple onedimensional example.

EXAMPLE 3.

Consider the polynomial

$$f(x) = x^5 + x^4 - 11x^3 - 3x^2 + 18x$$

which has only simple real zeros contained in the interval $[x]^0 = [-5, 6]$. Hence (7) can not be performed since $0 \in f'([x]^0)$. Using a modification of the Interval-Newton-Method described already in [3] one can compute disjoint subintervals of $[x]^0$ for which the interval arithmetic evaluation does not contain zero. Hence (7) can be performed for each of these intervals. If such a subinterval contains a zero then a) of Theorem 3 holds, otherwise b) is true. The following table contains the intervals which were obtained by applying the generalized Interval-Newton-Method until $0 \notin f'([x])$ for all computed subintervals of $[x]^0$ (for simplicity we only give three digits in the mantissa).

TABLE 1.

n	
1	$[-0.356 \times 10^1; \quad -0.293 \times 10^1]$
2	$[-0.141 \times 10^1; \quad -0.870]$
3	$[-0.977; \quad 0.499]$
4	$[0.501; \quad 0.633]$
5	$[0.140 \times 10^1; \quad 0.185 \times 10^1]$
6	$[0.188 \times 10^1; \quad 0.212 \times 10^1]$
7	$[0.265 \times 10^1; \quad 0.269 \times 10^1]$
8	$[0.297 \times 10^1; \quad 0.325 \times 10^1]$
9	$[0.327 \times 10^1; \quad 0.600 \times 10^1]$

TABLE 2.

n	1	2	3	4*	5*	6	7*	8	9*
	5	6	9	1	2	6	1	5	3

The subintervals which do not contain a zero of f are marked by a star in Table 2. The number in the second line exhibits the number of steps until the intersection becomes empty. For $n = 9$ we have a diameter of approx. 2.75, which is not small, and after only 3 steps the intersection becomes empty. The intervals $n = 1, 2, 3, 6, 8$ each contain a zero of f. In the second line the number of steps are given which have to be performed until the lower and upper bound can be no longer improved on the computer. These numbers confirm the quadratic convergence of the diameters of the enclosing intervals. (For $n = 3$ the enclosed zero is $x^* = 0$ and we are in the underflow range). □

For more details concerning the speed of divergence see [4].

The Interval-Newton-Method has the big disadvantage that even if the interval arithmetic evaluation $f'([x]^0)$ of the Jacobian contains no singular matrix its feasibility is not guaranteed, $IGA\left(f'([x]^0), f(m[x]^0)\right)$ can in general only be computed if $d[x]^0$ is sufficiently small. For this reason Krawczyk [12] had the idea to introduce a mapping which today is called the *Krawczyk-Operator:* Assume again that a mapping (10) with the corresponding properties is given. Then analogously to (14) we consider the so-called *Krawczyk-Operator*

$$K[x] = m[x] - C \cdot f(m[x]) + (I - C \cdot f'([x]))([x] - m[x]) \tag{25}$$

where C is a nonsingular real matrix. If we choose $C = m\left(f'([x])\right)^{-1}$ (= the inverse of the center of the interval arithmetic evaluation of the Jacobian) and $m[x]$ as the center of $[x]$ then for the so-called *Krawczyk-Method*

$$[x]^{k+1} = K[x]^k \cap [x]^k, \quad k = 0, 1, 2, \dots \tag{26}$$

the same result as formulated for the Interval-Newton-Method in Theorem 4 holds.

PROOF.

a) By (11) we have

$$f(x^*) - f(m[x]) = J(m[x])(x^* - m[x])$$

and since $f(x^*) = 0$ it follows

$$
\begin{aligned}
x^* &= m[x] - C \cdot f(m[x]) + (I - C \cdot J(m[x]))(x^* - m[x]) \\
&\in m[x] - C \cdot f(m[x]) + (I - C \cdot f'([x]))([x] - m[x]) \\
&= K[x].
\end{aligned}
$$

Hence if $x^* \in [x]^0$ then $x^* \in K[x]^0$ and therefore $x^* \in K[x]^0 \cap [x]^0 = [x]^1$. Mathematical induction proves $x^* \in [x]^k$, $k \geq 0$.
For the diameters of the sequence $\{[x]^k\}_{k=0}^{\infty}$ we have

$$
\begin{aligned}
d[x]^{k+1} &\leq d\,K[x]^k \\
&\leq |I - C_k \cdot f'([x]^k)|\, d[x]^k \\
&\leq |I - IGA(f'([x]^k)) \cdot f'([x]^k)|\, d[x]^k \\
&\leq A \cdot d[x]^k
\end{aligned}
$$

where A is defined by (16). Because of $\rho(A) < 1$ we have $\lim\limits_{k \to \infty} d[x]^k = 0$ and since $x^* \in [x]^k$ it follows $\lim\limits_{k \to \infty} [x]^k = x^*$. The proof for the quadratic convergence behaviour (18) follows from

$$
\begin{aligned}
d[x]^{k+1} &\leq |I - C_k \cdot f'([x]^k)| \, d[x]^k \\
&\leq |C_k| \cdot |C_k^{-1} - f'([x]^k)| \, d[x]^k \\
&\leq |IGA(f'([x]^0))| \cdot |f'([x]^k) - f'([x]^k)| \, d[x]^k \\
&= |IGA(f'([x]^0))| \cdot d\,f'([x]^k) \cdot d[x]^k
\end{aligned}
$$

by using (17).

b) Assume now that $K[x]^{k_0} \cap [x]^{k_0} = \emptyset$ for some $k_0 \geq 0$. Then $f(x) \neq 0$ for $x \in [x]^0$ since if $f(x^*) = 0$ for some $x^* \in [x]^0$ then Krawczyk's method is well defined and $x^* \in [x]^k$, $k \geq 0$.

If on the other hand $f(x) \neq 0$ and $N[x]^k \cap [x]^k \neq \emptyset$ then $\{[x]^k\}$ is well defined. Because of $\rho(A) < 1$ we have $d[x]^k \to 0$ and since we have a nested sequence it follows $\lim_{k \to \infty} [x]^k = \hat{x} \in \mathbb{R}^n$. Since the Krawczyk-Operator is continuous and since the same holds for forming intersections we obtain by passing to infinity in (26)

$$
\begin{aligned}
\hat{x} &= K\hat{x} \cap \hat{x} = K\hat{x} \\
&= \hat{x} - f'(\hat{x})^{-1} f(\hat{x}).
\end{aligned}
$$

From this it follows that $f(\hat{x}) = 0$ in contrast to the assumption that $f(x) \neq 0$ for $x \in [x]^0$.

This completes the proof of Theorem 4 for the Krawczyk-Method. $\quad\square$

In case b) of the Theorem 4, that is if $K[x]^{k_0} \cap [x]^{k_0} = \emptyset$ for some k_0, we speak again of divergence (of the Krawczyk-Method). Similar as for the Interval-Newton-Method k_0 is small if the diameter of $[x]^0$ is small. This will be demonstrated subsequently.

As for the Interval-Newton-Operator we can represent $K[x]$ using two real vectors k^1 and k^2 and we write $K[x] = [k^1, k^2]$. Now $K[x] \cap [x] = \emptyset$ iff

$$
(x^2 - k^1)_{i_0} < 0 \tag{27}
$$

or

$$
(k^2 - x^1)_{i_0} < 0 \tag{28}
$$

for at least one $i_0 \in \{1, 2, \ldots, n\}$. (Compare with (19) and (20)).
We first prove that for $K[x]$ defined by (25) we have the vector inequalities

$$
x^2 - k^1 \leq \mathcal{O}\left(\|d[x]\|^2\right) + C \cdot f(x^2) \tag{29}
$$

and

$$
k^2 - x^1 \leq \mathcal{O}\left(\|d[x]\|^2\right) - C \cdot f(x^1) \tag{30}
$$

where $[x] = [x^1, x^2]$ and $\mathcal{O}\left(\|d[x]\|^2\right)$ denotes a real vector with components all of order $\mathcal{O}\left(\|d[x]\|^2\right)$.

We prove (29). Let $f'([x]) = [f_1', f_2']$ where f_1', f_2' are real matrices. If $\frac{1}{2}(f_1' + f_2')$ is nonsingular then we set

$$C := \frac{1}{2}(f_1' + f_2')^{-1}.$$

An easy computation shows that

$$I - C \cdot f'([x]) = \left(\frac{f_1' + f_2'}{2}\right)^{-1} \left[\frac{f_1' - f_2'}{2}, \frac{f_2' - f_1'}{2}\right]$$

and therefore

$$K[x] = m[x] - C \cdot f(m[x]) + \left(\frac{f_1' + f_2'}{2}\right)^{-1} \left[\frac{f_1' - f_2'}{2}, \frac{f_2' - f_1'}{2}\right] \cdot \frac{d[x]}{2}.$$

Hence

$$
\begin{aligned}
k^2 - x^1 &= m[x] - x^1 - C \cdot f(m[x]) + \left(\frac{f_1' + f_2'}{2}\right)^{-1} \cdot \frac{f_2' - f_1'}{2} \cdot \frac{d[x]}{2} \\
&= \frac{x^2 - x^1}{2} - C \cdot f(m[x]) + \mathcal{O}\left(\|d[x]\|^2\right)
\end{aligned}
$$

where we have used (17).
Choosing $y := x^1$ in (11) we have

$$f(x) - f(x^1) = J(x) \cdot (x - x^1)$$

where now

$$J(x) = \int_0^1 f'(x^1 + t(x - x^1))dt, \quad x \in [x]. \tag{31}$$

For $x = m[x]$ we therefore have

$$f(m[x]) - f(x^1) = J(m[x]) \cdot (m[x] - x^1)$$

where $J(m[x])$ is defined by (31). It follows that

$$
\begin{aligned}
k^2 - x^1 &= \frac{1}{2}d[x] - C \cdot f(x^1) - C \cdot J(m[x]) \cdot \frac{d[x]}{2} + \mathcal{O}\left(\|d[x]\|^2\right) \\
&= \frac{1}{2}(I - C \cdot J(m[x])) - C \cdot f(x^1) + \mathcal{O}\left(\|d[x]\|^2\right).
\end{aligned}
$$

Since

$$
\begin{aligned}
I - C \cdot J(m[x]) &= C\left(C^{-1} - J(m[x])\right) \\
&\in IGA(f'([x])) \cdot (f'([x]) - f'([x]))
\end{aligned}
$$

the assertion follows by applying (17).
The second inequality can be shown in the same manner, hence (29) and (30) are proved.

If $f(x) \neq 0$, $x \in [x]$ and $d[x]$ is sufficiently small, then there exists an $i_0 \in \{1, 2, \ldots, n\}$ such that

$$(C \cdot f(x^1))_{i_0} \neq 0 \tag{32}$$

and

$$\text{sign}(C \cdot f(x^2))_{i_0} = \text{sign}(C \cdot f(x^1))_{i_0} . \tag{33}$$

This can be seen as follows: Since $x^1 \in [x]$ we have $f(x^1) \neq 0$ and since C is nonsingular it follows that $C \cdot f(x^1) \neq 0$ and therefore $(C \cdot f(x^1))_{i_0} \neq 0$ for at least one $i_0 \in \{1, 2, \ldots, n\}$ which proves (32). Using again (11) with $y = x^1$ we have

$$f(x) - f(x^1) = J(x) \cdot (x - x^1), \quad x \in [x]$$

where

$$J(x) = \int_0^1 f'(x^1 + t(x - x^1))dt, \quad x \in [x] . \tag{34}$$

By choosing $x = x^2$ we have

$$f(x^2) - f(x^1) = J(x^2) \cdot (x^2 - x^1)$$

where $J(x^2)$ is defined by (34). It follows

$$C \cdot f(x^2) = C \cdot f(x^1) + C \cdot J(x^2) \cdot (x^2 - x^1) .$$

Since the second term on the right hand side approaches zero if $d[x] \rightarrow 0$ we have (33) for sufficiently small diameter $d[x]$.
Using (29), (30) together with (32) and (33) we can now show that for sufficiently small diameters of $[x]$ the intersection $K[x] \cap [x]$ becomes empty. See the analogous conclusions for the Interval-Newton-Method using (23), (24) together with (21) and (22). By the same motivation as for the Interval-Newton-Method we denote this behaviour as 'quadratic divergence' of the Krawczyk-Method.

It is important that either using the Interval-Newton-Operator or the Krawczyk-Operator one can also *prove* the *existence of a solution* of $f(x) = 0$ in a given interval vector. We formulate this fact as a theorem.

THEOREM 5.

Let $f : D \subseteq \mathbb{R}^n \rightarrow \mathbb{R}^n$ be a continuously differentiable mapping and assume that the interval arithmetic evaluation $f'([x])$ of the Jacobian exists for some interval vector $[x] \subset D$.
a) Suppose that the Gaussian algorithm is feasible for $f'([x])$ and assume

$$y - IGA(f'([x]), f(y)) \subseteq [x]$$

where $y \in [x]$ is fixed. Then f has a unique zero x^* in $[x]$.
b) *Suppose that C is a nonsingular matrix and assume*

$$y - C \cdot f(y) + (I - C \cdot f'([x])) \cdot ([x] - y) \subseteq [x]$$

for some fixed $y \in [x]$. Then f has a zero x^* in $[x]$.

PROOF.

a) Since the Gaussian algorithm is feasible it follows that $f'([x])$ contains no singular matrices.
For fixed $y \in [x]$ we consider the equation (11) and $J(x)$ defined by (12). $J(x)$ is nonsingular because of (13).
Now consider the mapping

$$p : [x] \subseteq D \subseteq \mathbb{R}^n \to \mathbb{R}^n$$

where

$$p(x) = x - J(x)^{-1} f(x)$$

and $y \in [x]$ is fixed.

It follows, using the assumption,

$$
\begin{aligned}
p(x) &= x - J(x)^{-1} f(y) + J(x)^{-1}(f(y) - f(x)) \\
&= y - J(x)^{-1} f(y) \\
&\in y - IGA\left(f'([x]), f(y)\right) \\
&\subseteq [x], \quad x \in [x].
\end{aligned}
$$

Hence the continuous mapping p maps the nonempty convex and compact set $[x]$ into itself. Therefore, by the Brouwer fixed point theorem it has a fixed point x^* in $[x]$ from which it follows that f has a solution in $[x]$. The uniqueness follows from the fact that $f'([x])$ contains no singular matrices.

b) Consider for the nonsingular matrix C the continuous mapping

$$q : [x] \subseteq D \subseteq \mathbb{R}^n \to \mathbb{R}^n$$

defined by

$$q(x) = x - C \cdot f(x), \quad x \in \mathbb{R}.$$

It follows, using the assumption,

$$
\begin{aligned}
q(x) &= x - C \cdot f(x) \\
&= x - C \cdot (f(x) - f(y)) - C \cdot f(y) \\
&= y + (x - y) - C \cdot J(x) \cdot (x - y) - C \cdot f(y) \\
&\in y - C \cdot f(y) + (I - C \cdot f'([x])) \cdot ([x] - y) \\
&\subseteq [x], \quad x \in [x].
\end{aligned}
$$

By the same reasoning as before it follows $f(x^*) = 0$ for some $x^* \in [x]$. □

REMARK

It is easy to show that in case a) of the preceding theorem the unique zero of x^* is even in $y - IGA(f'([x]), f(y))$ and in case b) all zeros x^* of f in $[x]$ are even in $y - C \cdot f(y) + (I - C \cdot f'([x])) \cdot ([x] - y)$.

4. Verification of Solutions of Nonlinear Systems

The result of the last theorem in the preceding section can be used in a systematic manner for verifying the existence of a solution of a nonlinear system in an interval vector. Besides of the existence of a solution also componentwise error-bounds are delivered by such an interval vector. We are now going to discuss how such an interval vector can be constructed.

For a nonlinear mapping $f : D \subset \mathbb{R}^n \to \mathbb{R}^n$ we consider Newton's method

$$x^{k+1} = x^k - f'(x^k)^{-1} f(x^k), \quad k = 0, 1, \dots . \tag{35}$$

The Newton-Kantorowich theorem gives sufficient conditions for the convergence of Newton's method starting at x^0. Furthermore it contains an error estimation. A simple discussion of this estimation in conjunction with the quadratic convergence property (18) which we have also proved for the Krawczyk-Method will lead us to a test interval which can be computed using only iterates of Newton's method.

THEOREM 6. (See [19], Theorem 12.6.2)

Assume that $f : D \subseteq \mathbb{R}^n \to \mathbb{R}^n$ is differentiable in the ball $\{x \,|\, \|x - x^0\| \leq r\}$ and that

$$\|f'(x) - f'(y)\| \leq L\|x - y\|$$

for all x, y from this ball. Suppose that $f'(x^0)^{-1}$ exists and that $\|f'(x^0)^{-1}\| \leq B_0$. Let

$$\|x^1 - x^0\| = \|f'(x^0)^{-1} \cdot f(x^0)\| \leq \eta_0$$

and assume that

$$h_0 = B_0 \eta_0 L \leq \frac{1}{2}, \quad r_0 = \frac{1 - \sqrt{1 - 2h_0}}{h_0} \eta_0 \leq r .$$

Then the Newton iterates are well defined, remain in the ball $\{x \,|\, \|x - x^0\| \leq r_0\}$ and converge to a solution x^* of $f(x) = 0$ which is unique in $D \cap \{x \,|\, \|x - x^0\| < r_1\}$ where

$$r_1 = \frac{1 + \sqrt{1 - 2h_0}}{h_0} \eta_0$$

provided $r \geq r_1$. Moreover the error estimate

$$\|x^* - x^k\| \leq \frac{1}{2^{k-1}}(2h_0)^{2^k-1}\eta_0 \, , k \geq 0 \tag{36}$$

holds. □

Since $h_0 \leq \frac{1}{2}$, the error estimate (36) (for $k = 0, 1$ and the ∞-norm) leads to

$$\begin{aligned}\|x^* - x^0\|_\infty &\leq 2\eta_0 = 2\|x^1 - x^0\|_\infty \\ \|x^* - x^1\|_\infty &\leq 2h_0\eta_0 \leq \eta_0 = \|x^1 - x^0\|_\infty \, .\end{aligned}$$

This suggests a simple construction of an interval vector containing the solution x^*. The situation is illustrated in Figure 1.

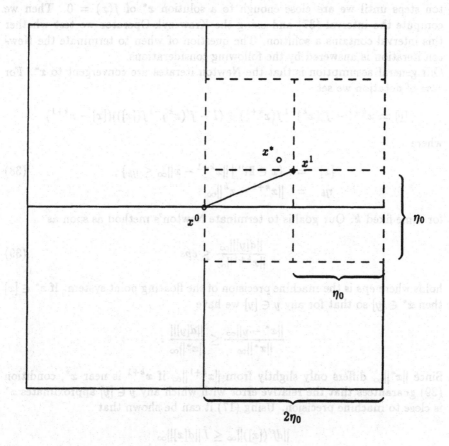

Figure 1: Error estimates (36) for $k = 1$ and the ∞-norm

If x^0 is close enough to the solution x^* then x^1 is much closer to x^* than x^0 since Newton's method is quadratically convergent. The same holds if we choose any vector $(\neq x^*)$ from the ball $\{x \mid \|x - x^1\|_\infty \leq \eta_0\}$ as starting vector for Newton's method. Because of (18) and since $x^* \in K[x]$ it is reasonable to assume that

$$K[x] = x^1 - f'(x^0)^{-1} \cdot f(x^1) + (I - f'(x^0)^{-1} \cdot f'([x])) \cdot ([x] - x^1) \subseteq [x]$$

for

$$[x] = \{x \mid \|x - x^1\|_\infty \leq \eta_0\}. \tag{37}$$

The important point is that this test interval $[x]$ can be computed without knowing B_0 and L. Of course all the preceding arguments are based on the assumption that the hypothesis of the Newton-Kantorowich theorem is satisfied, which may not be the case if x^0 is far away from x^*.

We try to overcome this difficulty by performing first a certain number of Newton steps until we are close enough to a solution x^* of $f(x) = 0$. Then we compute the interval (37) and using the Krawczyk-Operator we test whether this interval contains a solution. The question of when to terminate the Newton iteration is answered by the following considerations.

Our general assumption is that the Newton iterates are convergent to x^*. For ease of notation we set

$$[y] := x^{k+1} - f'(x^k)^{-1} f(x^{k+1}) + (I - f'(x^k)^{-1} f([x]))([x] - x^{k+1}).$$

where

$$\begin{aligned}
[x] &= \{x \in \mathbb{R}^n \mid \|x^{k+1} - x\|_\infty \leq \eta_k\}, \\
\eta_k &= \|x^{k+1} - x^k\|_\infty
\end{aligned} \tag{38}$$

for some fixed k. Our goal is to terminate Newton's method as soon as

$$\frac{\|d[y]\|_\infty}{\|x^{k+1}\|_\infty} \leq eps \tag{39}$$

holds where eps is the machine precision of the floating point system. If $x^* \in [x]$ then $x^* \in [y]$ so that for any $y \in [y]$ we have

$$\frac{\|x^* - y\|_\infty}{\|x^*\|_\infty} \leq \frac{\|d[y]\|}{\|x^*\|_\infty}.$$

Since $\|x^*\|_\infty$ differs only slightly from $\|x^{k+1}\|_\infty$ if x^{k+1} is near x^*, condition (39) guarantees that the relative error with which any $y \in [y]$ approximates x^* is close to machine precision. Using (17) it can be shown that

$$\|df'([x])\|_\infty \leq \hat{L}\|d[x]\|_\infty$$

and

$$\||d[y]|\||_\infty \leq \|f'(x^k)^{-1}\|_\infty \cdot \tilde{L}\||d[x]|\||_\infty^2$$

where $\tilde{L} = \max\{\hat{L}, L\}$, and since $\||d[x]|\||_\infty = 2\eta_k$ the inequality (39) holds if

$$4\frac{\|f'(x^k)^{-1}\|_\infty \tilde{L}\eta_k^2}{\|x^{k+1}\|_\infty} \leq eps \tag{40}$$

is true.

From Newton's method we have

$$x^{k+1} - x^k = f'(x^k)^{-1}\left\{f(x^k) - f(x^{k-1}) - f'(x^k)^{-1}(x^k - x^{k-1})\right\}$$

and by 3.2.12 in [19] it follows that

$$\eta_k \leq \frac{1}{2}\|f'(x^k)^{-1}\|_\infty \tilde{L}\eta_{k-1}^2.$$

Replacing the inequality sign by equality in this relation and eliminating $\|f'(x^k)^{-1}\|_\infty \tilde{L}$ in (40) we get the following stoping criterion for Newton's method:

$$\frac{8\eta_k^3}{\|x^{k+1}\|_\infty \eta_{k-1}^2} \leq eps. \tag{41}$$

Of course this is not a mathematical proof that if (41) is satisfied then the interval $[y]$ constructed as above will contain x^* and that the vectors in $[y]$ will approximate x^* with a relative error close to eps. However as has been shown in [5] the test based on the stopping criterion (41) works extremely well in practice.

The idea of this section has been generalized to nonsmooth mappings by X. Chen [9].

A very important point is also the fact that for the verification of solutions of nonlinear systems one can often replace the interval arithmetic evaluation of the Jacobian by an interval arithmetic enclosure of the slope-matrix of f. In this connection slopes have first been considered in [1]. See also [20].

Verification techniques have been applied to a series of fundamental problems by rewriting them as nonlinear systems. We mention the eigenvalue problem for matrices, the singular value problem, the generalized eigenvalue problem and the generalized singular value problem, e.g. . See [16] where one can find many references.

Interval arithmetic can also be applied in a systematic manner to bound the solution set of a given problem if the data are already contained in intervals. An interesting question in this context is how the solution set looks like. A couple of recent papers are concerned with the discussion of this question in the case of a linear system with interval entries. See [7], for example.

References

[1] Alefeld, G. (1981) Bounding the Slope of Polynomials and some Applications, Computing 26, 227-237

[2] Alefeld, G. (1994) Inclusion Methods for Systems of Nonlinear Equations - The Interval-Newton-Method and Modifications, in Herzberger, J.(ed.), Topics in Validated Computation, Elsevier

[3] Alefeld, G. (1968) Intervallrechnung über den komplexen Zahlen und einige Anwendungen, Ph. D. Thesis, University Karlsruhe

[4] Alefeld, G. (1991) Über das Divergenzverhalten des Intervall-Newton-Verfahrens, Computing 46, 289-294

[5] Alefeld, G., Gienger, A. and Potra, F. (1994) Efficient Numerical Validation of Solutions of Nonlinear Systems, SIAM J. Numer. Anal., 31(1), 252-260

[6] Alefeld, G. and Herzberger, J. (1983) Introduction to Interval Computations, Academic Press

[7] Alefeld, G., Kreinowich, V. and Mayer, G. (1998) The Shape of the Solution Set of Linear Interval Equations with Dependent Coefficients, Mathematische Nachrichten 192, 23-36

[8] Alefeld, G. and Lohner, R. (1985) On Higher Order Centered Forms, Computing 35, 177-184

[9] Chen, X. (1997) A Verification Method for Solutions of Nonsmooth Equations, Computing 58, 281-294

[10] Cornelius, H. and Lohner, R. (1984) Computing the Range of Values of Real Functions with Accuracy Higher than Second Order, Computing 33, 331-347

[11] Hertling, P. (Jan. 1998) A Lower Bound for Range Enclosure in Interval Arithmetic, Centre for Discrete Mathematics and Theoretical Computer Science Research Report Series, Department of Computer Science, University of Auckland

[12] Krawczyk, R. (1969) Newton-Algorithmen zur Bestimmung von Nullstellen mit Fehlerschranken, Computing 4, 187-201

[13] Krawczyk, R. and Neumaier, A. (1985) Interval Slopes for Rational Functions and Associated Centered Forms, SIAM J. Numer. Anal., 22(3), 604-616

[14] Kulisch, U. and Miranker, W. (1986) The Arithmetic of the Digital Computer: A New Approach, SIAM Review 28(1), 1-40

[15] Mayer, G. (1992) Old and New Aspects for the Interval Gaussian Algorithm, in Kaucher, E., Markov, S. M. and Mayer, G. (eds.), Computer Arithmetic, Scientific Computation and Mathematical Modelling, J. C. Baltzer AG Scientific Publishing Co., Basel

[16] Mayer, G. (1994) Result Verification for Eigenvectors and Eigenvalues, in Herzberger, J. (ed.), Topics in Validated Computations, Elsevier

[17] Moore, R. E. (1966) Interval Analysis, Prentice Hall, Englewood Cliffs, New Jersey

[18] Nguyen, H. T., Kreinowich, Y., Nesterov, V. and Nakumura, M. (1997) On Hardware Support for Interval Computations and for Soft Computing: Theorems, IEEE Transactions on Fuzzy Systems, Vol 5, No 1, 108-127

[19] Ortega, J. M. and Rheinboldt, W. C. (1970) Iterative Solution of Nonlinear Equations in Several Variables, Academic Press, New York and London

[20] Rall, L. B. (1969) Computational Solution of Nonlinear Operator Equations, Wiley, New York

[21] Ratschek, H. (1980) Centered Forms, SIAM J. Numer. Anal., 17, 656-662

[22] Schwandt, H. (1981) Schnelle fast global konvergente Verfahren für die Fünf-Punkte-Diskretisierung der Poissongleichung mit Dirichletschen Randbedingungen auf Rechteckgebieten, Thesis, Fachbereich Mathematik der TU Berlin

A NEW CONCEPT OF CONSTRUCTION OF ADAPTIVE CALCULATION MODELS FOR HYPERBOLIC PROBLEMS

A.M. BLOKHIN
Sobolev Institute of Mathematics
Koptyuga 4
Novosibirsk 630090, Russia
E-mail: Blokhin@Math.nsc.ru

Abstract. Rapid development of methods of mathematical simulation (especially in Continuum Mechanics), caused by the wide inculcation of computers into scientific practice, became a subject of various discussions in the recent years. Not going into details, by mathematical modelling in these lectures we mean the chain of necessary steps, namely:

Generation of the physic model of the considered event → Formulation and investigation of the mathematical model → Construction of the calculation algorithm to find an approximate solution to the mathematical model → Calculation of the approximate solution with the help of computer and comparison of the obtained results with experimental data.

We think that two cells in the middle of the chain are of particular importance. Moreover, we will suppose further that the mathematical and calculation models have to be considered simultaneously and properties of the mathematical model have to be taken into consideration while constructing the calculation model. We will call such calculation model **an adequate calculation model** (or an adaptive calculation model). We also intend to use only adequate calculation models for finding solutions to mathematical models which are described by Hyperbolic Systems.

This paper consists of two lectures only; the third lecture was not included because of lack of place. In the first lecture we discuss the situation when we violate the principle of adequacy. With this purpose we consider the question on stability of difference schemes on the example of linearized system of Gas Dynamics. Taken a well-known difference scheme, we show that it is stable if we consider a difference Cauchy problem. If we add boundary conditions (for example, if we consider a mixed problem on stability of shock waves in Gas Dynamics) we can show that, for any realization of

H. Bulgak and C. Zenger (eds.), Error Control and Adaptivity in Scientific Computing, 23–64.

boundary conditions, the difference scheme is unstable. So, when we construct the calculation model we must take into account boundary conditions without fail.

In the second lecture we consider the question of construction of adequate calculation models for quasilinear Gas Dynamic Equations.

Introduction

Present time is characterized by wide distribution of the so-called methods of mathematical simulation. By the mathematical simulation (for example, in physics) we mean the following chain of actions while investigating a phenomenon: **description of the physical model of the phenomenon — creation of the mathematical model of the phenomenon — construction of the calculation model of the phenomenon — creation of the calculating program - calculation with a computer**.

Long-standing experience of solution of practical problems from continuum mechanics enables us to conclude that conscious application of methods of mathematical simulation is not possible without detailed analysis of relations between the mathematical and calculation models of the phenomenon. It should be noted that the approach based on simultaneous investigation of the initial mathematical problem and its finite-difference analog has been used in theory of differential equations for a long time. It is worth to note here , for example, the fundamental work [1] and also monographs [2–4] in which this approach was applied and developed.

As a basis in construction and investigation of calculation models, we consider the requirement **of adequacy of the calculation model to the initial differential problem**. By adequacy we mean the following: the calculation model is constructed such that a theorem on existence of solution to the initial problem can be proved with its help. The latter seems to be important since, while calculating, we must be sure that the approximate solutions tends in the limit to the solution of the initial differential problem.

Some **initial mathematical models** will be considered in these lectures. First, **the mixed problem for symmetric t-hyperbolic system with dissipative boundary conditions** will be taken (see [2, Sec.15]).

Problem I. Find the solution $\mathbf{U}$ to a symmetric t-hyperbolic (by Friedrichs) system:

$$A \cdot \mathbf{U}_t + B \cdot \mathbf{U}_x + C \cdot \mathbf{U}_y = 0, \text{ in } R_{++}^3, \tag{0.1}$$

which satisfies boundary conditions at $x = 0$:

$$\mathbf{U^I} = S \cdot \mathbf{U^{II}} \qquad (0.2)$$

and initial data at $t = 0$:

$$\mathbf{U}(0, x, y) = \mathbf{U}_0(x, y). \qquad (0.3)$$

Here $A = \text{diag}(A^I, A^{II}, A^{III})$, $B = \text{diag}(I_{N_0}, -I_{N_1}, O_{N_2})$ are block-diagonal matrices;

$$A^I = \text{diag}(a_1, a_2, \ldots, a_{N_0}),$$

$$A^{II} = \text{diag}(a_{N_0+1}, \ldots, a_{N_0+N_1}),$$

$$A^{III} = \text{diag}(a_{N_0+N_1+1}, \ldots, a_N)$$

are diagonal matrices; $a_i > 0$, $i = \overline{1, N}$; $N_0 + N_1 + N_2 = N$; I_{N_0}, I_{N_1} are the unit matrices of order N_0, N_1, correspondingly; O_{N_2} is the zero square matrix of order N_2;

$$C = \begin{pmatrix} C_1 & C_2 & C_3 \\ C_2^* & C_4 & C_5 \\ C_3^* & C_4^* & C_6 \end{pmatrix},$$

and $C_1 = C_1^*$, $C_4 = C_4^*$, $C_6 = C_6^*$ are square matrices of order N_0, N_1, N_2, correspondingly; C_2, C_3, C_5 are matrices of order $N_0 \times N_1$, $N_0 \times N_2$, $N_1 \times N_2$, correspondingly;

$$\mathbf{U} = \mathbf{U}(t, x, y) = \begin{pmatrix} u_1 \\ \vdots \\ u_N \end{pmatrix}, \quad \mathbf{U}^I = \begin{pmatrix} u_1 \\ \vdots \\ u_{N_0} \end{pmatrix},$$

$$\mathbf{U}^{II} = \begin{pmatrix} u_{N_0+1} \\ \vdots \\ u_{N_0+N_1} \end{pmatrix}, \quad \mathbf{U}^{III} = \begin{pmatrix} u_{N_0+N_1+1} \\ \vdots \\ u_N \end{pmatrix};$$

S is a real constant matrix of order $N_0 \times N_1$,

$$R_{++}^3 = \{(t, x, y); t, x > 0, y \in R^1\}.$$

Remark 0.1. Symmetric system (0.1) with real constant coefficients is given in the so-called **canonical form** (for details see [2, Sec.11]).

Let boundary conditions (0.2) be **strictly dissipative** (see. [2, Sec.15]), i.e.,

$$-(B \cdot \mathbf{U}, \mathbf{U})|_{x=0} = (\mathbf{U}^{II}, [I_{N_1} - S^* \cdot S] \cdot \mathbf{U}^{II})|_{x=0} \geq$$

$$\geq k_0 \cdot (\mathbf{U}^{II}, \mathbf{U}^{II})|_{x=0}, \qquad (0.4)$$

where $k_0 > 0$ is a constant. Then, by [2, Sec.16], an energy integral identity on a smooth solution of system (0.1) can be written down (under fulfilment of (0.4) the identity is called **a dissipative energy integral**):

$$\frac{\partial}{\partial t}(A \cdot \mathbf{U}, \mathbf{U}) + \frac{\partial}{\partial x}(B \cdot \mathbf{U}, \mathbf{U}) + \frac{\partial}{\partial y}(C \cdot \mathbf{U}, \mathbf{U}) = 0. \qquad (0.5)$$

With its help, for the case of dissipative boundary conditions (0.2), the following **a priori estimation** for **problem I** can be derived:

$$J(t) \leq J(0), \quad t > 0, \qquad (0.6)$$

where

$$J(t) = \iint_{R_+^2} (A \cdot \mathbf{U}, \mathbf{U}) dx dy, \quad R_+^2 = \{(x, y); x > 0, y \in R^1\}.$$

Coming back to **problem I**, we will construct difference models such that they permit existence of **a difference analog** of dissipative energy integral (0.5). This remark stays valid for other mathematical models. Finally, it will bring a possibility to obtain an energy estimation (a difference analog of a priori estimate (0.6))) and, thus, to prove stability of the considered scheme. That will show adequacy of the difference model of the initial **problem I** since, with the known energy estimation, a theorem on existence of a sufficiently smooth solution can be proved by standard arguments (which the reader can find in [2]).

Remark 0.2. If the matrix C is diagonal the constant k_0 from (0.4) can be considered close to the unit.

As the second model, we consider the mixed problem for the system of acoustics equations with boundary conditions on a shock wave taken from [4].

Problem II. Find the solution $\mathbf{U}$ to the system of acoustics equations

$$A \cdot \mathbf{U}_t + B \cdot \mathbf{U}_x + (C_0 + \omega \cdot A) \cdot \mathbf{U}_y = 0 \text{ in } R_{++}^3, \qquad (0.7)$$

which satisfies the boundary conditions at $x = 0$:

$$u + d \cdot p = 0, \quad v_t + \omega \cdot v_y - \lambda \cdot p_y = 0, \qquad (0.8)$$

and the initial data at $t = 0$:

$$\mathbf{U}(0, x, y) = \mathbf{U}_0(x, y). \qquad (0.9)$$

Here

$$A = \begin{pmatrix} 1 & 0 & 0 \\ 0 & M^2 & 0 \\ 0 & 0 & M^2 \end{pmatrix}, \quad B = \begin{pmatrix} 1 & 1 & 0 \\ 1 & M^2 & 0 \\ 0 & 0 & M^2 \end{pmatrix},$$

$$C_0 = \begin{pmatrix} 0 & 0 & 1 \\ 0 & 0 & 0 \\ 1 & 0 & 0 \end{pmatrix}, \quad \mathbf{U} = \begin{pmatrix} p \\ u \\ v \end{pmatrix};$$

$M(0 < M < 1)$, $\omega \geq 0$, d, λ are constants. System (0.7) is obtained by linearization of the gas dynamics equations (see below) with respect to a constant fundamental solution behind the oblique jump of sealing with special scaling of the linearized system, boundary conditions (0.8) are linearization of the Rankine-Hugoniot conditions on the shock wave; u, v, p are small perturbations of the vector of velocity, pressure (see [4]). In some cases, together with **problem II**, we will consider the relations:

$$S_t + S_x + \omega \cdot S_y = 0 \text{ in } R^3_{++},$$

$$S + p = 0 \text{ at } x = 0,$$

$$S(0, x, y) = S_0(x, y) \text{ at } t = 0.$$

Here S are small perturbations of the entropy (see [4]).

Problem II is studied in detail in [4]. The basic moment in the investigation is the fact that **prolem II** can be reduced to the following auxiliary problem.

Problem III. We seek the solution to the wave equation

$$\{M^2 \cdot (\hat{\tau} + \xi)^2 - \xi^2 - \eta^2\}p = 0 \text{ in } R^3_{++}, \tag{0.10}$$

which satisfies the boundary condition at $x = 0$

$$\{M^2 \cdot (1 + d) \cdot \hat{\tau}^2 - \beta^2 \cdot \hat{\tau}\xi + M^2 \cdot \lambda\eta^2\}p = 0 \tag{0.11}$$

and the initial data at $t = 0$

$$p(0, x, y) = p_0(x, y),$$

$$\frac{\partial p}{\partial t}(0, x, y) = -\omega \cdot \frac{\partial p_0}{\partial y}(x, y) - \frac{\partial p_0}{\partial x}(x, y) - \frac{\partial u_0}{\partial x}(x, y) - \frac{\partial v_0}{\partial y}(x, y). \tag{0.12}$$

Here $\tau = \frac{\partial}{\partial t}$, $\xi = \frac{\partial}{\partial x}$, $\eta = \frac{\partial}{\partial y}$, $\hat{\tau} = \tau + \omega \cdot \eta$, $\beta^2 = 1 - M^2$.

To describe domains of ill-posedness of **problem III** we construct examples of ill-posedness of the Hadamard type (see [2]). For this purpose we seek solutions to **problem III** in the form:

$$p = p^0 \cdot \exp\{i \cdot (-\hat{\omega} \cdot t + \hat{k} \cdot y + \hat{l} \cdot x)\}, \tag{0.13}$$

where p^0, $\hat{\omega}$, $\hat{k}$, $\hat{l}$ are constants, $i = \sqrt{-1}$. The constants $\hat{\omega}$, $\hat{k}$, $\hat{l}$ are related by the algebraic equations:

$$M^2(\hat{l} - \hat{\omega} + \omega \cdot \hat{k})^2 = \hat{l}^2 + \hat{k}^2,$$

$$M^2(1 + d)(-\hat{\omega} + \omega \cdot \hat{k})^2 - \beta^2(-\hat{\omega} + \omega \cdot \hat{k})\hat{l} + M^2 \cdot \lambda \cdot \hat{k}^2 = 0. \tag{0.14}$$

Detailed analysis of (0.14) leads to the following conclusion:

a) In domain $\mathcal{K}$: $\lambda < 0$, $d \cdot \beta^2 + M^2 \cdot \lambda > 0$ **problem III** has a solution of the type (0.13) with $\operatorname{Im}\hat{\omega} \leq 0$, $\operatorname{Im}\hat{l} \geq 0$, $\operatorname{Im}\hat{k} = 0$.

It was shown in [4] that the domain $\mathcal{K}$ is a domain of well-posedness of **problem III** (and **problem II** simultaneously). Moreover, the domain $\mathcal{K}$ is a domain where the **uniform Lopatinskii condition** is fulfilled. The latter means well-posedness of **problem II** and problems close to the initial one.

b) At the rest points of the plane neither the uniform Lopatinskii condition nor the Lopatinskii condition is fulfilled. They are those points of the plane d, λ for which **problem III** has solutions of the type (0.13) with $\operatorname{Im}\hat{\omega} > 0$, $\operatorname{Im}\hat{l} < 0$, $\operatorname{Im}\hat{k} = 0$. For the last case ill-posedness of **problem III** (and initial **problem II** also) is established by construction of the following Hadamard example

$$p_{\hat{k}} = p^0 \cdot \exp\{\sqrt{|\hat{k}|} + i \cdot (-\hat{\omega} \cdot t + \hat{k} \cdot y + \hat{l} \cdot x)\}, \quad |\hat{k}| = 1, 2, \cdots.$$

Break-down of the uniform Lopatinskii condition means that there exists a problem close to the initial **problem II** such that a Hadamard example can be constructed ([4]).

Well-posedness of mixed **problem II** in the domain $\mathcal{K}$ is proved with the help of technique of dissipative energy integrals in its more complicated variant as compared with **problem I**. The matter is that boundary conditions (0.8) are not strictly dissipative, i.e., for **problem II**

$$-(BU, U)|_{x=0} > 0$$

is not true.

By this reason, this simple idea (obtaining of a priori estimation) does not work for **problem II**. In order to obtain a priori estimation for **problem II**, an extended system (i.e., a system for the vector U and derivatives of U) with strictly dissipative boundary conditions is constructed.

The extended system and boundary conditions for **problem II** were constructed in [4]. Therewith, if $(d, \lambda) \in \mathcal{K}$ then the following a priori estimation takes place:

$$\|U(t)\|_{W_2^2(R_+^2)} \leq C_1(T) \cdot \|U_0\|_{W_2^2(R_+^2)}, 0 < t \leq T < \infty. \tag{0.15}$$

Here $C_1(T) > 0$ is a constant dependent on T; W_2^2 is the Sobolev's space.

We note that if the point (d, λ) belongs to the domain where the uniform Lopatinskii condition does not hold, the following inequality is a priori estimation for **problem II**:

$$\|\mathbf{U}(t)\|_{W_2^2(R_+^2)} \leq C_2(T) \cdot \|\mathbf{U}_0\|_{W_2^3(R_+^2)}, \quad 0 < t \leq T < \infty. \tag{0.16}$$

Here $C_2(T) > 0$ is a constant. Estimation of the form (0.16) is called the estimation with the loss of smoothness ((0.15) is the estimation without the loss of smoothness).

The important moment, while constructing the extended system, is reduction of **mixed problem III** to a mixed problem for a symmetric t-hyperbolic (by Friedrichs) system. The idea of reduction is as follows. If p satisfies (0.10) then the vector

$$\mathbf{Y} = \begin{pmatrix} \tilde{L}_1 \\ L_2 \\ \tilde{L}_3 \end{pmatrix} p$$

satisfies the system (see [4]):

$$\{D \cdot \hat{\tau} - Q \cdot \xi - R \cdot \eta\}\mathbf{Y} = 0, \tag{0.17}$$

Here

$$L_1 = \frac{1}{\beta^2}\hat{\tau}, \quad L_2 = \xi - \frac{M^2}{\beta^2}\hat{\tau}, \quad L_3 = \eta,$$

$$\tilde{L}_1 = ML_1, \quad \tilde{L}_3 = \frac{1}{\beta}L_3, \quad D = \frac{1}{\beta^2}(E + M^2Q).$$

$$E = M\begin{pmatrix} 1 & m_1 & 0 \\ m_1 & 1 & 0 \\ 0 & 0 & 1 \end{pmatrix}, \quad Q = \begin{pmatrix} m_1 & 1 & 0 \\ 1 & m_1 & 0 \\ 0 & 0 & -m_1 \end{pmatrix},$$

$$R = \begin{pmatrix} 0 & 0 & 1 \\ 0 & 0 & m_1 \\ 1 & m_1 & 0 \end{pmatrix},$$

m_1 is a constant. We note that $D > 0$ if $|m_1| < 1$.

Compose the extended system from two systems of the form (0.17):

$$\left\{\begin{pmatrix} D & O_3 \\ O_3 & D \end{pmatrix}\hat{\tau} - \begin{pmatrix} Q & O_3 \\ O_3 & Q \end{pmatrix}\xi - \begin{pmatrix} R & O_3 \\ O_3 & R \end{pmatrix}\eta\right\}\mathbf{X} = 0, \tag{0.18}$$

where

$$\mathbf{X} = \{\tilde{L}_1 L_+, L_2 L_+, \tilde{L}_3 L_+, \tilde{L}_1 L_-, L_2 L_-, \tilde{L}_3 L_-\}^*,$$

$$L_{\pm} = (a_1 L_1 + a_2^{\pm} L_2)p,$$

$$a_1 = \frac{m}{M}, \quad a_2^{\pm} = -\frac{n}{a_{\pm}}, \quad \gamma = \frac{\beta}{M^2}, \quad n = -\frac{\lambda}{\beta},$$

$$a_{\pm} = \frac{M(\gamma \pm \sqrt{\gamma^2 - 4mn})}{2m}, \quad m = \beta d + \frac{\lambda M^2}{\beta}, \quad \gamma > 4mn.$$

We note that $m, n > 0$ in the domain $\mathcal{K}$.

Boundary condition (0.11) is presented in the form:

$$\{\tilde{L}_1 - a_{\pm} L_2\} L_{\pm} = 0, \quad x = 0. \tag{0.19}$$

Let $m_1 = a_+/(1 + a_+^2)$ for the first system of the form (0.17) and $m_1 = a_-/(1 + a_-^2)$ for the second system of the form (0.17). Then boundary conditons (0.19) are strictly dissipative because

$$\left(\begin{bmatrix} Q & 0_3 \\ 0_3 & Q \end{bmatrix} \mathbf{X}, \mathbf{X} \right) \Big|_{x=0} = a_+ (a_1 L_1 L_2 p + a_2^+ L_2^2 p)^2 +$$

$$+ \frac{a_+}{1 + a_+^2} (a_1 L_1 \tilde{L}_3 p + a_2^+ L_2 \tilde{L}_3 p)^2 + a_- (a_1 L_1 L_2 p + a_2^- L_2^2 p)^2 +$$

$$+ \frac{a_-}{1 + a_-^2} (a_1 L_1 \tilde{L}_3 p + a_2^- L_2 \tilde{L}_3 p)^2 > 0.$$

In [4] one can find another descriptions of systems (0.17), (0.18) which enable omitting the restriction $\gamma^2 > 4mn$.

As it was mentioned above for **problem II** the domain $\mathcal{K}$, $\lambda < 0$, $d \cdot \beta^2 + M^2 \cdot \lambda > 0$, is a domain of fulfilment of the uniform Lopatinskii condition. So, the estimations obtained for problems with constant coefficients are valid for problems with variable coefficients. Then, with the help of a certain technique, it is possible to prove a local theorem on existence and uniqueness of the classic solution for quasilinear equaitons of gas dynamics behind a curvilinear shock wave under the assumption that at every point of the curvilinear shock wave at the moment $t = 0$ the following inequalities take place:

$$\lambda < 0, \quad d \cdot \beta^2 + M^2 \cdot \lambda > 0.$$

We note that the basic results obtained for the case of constant coefficients do not take place for the case of variable coefficients (and, consequently, for the quasilinear case) if there exist estimations of the type (0.16) (with the loss of smoothness). All these facts are described in [4].

We will construct differential schemes for **problem II** such that they will admit existence of difference analogs of a priori estimations (0.15).

Up to this moment we considered only linear mathematical models since adequacy of a difference scheme to an initial nonlinear differential problem is difficult to establish by many reasons. Nevertheless, we will give an example of an initial quasilinear differential problem.

We consider a system of equations which describe three-dimensional gas flow under assumption that the gas is nonviscous, non thermoconductive, and is in the state of local thermodynamical equilibrium, i.e., there exsists a state equation. In the Cartesian coordinate system $\mathbf{x} = (x_1, x_2, x_3)$ the gas dynamics equations can be presented in the divergent form:

$$\rho_t + \mathrm{div}\,(\rho \cdot \mathbf{v}) = 0,$$

$$(\rho \cdot v_i)_t + \sum_{k=1}^{3}(\Pi_{ik})_{x_k} = 0, \quad i = 1, 2, 3, \qquad (0.20)$$

$$\{\rho \cdot (E + |\mathbf{v}|^2/2)\}_t + \mathrm{div}\,\{\rho \cdot \mathbf{v}(E + |\mathbf{v}|^2/2 + p \cdot V)\} = 0.$$

Here ρ is the gas density, $\mathbf{v} = (v_1, v_2, v_3)^*$ is the gas velocity, $|v|^2 = (v, v)$, $\Pi_{ik} = \rho \cdot v_i \cdot v_k + p \cdot \delta_{ik}$ is the tensor of impuls flux density, p is the pressure, E is the internal energy, $V = 1/\rho$ is the specific volume.

The Cauchy problem (0.20) was considered in [5], and a theorem on local existence of the classical solution to this problem was proved.

1. Lecture I. Stability of Numerical Boundary Treatments for Finite-Difference Schemes for the Acoustics Equations System

1.1. INVESTIGATION OF STABILITY OF A CERTAIN EXPLICIT TWO-LAYERS DIFFERENCE SCHEME

It was mentioned in **Introduction** that investigation of stability for difference schemes to **mixed problem I** will be carried out by the technique of finite-difference analogs of dissipative energy integrals. In this section we will present its basic elements.

In the domain R^3_{++} we construct a difference scheme with the steps $\Delta t = \Delta, \Delta x = h_x, \Delta y = h_y$. Introduce the following notation:

$$\mathbf{U}^k_{ij} = \mathbf{U}_i = \mathbf{U} = \mathbf{U}(k \cdot \Delta, i \cdot h_x, j \cdot h_y), \quad k, i, |j| = 0, 1, \cdots;$$

$$\varphi \mathbf{U}^k_{ij} = \mathbf{U}^{k+1}_{ij} = \hat{\mathbf{U}}_i = \hat{\mathbf{U}}, \quad \psi^{\pm 1}\mathbf{U}^k_{ij} = \mathbf{U}^k_{i\pm1,j},$$

$$\theta^{\pm 1}\mathbf{U}^k_{ij} = \mathbf{U}^k_{i,j\pm1}, \quad \tau = \varphi - 1, \quad \xi = \psi - 1, \quad \eta = \theta - 1,$$

$$\bar{\xi} = 1 - \psi^{-1}, \quad \bar{\eta} = 1 - \theta^{-1},$$

$$\xi_0 = \xi + \bar{\xi}; \quad \hat{\xi} = \xi \cdot \bar{\xi} = \xi - \bar{\xi};$$

$$\eta_0 = \eta + \bar{\eta}; \quad \hat{\eta} = \eta \cdot \bar{\eta} = \eta - \bar{\eta}; \quad r_x = \frac{\Delta}{h_x}, \quad r_y = \frac{\Delta}{h_y}.$$

Following [4,6,7], we formulate the difference model of **mixed problem I**:

$$\left\{ A \cdot \tau + \frac{1}{2} r_x \cdot B \cdot \xi_0 + \frac{1}{2} r_y \cdot C \cdot \eta_0 - \right.$$

$$\left. -A \cdot (a \cdot \hat{\xi} + b \cdot \hat{\eta}) \right\} \mathbf{U} = 0, \quad k, |j| = 0, 1, \cdots, \quad i = 1, 2, \cdots, \qquad (1.1)$$

$$\mathbf{U}_0^I = S\mathbf{U}_0^{II}, \quad k, |j| = 0, 1, \cdots, \qquad (1.2)$$

$$\mathbf{U}_{ij}^0 = \mathbf{U}_0(ih_x, jh_y), \quad i, |j| = 0, 1, \cdots. \qquad (1.3)$$

Here $a, b > 0$ are constants. It is shown from (1.2) that, in order to determine the components of the vector $\hat{\mathbf{U}}_0$, it is necessary to pose **additional boundary conditions**. In our case they take the form:

$$\hat{\mathbf{U}}_0^{II} = \mathbf{U}_1^{II}, \quad \hat{\mathbf{U}}_0^{III} = \mathbf{U}_1^{III}, \quad k, |j| = 0, 1, \cdots. \qquad (1.4)$$

We will prove stability of difference model (1.1)–(1.4).

Theorem 1.1. Let

$$r_x \leq \left[-\frac{1}{4} - 2a + \sqrt{\frac{1}{16} + 2a} \right] \cdot \min_{i=1, N_0 + N_1} a_i,$$

$$r_y \leq (\sqrt{b} - 2 \cdot b) \cdot \frac{\nu_0}{\nu_C}, \qquad (1.5)$$

$\nu_0 = \min_{i=\overline{1,N}} a_i$, $\nu_C = \|C\|$ is the operator norm of the matrix C,

$$a, b < 1/4.$$

Then difference model (1.1)–(1.4) is stable in the energy norm $\sqrt{J_k}$, here

$$J_k = \tilde{J}_k + h_x \cdot h_y \cdot \sum_{j=-\infty}^{\infty} \{ a \cdot (\mathbf{U}_0^{III}, A^{III} \cdot \mathbf{U}_0^{III}) + \frac{1}{2} \cdot (\mathbf{U}_0^{II}, B_+^{II} \cdot \mathbf{U}_0^{II}) \},$$

$$B_+^{II} = 2 \cdot a \cdot A^{II} + r_x \cdot I_{N_1}, \quad \tilde{J}_k = h_x \cdot h_y \cdot \sum_{i=1}^{\infty} \sum_{j=-\infty}^{\infty} (\mathbf{U}, A\mathbf{U}).$$

Proof. Multiply (1.1) scalarly by $2\mathbf{U}$ and obtain a difference analog of the energy integral identity (see **Introduction**):

$$\tau(\mathbf{U}, A \cdot \mathbf{U}) + \frac{r_x}{2} \cdot \xi_0(\mathbf{U}, B \cdot \mathbf{U}) + \frac{r_y}{2} \cdot \eta_0(\mathbf{U}, C \cdot \mathbf{U}) -$$

$$-a \cdot \hat{\xi}(\mathbf{U}, A \cdot \mathbf{U}) - b \cdot \hat{\eta}(\mathbf{U}, A \cdot \mathbf{U}) - \frac{r_x}{2} \cdot \xi(\bar{\xi}\mathbf{U}, B \cdot \bar{\xi}\mathbf{U}) -$$

$$-\frac{r_y}{2} \cdot \eta(\bar\eta\mathbf{U}, C \cdot \eta\mathbf{U}) - (\tau\mathbf{U}, A \cdot \tau\mathbf{U}) + a \cdot (\bar\xi\mathbf{U}, A \cdot \bar\xi\mathbf{U}) +$$

$$+ a \cdot (\xi\mathbf{U}, A \cdot \xi\mathbf{U}) + b \cdot (\bar\eta\mathbf{U}, A \cdot \bar\eta\mathbf{U}) + b \cdot (\eta\mathbf{U}, A \cdot \eta\mathbf{U}) = 0. \tag{1.6}$$

The following evident relations were used while deducing (1.6):

$$\begin{cases} (2\mathbf{U}, B \cdot \xi\mathbf{U}) = \xi(\mathbf{U}, B \cdot \mathbf{U}) - (\xi\mathbf{U}, B \cdot \xi\mathbf{U}), \\ (2\mathbf{U}, B \cdot \bar\xi\mathbf{U}) = \bar\xi(\mathbf{U}, B \cdot \mathbf{U}) + (\bar\xi\mathbf{U}, B \cdot \bar\xi\mathbf{U}), \\ (2\mathbf{U}, A \cdot \bar\xi\mathbf{U}) = \hat\xi(\mathbf{U}, A \cdot \mathbf{U}) - (\bar\xi\mathbf{U}, A \cdot \bar\xi\mathbf{U}) - (\xi\mathbf{U}, A \cdot \xi\mathbf{U}), \end{cases}$$

an so on. Multiply (1.6) by $h_x h_y$ and sum up this expression with respect to i (from 1 to $+\infty$) and j (from $-\infty$ to $+\infty$):

$$\tilde{J}_{k+1} - \tilde{J}_k + h_x \cdot h_y \cdot \sum_{j=-\infty}^{\infty} \{\frac{1}{2}r_x[-(\mathbf{U}_0, B \cdot \mathbf{U}_0) - (\mathbf{U}_1, B \cdot \mathbf{U}_1)]+$$

$$+ a[-(\mathbf{U}_0, A \cdot \mathbf{U}_0) + (\mathbf{U}_1, A \cdot \mathbf{U}_1)] + \frac{1}{2}r_x(\xi\mathbf{U}_0, B \cdot \xi\mathbf{U}_0)\}+$$

$$+ h_x \cdot h_y \cdot \sum_{i=1}^{\infty} \sum_{j=-\infty}^{\infty} \{-(\tau\mathbf{U}, A \cdot \tau\mathbf{U}) + a \cdot (\bar\xi\mathbf{U}, A \cdot \bar\xi\mathbf{U})+$$

$$+ a \cdot (\xi\mathbf{U}, A \cdot \xi\mathbf{U}) + b \cdot (\bar\eta\mathbf{U}, A \cdot \bar\eta\mathbf{U}) + b \cdot (\eta\mathbf{U}, A \cdot \eta\mathbf{U})\} = 0. \tag{1.7}$$

Transform (1.7) in the following way: since

$$-(\tau\mathbf{U}, A \cdot \tau\mathbf{U}) = -(A^{-1} \cdot [(r_x/2) \cdot B \cdot \xi_0 + (r_y/2) \cdot C \cdot \eta_0-$$

$$- A \cdot (a \cdot \hat\xi + b \cdot \hat\eta)]\mathbf{U}, [(r_x/2) \cdot B \cdot \xi_0 + (r_y/2) \cdot C \cdot \eta_0-$$

$$- A \cdot (a \cdot \hat\xi + b \cdot \hat\eta)]\mathbf{U}) \geq -\{(\xi\mathbf{U}, B_- \cdot A^{-1} \cdot B_- \cdot \xi\mathbf{U})+$$

$$+ (\bar\xi\mathbf{U}, B_+ \cdot A^{-1} \cdot B_+ \cdot \bar\xi\mathbf{U}) + (\eta\mathbf{U}, C_- \cdot A^{-1} \cdot C_- \cdot \eta\mathbf{U})+$$

$$+ (\bar\eta\mathbf{U}, C_+ \cdot A^{-1} \cdot C_+ \cdot \bar\eta\mathbf{U})\},$$

(1.7) together with (1.2) gives

$$\tilde{J}_{k+1} - \tilde{J}_k + h_x \cdot h_y \cdot \sum_{j=-\infty}^{\infty} \{(r_x/2)(\mathbf{U}_0^{II}, [I_{N_1} - S^* \cdot S]\mathbf{U}_0^{II})-$$

$$- (r_x/2)(\mathbf{U}_1, B \cdot \mathbf{U}_1) + a[-(\mathbf{U}_0, A \cdot \mathbf{U}_0) + (\mathbf{U}_1, A \cdot \mathbf{U}_1)]\}+$$

$$+ h_x \cdot h_y \cdot \sum_{i=1}^{\infty} \sum_{j=-\infty}^{\infty} \{(\xi\mathbf{U}, K_1 \cdot \xi\mathbf{U}) + (\bar\xi\mathbf{U}, K_2 \cdot \bar\xi\mathbf{U})$$

$$+(\eta \mathbf{U}, K_3 \cdot \eta \mathbf{U}) + (\bar{\eta}\mathbf{U}, K_4 \cdot \bar{\eta}\mathbf{U})\} \leq 0. \tag{1.8}$$

Here

$$K_1 = a \cdot A - B_- \cdot A^{-1} \cdot B_- - \frac{1}{2}r_x \cdot D_1,$$

$$K_2 = a \cdot A - B_+ \cdot A^{-1} \cdot B_+,$$

$$K_3 = a \cdot A - C_- \cdot A^{-1} \cdot C_-, K_4 = a \cdot A - C_+ \cdot A^{-1} \cdot C_+,$$

$$B_\pm = 2a \cdot A \pm r_x \cdot B, C_\pm = 2b \cdot A \pm r_y \cdot C,$$

$$D_1 = \text{diag}\,(O_{N_0}, I_{N_1}, O_{N_2}).$$

If (1.5) are true, then

$$K_{1,2,3,4} \geq 0.$$

So, (1.8) transformes into

$$J_{k+1} - J_k + h_x \cdot h_y \cdot \sum_{j=-\infty}^{\infty} \{(\mathbf{U}_0^{II}, [r_x \cdot I_{N_1} - (r_x/2 + a \cdot \nu)S^* \cdot S] \cdot \mathbf{U}_0^{II}) +$$

$$+(1/2)(\mathbf{U}_1^I, B_-^I \cdot \mathbf{U}_1^I)\} \leq 0, \tag{1.8'}$$

where

$$B_-^I = 2a \cdot A^I - r_x \cdot I_{N_0}, \nu = \max_{i=\overline{1,N}} a_i.$$

By (1.5), $B_-^I \geq 0$. Besides,

$$r_x \cdot I_{N_1} - ((r_x/2) + a \cdot \nu)S^* \cdot S \geq [r_x - (r_x/2 + a \cdot \nu)(1 - k_0)] \cdot I_{N_1} \geq 0,$$

at

$$k_0 \geq \frac{a\nu - (r_x/2)}{a\nu + (r_y/2)}.$$

(if C is a diagonal matrix, this equation is fulfilled a fortiori by **remark 0.2** from **Introduction**). With account of this inequality, the desired a priori estimation (a difference analog of the a priori estimation (0.6) from **Introduction**) follows from (1.8'):

$$J_{k+1} \leq J_k, \text{ i.e. } J_k \leq J_0, \quad k = 1, 2, \cdots. \tag{1.9}$$

It is known that exactly inequality (1.9) points at stability of scheme (1.1) (in a view of the boundary conditions) in the energy norm $\sqrt{J_k}$.

Remark 1.1. The technique how to construct finite-difference analogs of dissipative energy integrals for a certain class of explicit two-layers difference schemes is given in detail in the survey [8] also. The case of variable coefficients is studied in [4, 6].

1.2. SPECTRUM METHOD FOR INVESTIGATION OF STABILITY OF DIFFERENCE SCHEMES

We start now investigation of difference schemes for **mixed problem II**. We note that **problem II** differs in principle from **problem I** since boundary conditions (0.8) are not dissipative any more. In this connection, a more complicated investigation is to be carried out both for **mixed problem** and difference models of this problem. Before proceeding to construction of adequate difference models for **mixed problem II**, we will make a brief acquaintance with the **spectrum method** and its application to investigation of difference models for **problem II** (for more details see [9]).

Chosen the difference scheme from section 1, we formulate a difference model for **problem II** (system (0.7) is considered simultaneously with the equation for S, see **Introduction**):

$$\{A \cdot \tau + \frac{1}{2}r_x B \cdot \xi_0 + \frac{1}{2}r_y C \cdot \eta_0 - A \cdot (a \cdot \hat{\xi} + b \cdot \hat{\eta})\}U = 0,$$

$$l = 1, 2, \cdots; k, |j| = 0, 1, 2, \cdots; \tag{2.1}$$

$$U_{lj}^0 = U_0(l \cdot h_x, j \cdot h_y), l, |j| = 0, 1, 2, \cdots; \tag{2.2}$$

$$\begin{cases} u_{0j}^k + d \cdot p_{0j}^k = 0, \quad S_{0j}^k + p_{0j}^k = 0, \\ \\ \tau v_{0j}^k + \frac{r_y}{2}\omega\eta_0 v_{0j}^k - \lambda\frac{r_y}{2}\eta_0 p_{0j}^k - b\hat{\eta}v_{0j}^k = 0, \\ \\ k, |j| = 0, 1, 2, \cdots. \end{cases} \tag{2.3}$$

Here $a, b > 0$ are arbitrary as yet constants; $C = C_0 + \omega A$ (see **Introduction**), $U = U_{ij}^k$. It is seen from (2.3) that an additional condition is required in order to determine $\hat{U}_0$. In this connection, there appears the question on stability of scheme (2.1) with boundary conditions (2.3) and the additional boundary condition. Below it becomes clear that the answer to this question is not evident.

First we will carry out the spectrum analysis of the difference scheme for Cauchy problem (2.1), (2.2) (therewith $k, |l|, |j| = 0, 1, ...$). We seek particular solutions to (2.1) in the form:

$$U = q^k \cdot \chi^l \cdot e^{i \cdot j \cdot \varphi_0} \cdot V_0, \tag{2.4}$$

where V_0 is a constant vector, φ_0 is a real parameter; χ, q are complex numbers, $i = \sqrt{-1}$.

Substituting (2.4) into (2.1), we obtain

$$\left\{(q-1) \cdot I_4 + \frac{r}{2}A^{-1}[B \cdot \xi_0(\chi) + 2 \cdot i \cdot \sin\varphi_0 \cdot C]-\right.$$

$$-\left[a\cdot\hat{\xi}(\chi)-4\sin^2\frac{\varphi_0}{2}\cdot b\right]I_4\right\}\cdot\mathbf{V}_0=0, \tag{2.5}$$

where $\xi_0(\chi)=(\chi-\chi^{-1})$, $\hat{\xi}(\chi)=\chi-2+\chi^{-1}$, $r=r_x=r_y(h=h_x=h_y)$, $\chi=e^{i\psi_0}$, ψ_0 is a real parameter. Therewith system (2.5) has a nontrivial solution $\mathbf{V}_0$ for $q=q(\varphi_0,\psi_0)$ such that the determinant of (2.5) turns into zero. The necessary condition of stability of (2.1) is validity of the following inequality for any φ_0, ψ_0

$$|q(\varphi_0,\psi_0)|\leq 1, \tag{2.6}$$

i.e., the spectrum of the difference Cauchy problem must lie in the unit circle (see also [9]). We check validity of (2.6) at $\varphi_0=\psi_0$. In this case equality to zero of the determinant of (2.5) means that

$$q=q_1=1-4(a+b)\sin^2\frac{\varphi_0}{2}-r(\omega+1)\sin\varphi_0 i,$$

$$q=q_{2,3}=1-4(a+b)\sin^2\frac{\varphi_0}{2}-r(\omega+1\pm\sqrt{2}\kappa)\sin\varphi_0 i,$$

$$\kappa=M^{-1}.$$

In view of $|q_{1,2,3}|\leq 1$, we come to the following restriction on a,b,r:

$$0<a+b<\frac{1}{2},$$

$$r\leq\frac{\sqrt{2(a+b)}}{1+\omega+\sqrt{2}\kappa}. \tag{2.7}$$

We consider now a one-dimensional variant of **mixed problem II** . In the domain $t>0$, $x>0$ we seek the solution to the acoustics equations system

$$\mathbf{A}_0\mathbf{W}_t+\mathbf{B}_0\mathbf{W}_x=0, \tag{2.8}$$

which satisfies the boundary conditions at $x=0,t>0$

$$u+d\cdot p=0,\quad S+p=0 \tag{2.9}$$

and the initial data at $t=0,x>0$

$$\mathbf{W}(0,x)=\mathbf{W}_0(x). \tag{2.10}$$

Here

$$A_0=\begin{pmatrix} M^2 & 0 & 0 \\ 0 & 1 & 0 \\ 0 & 0 & 1 \end{pmatrix},\quad B_0=\begin{pmatrix} M^2 & 1 & 0 \\ 1 & 1 & 0 \\ 0 & 0 & 1 \end{pmatrix},$$

$$\mathbf{W} = \begin{pmatrix} u \\ p \\ S \end{pmatrix}.$$

In this case difference model (2.1)–(2.3) will transorm into

$$\left\{ A_0 \cdot \tau + \frac{1}{2} r \cdot B_0 \cdot \xi_0 - a \cdot A_0 \cdot \hat{\xi} \right\} \cdot \mathbf{W} = 0,$$

$$l = 1, 2, \ldots, \quad k = 0, 1, 2, \ldots; \tag{2.11}$$

$$\mathbf{W}_l^0 = \mathbf{W}_0(l \cdot h), \quad l = 0, 1, 2, \ldots; \tag{2.12}$$

$$u_0^k + d \cdot p_0^k = 0, \quad s_0^k + p_0^k = 0, \quad k = 0, 1, 2, \ldots. \tag{2.13}$$

Now we carry out the spectrum analysis for problem (2.11)–(2.13). We seek its solution in the form:

$$\mathbf{W} = q^k \cdot \chi^l \cdot \mathbf{Q}_0, \tag{2.14}$$

where

$$\mathbf{Q}_0 = \begin{pmatrix} u_0 \\ p_0 \\ S_0 \end{pmatrix} \text{ is a constant vector.}$$

Substituting (2.14) into (2.11), we obtain

$$\begin{pmatrix} \sigma & g \cdot \kappa^2 & 0 \\ g & \sigma & 0 \\ 0 & 0 & \sigma \end{pmatrix} \cdot \mathbf{Q}_0 = 0, \tag{2.15}$$

where $\sigma = q - 1 - a \cdot \hat{\xi}(\chi) + g$, $g = \frac{r}{2} \cdot \xi_0(\chi)$.

It follows from (2.15) that two types of solutions (2.14) are possible:

$$1) \quad S_0 \neq 0, \quad u_0 = p_0 = 0, \quad \sigma(\chi_0) = 0,$$

$$\mathbf{Q}_0 = \begin{pmatrix} 0 \\ 0 \\ S_0 \end{pmatrix}, \tag{2.16}$$

$$2) \quad \begin{cases} S_{1,2} = 0, \sigma(\chi_1) = -\kappa g(\chi_1), \sigma(\chi_2) = \kappa g(\chi_2), \\ \mathbf{Q}_1 = \begin{pmatrix} u_1 \\ p_1 \\ 0 \end{pmatrix}, \mathbf{Q}_2 = \begin{pmatrix} u_2 \\ p_2 \\ 0 \end{pmatrix}, \\ u_1 - \kappa p_1 = 0, u_2 + \kappa p_2 = 0. \end{cases} \tag{2.17}$$

38

While deducing (2.16), (2.17), we suppose $g(\chi_0) \neq 0$, $g(\chi_{1,2}) \neq 0$. With account of (2.16), (2.17), solutions to (2.11) are sought in the form:

$$\mathbf{W} = q^k \cdot [\chi_0^l \cdot \mathbf{Q}_0 + \chi_1^l \cdot \mathbf{Q}_1 + \chi_2^l \cdot \mathbf{Q}_2], \qquad (2.18)$$

$|\chi_{0,1,2}| < 1, \chi_{1,2}$ are simple roots.

The following Lemma will be used in the sequel.

Lemma 2.1. The roots of each equations

$$\sigma(\chi) = 0,$$

$$\sigma(\chi) = -\kappa g(\chi),$$

$$\sigma(\chi) = \kappa g(\chi)$$

satisfies the condition: at $|q| > 1$ one root lies inside the unit circle, another one lies outside of the unit circle.

Proof. With the help of the linear-fractional transformations

$$q = \frac{\rho + 1}{\rho - 1}, \chi = \frac{\zeta + 1}{\zeta - 1}.$$

we rewrite $\sigma(\chi) = 0$ as

$$\zeta^2 + r(\rho - 1)\zeta - 1 - 2a(\rho - 1) = 0.$$

The roots of the equation are:

$$\zeta_{1,2} = \frac{-r(\rho - 1) \pm \sqrt{r^2(\rho - 1)^2 + 4 + 8a(\rho - 1)}}{2}.$$

Since

$$r^2(\rho - 1)^2 + 4 + 8a(\rho - 1) = [r(\rho - 1) + 4a/r]^2 + 4(1 - 4a^2/r^2)$$

then

$$\mathrm{Re}\left(\sqrt{r^2(\rho - 1)^2 + 4 + 8a(\rho - 1)}\right) \geq r\,\mathrm{Re}\,(\rho - 1) + 4a/r.$$

Here we suppose that a, r meet restrictions which were inferred while analysing the spectrum of the one-dimensional Cauchy problem:

$$0 < a < \frac{1}{2}, \quad r \leq \frac{\sqrt{2a}}{k_1}, \quad k_1 = 1 + \kappa. \qquad (2.19)$$

Indeed, by (2.19),

$$\frac{4a}{r} - r > 0, \quad 1 - \frac{4a^2}{r^2} \geq 0.$$

Consequently, at $|q| > 1 (\mathrm{Re}\rho > 0)$:

$$\mathrm{Re}\,\zeta_1 > 0 \quad (\text{i.e. } |\chi_1| > 1),$$

$$\mathrm{Re}\,\zeta_2 < 0 \quad (\text{i.e. } |\chi_2| < 1),$$

Equations

$$\sigma(\chi) = -\kappa g(\chi), \quad \sigma(\chi) = \kappa g(\chi)$$

are analysed by analogy.

Remark 2.1. At $q = 1$ the roots of equations are: for $\sigma(\chi) = 0$

$$\chi_1 = 1, \quad \chi_2 = \frac{2a + r}{2a - r};$$

for $\sigma(\chi) = -\kappa g(\chi)$

$$\chi_1 = 1, \quad \chi_2 = \frac{2Ma + r(1 + M)}{2Ma - r(1 + M)};$$

for $\sigma(\chi) = \kappa g(\chi)$

$$\chi_1 = 1, \quad \chi_2 = \frac{2Ma - r(1 - M)}{2Ma + r(1 - M)} = \hat{\chi}.$$

By virtue of **Lemma 2.1**, as χ_0, χ_1, χ_2 in (2.18) we take

$$\begin{cases} \chi_0 = \dfrac{2a + q - 1 - \sqrt{r^2 + (q - 1)(4a + q - 1)}}{2a - r}, \\[3mm] \chi_1 = \dfrac{2a + q - 1 - \sqrt{r^2 k_1^2 + (q - 1)(4a + q - 1)}}{2a - rk_1}, \\[3mm] \chi_2 = \dfrac{2a + q - 1 - \sqrt{r^2 k_2^2 + (q - 1)(4a + q - 1)}}{2a + rk_2}, \end{cases} \qquad (2.20)$$

where $k_2 = \kappa - 1$.

We already know that an additional condition is necessary in order to determine $\hat{W}_0$. To derive it, we, first, write a differential relation, a consequence of system (2.8),

$$w_t + w_x + \kappa^2 p_x + \hat{\beta} u_x = 0, \qquad (2.21)$$

$$w = u + \hat{\beta} p,$$

where $\hat{\beta}$ is a constant , and, then, an approximation of (2.21) at $x = 0$:

$$\tau w_0^k + (1 + \hat{\beta})\gamma u_0^k + (\kappa^2 + \hat{\beta})\mu p_0^k = 0, \qquad (2.22)$$

$\gamma(\psi), \mu(\psi)$ are difference operators approximating the operator $\Delta \cdot \dfrac{\partial}{\partial x}$ at $x = 0$ (for example, $\gamma(\psi) = r(\psi - 1)$ an so on). With account of (2.17), (2.22), we obtain a linear algebraic system to determine u_1, p_1, u_2, p_2. This system must have a nontrivial solution since the determinant of the system is not the identical zero. From this condition we derive the determinant relation for q:

$$(d - \kappa)(a_2 + \kappa a_0) = (d + \kappa)(a_3 - \kappa a_1), \qquad (2.23)$$

where

$$a_0 = q - 1 + (1 + \hat{\beta}) \cdot \gamma(\chi_1),$$

$$a_1 = q - 1 + (1 + \hat{\beta}) \cdot \gamma(\chi_2),$$

$$a_2 = \hat{\beta}(q - 1) + (\kappa^2 + \hat{\beta}) \cdot \mu(\chi_1),$$

$$a_3 = \hat{\beta}(q - 1) + (\kappa^2 + \hat{\beta}) \cdot \mu(\chi_2).$$

Let $\hat{\beta} = -\kappa$, $\gamma = \mu = r(\psi - 1)$. Then (2.23) transforms into $(d \neq -\kappa)$:

$$q - 1 = k_2 r(\chi_2 - 1). \qquad (2.23')$$

However, (2.23′) does not have solutions (under $\dfrac{4a^2}{r^2 k_2^2} \neq 1$). Consequently, in the one-dimensional case it is possibble to find a boundary condition such that (2.11)–(2.13) does not admit existence of the Hadamard example, in other words, it does not have solutions of the form (2.18) with $|q| > 1$.

Now turn back to (2.1)–(2.3). We supplement (2.3) with one more condition. For this purpose we write down the differential relation

$$w_t + w_x + \omega w_y + \beta_2 u_x + \beta_0 \kappa^2 p_x + \beta_2 v_y + \beta_1 \kappa^2 p_y = 0, \qquad (2.24)$$

$$w = \beta_0 \cdot u + \beta_1 \cdot v + \beta_2 \cdot p,$$

which is a consequence of (0.7), and an approximation of (2.24) at $x = 0$:

$$\tau w_{0j}^k + (\beta_0 + \beta_2)\gamma u_{0j}^k + \beta_1 \delta v_{0j}^k + (\beta_0 \kappa^2 + \beta_2)\mu p_{0j}^k + \beta_0 \omega \frac{r}{2}\eta_0 u_{0j}^k +$$

$$+ (\omega\beta_1 + \beta_2)\frac{r}{2}\eta_0 v_{0j}^k + + (\beta_1 \kappa^2 + \omega\beta_2)\frac{r}{2}\eta_0 p_{0j}^k - b\hat{\eta} w_{0j}^k = 0. \qquad (2.25)$$

Here $\beta_{0,1,2}$ are constants; $\gamma(\psi)$, $\delta(\psi)$, $\mu(\psi)$ are difference operators approximating the operator $\Delta \cdot \dfrac{\partial}{\partial x}$ at $x = 0$ (for example, $\gamma = \dfrac{r}{2}(4\psi - \psi^2 - 3)$ and so on); $r = r_x = r_y(h = h_x = h_y)$.

The following theorem is true.

Theorem 2.1. *Difference scheme* (2.1) *with boundary conditions* (2.3) *and* (2.25) *is unstable for any* $\beta_0, \beta_1, \beta_2$ *and for any* $\gamma(\psi)$, $\delta(\psi)$, $\mu(\psi)$, *approximating the differential operator* $\Delta \cdot \frac{\partial}{\partial x}$ *at* $x = 0$.

Proof. Rewrite (2.5) as follows:

$$\begin{pmatrix} \sigma & 0 & g\kappa^2 & 0 \\ 0 & \sigma & \nu\kappa^2 & 0 \\ g & \nu & \sigma & 0 \\ 0 & 0 & 0 & \sigma \end{pmatrix} \cdot \mathbf{V}_0 = 0, \qquad (2.5')$$

where

$$\sigma = q - 1 - a \cdot \hat{\xi}(\chi) + g + 4b \cdot \sin^2(\frac{\varphi_0}{2}) + \omega\nu,$$

$$g = \frac{r}{2} \cdot \xi_0(\chi), \nu = i \cdot r \cdot \sin \varphi_0.$$

The solution to (2.1) is sought in the form :

$$\mathbf{U} = q^k \cdot exp\{j\varphi_0 i\}[\chi_0^l \cdot \mathbf{V}_0 + \chi_1^l \cdot \mathbf{V}_1 + \chi_2^l \cdot \mathbf{V}_2], \qquad (2.26)$$

where φ_0 is a real parameter; χ_0 is a root of $\sigma(\chi) = 0$; χ_1, χ_2 are roots of $M^2\sigma^2(\chi) = g^2(\chi) + \nu^2$, and χ_1, χ_2 are simple roots, and $|\chi_{0,1,2}| < 1$;

$$\mathbf{V}_0 = \begin{pmatrix} u_0 \\ v_0 \\ 0 \\ S_0 \end{pmatrix}, \quad \mathbf{V}_1 = \begin{pmatrix} u_1 \\ v_1 \\ p_1 \\ 0 \end{pmatrix}, \quad \mathbf{V}_2 = \begin{pmatrix} u_2 \\ v_2 \\ p_2 \\ 0 \end{pmatrix},$$

relations on $\mathbf{V}_{0,1,2}$ are (see (2.5')):

$$\begin{cases} g(\chi_0) \cdot u_0 + \nu v_0 = 0, \\ \sigma(\chi_1) \cdot u_1 + g(\chi_1)\kappa^2 \cdot p_1 = 0, \\ \sigma(\chi_1) \cdot v_1 + \nu\kappa^2 \cdot p_1 = 0, \\ \sigma(\chi_2) \cdot u_2 + g(\chi_2)\kappa^2 \cdot p_2 = 0, \\ \sigma(\chi_2) \cdot v_1 + \nu\kappa^2 \cdot p_2 = 0. \end{cases} \qquad (2.27)$$

Since, at $\varphi_0 = 0$ and $q = 1, \chi_0 = 1, \chi_1 = 1$, $\chi_2 = \hat{\chi}$ (see **Remark 2.1**) we will investigate the formulated difference model on "long waves" under the assumption $\varphi_0 = \varepsilon\hat{\varphi}$, where ε is a small parameter, $\hat{\varphi} = O(1)$. We assume the relations

$$\begin{cases} q - 1 = \varepsilon\hat{q} + O(\varepsilon^2), \\ \chi_0 - 1 = \varepsilon\hat{\chi}_0 + O(\varepsilon^2), \\ \chi_1 - 1 = \varepsilon\hat{\chi}_1 + O(\varepsilon^2), \\ \chi_2 - \hat{\chi} = \varepsilon\hat{\chi}_2 + O(\varepsilon^2). \end{cases} \tag{2.28}$$

are valid. Relations (2.27), boundary conditions (2.3), (2.25) form a linear algebraic system to determine u_0, v_0, u_1, v_1, p_1, u_2, v_2, p_2. The determinant condition on q is derived from the requirement that the determinant of this system is equal to zero:

$$\hat{\sigma} \cdot \sigma(\chi_1) \cdot \sigma(\chi_2) \cdot g(\chi_0) \cdot d(a_6 - a_7) -$$

$$-\sigma(\chi_1) \cdot \sigma(\chi_2) \cdot \lambda \cdot \nu^2 \cdot (a_6 - a_7) +$$

$$+\sigma(\chi_2) \cdot g(\chi_0) \cdot g(\chi_1) \cdot \kappa^2 (a_7\hat{\sigma} - a_1\hat{\sigma}d + a_3\lambda\nu) +$$

$$+g(\chi_1) \cdot \sigma(\chi_2) \cdot \lambda \cdot \kappa^2 \cdot \nu^2 \cdot (a_1 - a_0) +$$

$$+\sigma(\chi_2) \cdot \nu^2 \cdot \kappa^2 (a_7\hat{\sigma} - a_0\hat{\sigma}d + a_4\lambda\nu) -$$

$$-\sigma(\chi_1) \cdot g(\chi_0) \cdot g(\chi_2) \cdot \kappa^2 (a_6\hat{\sigma} - a_2\hat{\sigma}d + a_3\lambda\nu) -$$

$$-g(\chi_2) \cdot \sigma(\chi_1) \cdot \lambda \cdot \kappa^2 \cdot \nu^2 \cdot (a_2 - a_0) +$$

$$+\hat{\sigma} \cdot g(\chi_1) \cdot g(\chi_2) \cdot g(\chi_0) \cdot \kappa^4 (a_1 - a_2) +$$

$$+\hat{\sigma} \cdot g(\chi_2) \cdot \nu^2 \cdot \kappa^4 (a_0 - a_2) -$$

$$-\sigma(\chi_1) \cdot \nu^2 \cdot \kappa^2 (a_6\hat{\sigma} - a_0\hat{\sigma}d + a_5\lambda\nu) -$$

$$-\hat{\sigma} \cdot g(\chi_1) \cdot \nu^2 \cdot \kappa^4 (a_0 - a_1) +$$

$$+\sigma(\chi_2) \cdot \nu \cdot \kappa^2 \cdot d \cdot g(\chi_0) \cdot \hat{\sigma}(a_3 - a_4) +$$

$$+g(\chi_2) \cdot \nu \cdot \kappa^4 \cdot g(\chi_0) \cdot \hat{\sigma}(a_4 - a_3) +$$

$$+\sigma(\chi_1) \cdot \nu \cdot \kappa^2 \cdot d \cdot g(\chi_0) \cdot \hat{\sigma}(a_5 - a_3) -$$

$$-g(\chi_1) \cdot \nu \cdot \kappa^4 \cdot g(\chi_0) \cdot \hat{\sigma}(a_5 - a_3) -$$

$$-\nu^3 \cdot \kappa^4 \cdot \hat{\sigma}(a_5 - a_4) = 0. \tag{2.29}$$

Here

$$\hat{\sigma} = q - 1 + \omega\nu + 4b \cdot \sin^2 \frac{\varphi_0}{2},$$

$$a_0 = \beta_0 \cdot \hat{\sigma} + (\beta_0 + \beta_2) \cdot \gamma(\chi_0),$$

$$a_1 = \beta_0 \cdot \hat{\sigma} + (\beta_0 + \beta_2) \cdot \gamma(\chi_1),$$

$$a_2 = \beta_0 \cdot \hat{\sigma} + (\beta_0 + \beta_2) \cdot \gamma(\chi_2),$$

$$a_3 = \beta_1 \cdot \hat{\sigma} + \beta_2 \cdot \nu + \beta_1 \cdot \delta(\chi_0),$$

$$a_4 = \beta_1 \cdot \hat{\sigma} + \beta_2 \cdot \nu + \beta_1 \cdot \delta(\chi_1),$$

$$a_5 = \beta_1 \cdot \hat{\sigma} + \beta_2 \cdot \nu + \beta_1 \cdot \delta(\chi_2),$$

$$a_6 = \beta_2 \cdot \hat{\sigma} + \beta_1 \cdot \nu\kappa^2 + (\beta_0\kappa^2 + \beta_2) \cdot \mu(\chi_1),$$

$$a_7 = \beta_2 \cdot \hat{\sigma} + \beta_1 \cdot \nu\kappa^2 + (\beta_0\kappa^2 + \beta_2) \cdot \mu(\chi_2).$$

We substitute (2.28) into (2.29) and into

$$\sigma(\chi_0) = 0,$$

$$M^2\sigma^2(\chi_{1,2}) = g^2(\chi_{1,2}) + \nu^2,$$

and obtain

$$\hat{\chi}_0 = -\frac{\bar{q}}{r}, \quad \hat{q} = \bar{q} - ir\omega\hat{\varphi},$$

$$\hat{\chi}_1 = \hat{\varphi}\frac{1 + z^2}{2z}, \quad \bar{q} = r\hat{\varphi}\frac{\kappa(1 - z^2) - 1 - z^2}{2z},$$

$$(Mm_1 - m_2)[(z^2k_1 - k_2)^2(dM(1 - z^2) - 1 - z^2) +$$

$$+ 4z^2(z^2k_1 - k_2) - 4z^2(1 - z^2) \cdot \lambda M] = 0, \tag{2.30}$$

where

$$m_1 = (\kappa^2\beta_0 + \beta_2) \cdot \mu(\hat{\chi}), \quad m_2 = (\beta_0 + \beta_2) \cdot \gamma(\hat{\chi}).$$

While deriving (2.30) we assume that

$$\gamma(\chi_0) = r\hat{\chi}_0\varepsilon + O(\varepsilon^2), \quad \delta(\chi_0) = r\hat{\chi}_0\varepsilon + O(\varepsilon^2),$$

$$\gamma(\chi_1) = r\hat{\chi}_1\varepsilon + O(\varepsilon^2), \quad \delta(\chi_1) = r\hat{\chi}_1\varepsilon + O(\varepsilon^2),$$

$$\mu(\chi_1) = r\hat{\chi}_1\varepsilon + O(\varepsilon^2). \tag{2.31}$$

It becomes clear that, for any reasonable approximations of $\Delta \cdot \dfrac{\partial}{\partial x}$ at $x = 0$, (2.31) are valid. We also assume that

$$\gamma(\chi_2) = \gamma(\hat{\chi}) + O(\varepsilon), \quad \delta(\chi_2) = \delta(\hat{\chi}) + O(\varepsilon), \quad \mu(\chi_2) = \mu(\hat{\chi}) + O(\varepsilon),$$

therewith $\gamma(\hat{\chi}), \mu(\hat{\chi}) \neq 0$.

It is easy to see that the last relation from (2.30) has the root $z^2 = 1$. Therewith

$$\bar{q} = -\frac{r\hat{\varphi}}{z}, \quad \hat{\chi}_0 = \frac{\hat{\varphi}}{z}, \quad \hat{\chi}_1 = \frac{\hat{\varphi}}{z}.$$

Let $z = -\text{sign}(\hat{\varphi})$. Then

$$\text{Re}\,\bar{q} = \text{Re}\,\hat{q} > 0, \quad \text{Re}\,\hat{\chi}_0 = \text{Re}\,\hat{\chi}_1 < 0$$

and

$$|\chi_0|^2 = 1 + 2\varepsilon\text{Re}\hat{\chi}_0 + O(\varepsilon^2) < 1,$$

$$|\chi_1|^2 = 1 + 2\varepsilon\text{Re}\hat{\chi}_1 + O(\varepsilon^2) < 1,$$

$$|\chi_2|^2 = \hat{\chi}^2 + 2\varepsilon\hat{\chi}\text{Re}\hat{\chi}_2 + O(\varepsilon^2) < 1,$$

$$|q|^2 = 1 + 2\varepsilon\text{Re}\hat{q} + O(\varepsilon^2) > 1.$$

The last three inequalities mean that we constructed a Hadamard example which confirms instability of (2.1) (with account of the boundary conditions).

Remark 2.2. Theorem 2.1 holds true also for the cases when we either use another approximations of $\dfrac{\partial}{\partial y}$ in (2.1)–(2.3), (2.25) or add terms of the form $\hat{b}\hat{\eta}\mathbf{U}$, $\hat{b}\hat{\eta}\xi\mathbf{U}$ and so on ($\hat{b}$ is a constant) into (2.1), (2.25), and the third condition from (2.3). This fact is evident since the difference Hadamard example was constructed by the first approximation, i.e., the first terms from (2.28) were used (see Proof of **Theorem 2.1**).

An analogous conclusion is true if we add a term of the type $\hat{a}\cdot\gamma^2(\psi)w_{0j}^k$, $\hat{a}\cdot\gamma(\psi)\delta(\psi)w_{0j}^k$ (is a constant) and so on into additional boundary condition (2.25) or a term of the type $\hat{a}\cdot\gamma^2(\psi)v_{0j}^k$, $\hat{a}\cdot\gamma(\psi)\delta(\psi)v_{0j}^k$ and so on into the third boundary condition (2.3).

Remark 2.3. The spectrum analysis carried out in this section shows that stability depends on the type of the problem (is it the difference Cauchy problem or the difference boundary value problem) and its dimension. In this connection, recommendations obtained for either the difference Cauchy problem or the one-dimensional difference boundary value problem can not be automatically applied to multi-dimensional cases of the difference boundary-value problem.

Remark 2.4. This section is written on the basis of papers [10], [11] (see [12] too).

2. Lecture II. Construction of adequate calculation models for Gas Dynamic Equations

2.1. INVESTIGATION OF SOME DIFFERENCE-DIFFERENTIAL MODELS FOR PROBLEM II

Results from subsections 1.2, 1.3 (see Lecture I) confirm that construction of stable difference models for **mixed problem II** is a complicated problem. The main point is approximation of boundary conditions such that the difference model in whole stays stable. In this connection the idea to avoid discretization with respect to x (at least theoretically) seems to be interesting. As a result we come to the so-called difference-differential schemes.

In the domain R^3_{++} we carry out **discretization** over t and y with the steps $\Delta t = \Delta$, $\Delta y = h_y$. We introduce the notation:

$$\mathbf{U}(k \cdot \Delta, x, j \cdot h_y) = \mathbf{U}^k_j(x) = \mathbf{U}, \quad \varphi \mathbf{U}^k_j(x) = \varphi \mathbf{U} = \mathbf{U}^{k+1}_j(x),$$

$$\theta \mathbf{U}^k_j(x) = \theta \mathbf{U} = \mathbf{U}^k_{j+1}(x), \quad \theta^{-1} \mathbf{U}^k_j(x) = \theta^{-1}\mathbf{U} = \mathbf{U}^k_{j-1}(x),$$

$$L_t \mathbf{U}^k_j(x) = L_t \mathbf{U} = \tilde{\mathbf{U}}, \quad L_t = \alpha\varphi + \delta, \quad \hat{\tau} = \frac{\varphi - 1}{\Delta}, \quad \xi = \frac{\partial}{\partial x},$$

$$\hat{\xi} = L_t \xi, \eta = \theta - 1, \quad \bar{\eta} = 1 - \theta^{-1}, \quad \hat{\eta}_0 = \frac{\eta + \bar{\eta}}{2h_y}, \hat{\eta} = L_t \hat{\eta}_0,$$

where α, $\delta \geq 0$ $(\alpha + \delta = 1)$ are constants.

Now we formulate a difference-differential model to the initial mixed problem

$$A\hat{\tau}\mathbf{U} + B\hat{\xi}\mathbf{U} + C \cdot \hat{\eta}\mathbf{U} = 0, \quad x > 0, k, |j| = 0, 1, \dots; \qquad (1.1)$$

$$\begin{cases} u^k_j(0) + dp^k_j(0) = 0, \\ \hat{\tau}v^k_j(0) + \omega\hat{\eta}v^k_j(0) - \lambda\hat{\eta}p^k_j(0) = 0, \end{cases} \quad k, |j| = 0, 1, \dots; \qquad (1.2)$$

$$\mathbf{U}^0_j(x) = \mathbf{U}_0(x, j \cdot h_y), \quad x \geq 0, \quad |j| = 0, 1, \dots. \qquad (1.3)$$

Carried out discretization over t in R^3_{++}, we have one more difference-differential model

$$A\hat{\tau}\mathbf{U}^k + B\hat{\xi}\mathbf{U}^k + C \cdot \hat{\eta}\mathbf{U}^k = 0, \quad (x, y) \in R^2_+, \quad k = 0, 1, \dots; \qquad (1.4)$$

$$\begin{cases} u^k + dp^k = 0, \\ \hat{\tau}v^k + \omega\hat{\eta}v^k - \lambda\hat{\eta}p^k = 0, \end{cases} \quad x = 0, \quad y \in R^1, \quad k = 0, 1, \dots; \qquad (1.5)$$

$$\mathbf{U}^0(x, y) = \mathbf{U}_0(x, y), \quad x \geq 0, \quad y \in R^1. \qquad (1.6)$$

Here

$$\mathbf{U}(k \cdot \Delta, x, jh_y) = \mathbf{U}^k(x, y) = \mathbf{U}^k, \quad \hat{\eta} = L_t \eta, \quad \eta = \frac{\partial}{\partial y}.$$

We will call **difference-differential model** (1.1)–(1.3) by **model I** and **difference-differential model** (1.4)–(1.6) by **model II**.

The following facts are valid for the difference-differential **models I and II** (see [4, ch. III, sec 1], **Introduction**). Multiply (1.1) (or (1.4)) scalarly by

$$\begin{pmatrix} M^2(\hat{T}+\hat{\xi}) \\ -\hat{\xi} \\ -\hat{\eta} \end{pmatrix},$$

where $\hat{T} = \hat{\tau} + \omega\hat{\eta}$. After some transformations we obtain that the component p of the vector $\mathbf{U}$ satisfies the difference-differential wave equation:

$$\{M^2(\hat{T}+\hat{\xi})^2 - \hat{\xi}^2 - \hat{\eta}^2\}p = 0. \tag{1.7}$$

Multiply (1.1) scalarly by

$$\begin{pmatrix} M^2\hat{T} \\ -\hat{T} \\ 0 \end{pmatrix}.$$

Let $x = 0$ in the expression:

$$\{M^2(1+d)\hat{T}^2 - \beta^2\hat{T}\hat{\xi} + M^2\lambda\hat{\eta}^2\}p_j^k(0) = 0. \tag{1.8}$$

Then, instead of problem (1.1)–(1.3), we come to another difference-differential model:

$$\begin{cases} \{M^2(\hat{T}+\hat{\xi})^2 - \hat{\xi}^2 - \hat{\eta}^2\}p = 0, \\ M^2(\hat{T}+\hat{\xi})u = -\hat{\xi}p, \\ M^2(\hat{T}+\hat{\xi})v = -\hat{\eta}p, x > 0, k, |j| = 0, 1, \ldots; \end{cases} \tag{1.9}$$

$$\begin{cases} \{M^2(1+d)\hat{T}^2 - \beta^2\hat{T}\hat{\xi} + M^2\lambda\hat{\eta}^2\}p_j^k(0) = 0, \\ u_j^k(0) = -dp_j^k(0), \\ \hat{T}v_j^k(0) = \lambda\hat{\eta}p_j^k(0) = 0, \quad k, |j| = 0, 1, \ldots; \end{cases} \tag{1.10}$$

$$\begin{cases} \mathbf{U}_j^0(x) = \mathbf{U}_0(x, j \cdot h_y), \\ (\hat{T}+\hat{\xi})p_j^0(x) + \hat{\xi}u_j^0(x) + \hat{\eta}v_j^0(x) = 0, \quad x \geq 0, |j| = 0, 1, \ldots. \end{cases} \tag{1.11}$$

We will show that, under certain conditions, problems (1.1)–(1.3) and (1.9)–(1.11) are equivalent. Indeed, it follows from (1.9) that

$$\{M^2(\hat{T}+\hat{\xi})^2 - \hat{\xi}^2 - \hat{\eta}^2\}p = M^2(\hat{T}+\hat{\xi})\mathcal{P} = 0,$$

where

$$\mathcal{P} = \mathcal{P}_j^k(x) = (\hat{T}+\hat{\xi})p_j^k(x) + \hat{\xi}u_j^k(x) + \hat{\eta}v_j^k(x).$$

By analogy, from (1.10)

$$\{M^2(1+d)\hat{T}^2 - \beta^2\hat{T}\hat{\xi} + M^2\lambda\hat{\eta}^2\}p_j^k(0) = M^2\hat{T}P_j^k(0) = 0.$$

Thus, we formulate a problem for the aggregate $\mathcal{P}$:

$$(\hat{T} + \hat{\xi})\mathcal{P}_j^k(x) = 0, \quad x > 0, \quad k, |j| = 0, 1, \ldots; \tag{1.12}$$

$$\hat{T}\mathcal{P}_j^k(0) = 0, \quad k, |j| = 0, 1, \ldots; \tag{1.13}$$

$$\mathcal{P}_j^0(x) = 0, \quad x \geq 0, \quad |j| = 0, 1, \ldots. \tag{1.14}$$

It follows from (1.13) that

$$\hat{T}\mathcal{P}_j^k(0) = \hat{\tau}\mathcal{P}_j^k(0) + \omega\hat{\eta}_0\tilde{\mathcal{P}}_j^k(0) = 0.$$

Multiplying this relation by $\hat{\mathcal{P}}_j^k(0)$ and taking into consideration the evident relation

$$2\mathcal{P}_j^k(0)\hat{\tau}\mathcal{P}_j^k(0) = \hat{\tau}(\mathcal{P}_j^k(0))^2 - \Delta(\hat{\tau}\mathcal{P}_j^k(0))^2,$$

$$2\hat{\mathcal{P}}_j^k(0)\hat{\eta}_0\hat{\mathcal{P}}_j^k(0) = \hat{\eta}_0(\hat{\mathcal{P}}_j^k(0))^2 - \frac{\eta}{2h_y}(\bar{\eta}\hat{\mathcal{P}}_j^k(0))^2,$$

we finally obtain

$$\hat{\tau}(\mathcal{P}_j^k(0))^2 + \omega\hat{\eta}_0(\tilde{\mathcal{P}}_j^k(0))^2 + (\alpha - \delta)\Delta(\hat{\tau}\mathcal{P}_j^k(0))^2 - \omega\frac{\eta}{2h_y}(\bar{\eta}\hat{\mathcal{P}}_j^k(0))^2 = 0.$$

Thus at $\alpha \geq \delta$ the inequality

$$\|\mathcal{P}^{k+1}\|_0^2 \leq \|\mathcal{P}^k\|_0^2,$$

is true, where $\|\mathcal{P}^k\|_0^2 = h_y \sum_{j=-\infty}^{\infty} (\mathcal{P}_j^k(0))^2$. By (1.14), $\|\mathcal{P}^0\|_0^2 = 0$, then $\mathcal{P}_j^k(x_0) = 0$, $k, |j| = 0, 1, \ldots$. By analogy, presenting (1.12) in the form

$$\hat{\tau}\mathcal{P} + \xi\tilde{\mathcal{P}} + \omega\hat{\eta}_0\tilde{\mathcal{P}} = 0,$$

and multiplying it by $2\tilde{\mathcal{P}}$, we come to the expression

$$\hat{\tau}(\mathcal{P})^2 + \xi(\tilde{\mathcal{P}})^2 + \omega\hat{\eta}_0(\hat{\mathcal{P}})^2 + (\alpha - \delta)\Delta(\hat{\tau}\mathcal{P})^2 - \omega\frac{\eta}{2h_y}(\bar{\eta}\tilde{\mathcal{P}})^2 = 0.$$

Hence, under the assumption $\alpha \geq \delta$, we obtain

$$\|\mathcal{P}^{k+1}\|^2 \leq \|\mathcal{P}^k\|^2,$$

where $\|\mathcal{P}^k\|^2 = h_y \sum_{j=-\infty}^{\infty} \int_0^\infty (\mathcal{P}_j^k(x))^2 dx$. By (1.14), $\|\mathcal{P}^0\|_0^2 = 0$, then $\mathcal{P}_j^k(x) \equiv 0$, $x \geq 0$, $k, |j| = 0, 1, \ldots$. Thus, the first equation from (1.1) holds everywhere. Consequently, problems (1.1)–(1.3) and (1.9)–(1.11) are equivalent.

If p satisfies (1.7) then the vector

$$\mathbf{Y}_j^k(x) = Y = \begin{bmatrix} \tilde{L}_1 \\ L_2 \\ \tilde{L}_3 \end{bmatrix} p$$

satisfies the system [4]

$$\{D \cdot \hat{T} - Q \cdot \hat{\xi} - R \cdot \hat{\eta}\} \mathbf{Y} = 0, \tag{1.15}$$

Here

$$L_1 = \frac{1}{\beta^2}\hat{T}, \quad L_2 = \hat{\xi} - \frac{M^2}{\beta^2}\hat{T}, \quad L_3 = \hat{\eta},$$

$$\tilde{L}_1 = ML_1, \quad \tilde{L}_3 = \frac{1}{\beta}L_3,$$

$$E = M\begin{pmatrix} 1 & m_1 & 0 \\ m_1 & 1 & 0 \\ 0 & 0 & 1 \end{pmatrix}, \quad Q = \begin{pmatrix} m_1 & 1 & 0 \\ 1 & m_1 & 0 \\ 0 & 0 & -m_1 \end{pmatrix},$$

$$R = \begin{pmatrix} 0 & 0 & 1 \\ 0 & 0 & m_1 \\ 1 & m_1 & 0 \end{pmatrix}, \quad D = \frac{1}{\beta^2}(E + M^2 Q),$$

m_1 is a constant. We note that $D > 0$ if $|m_1| < 1$.

Compose an extended system from the remained two systems of the type (1.15):

$$\left\{\begin{pmatrix} D & O_3 \\ O_3 & D \end{pmatrix}\hat{T} - \begin{pmatrix} Q & O_3 \\ O_3 & Q \end{pmatrix}\hat{\xi} - \begin{pmatrix} R & O_3 \\ O_3 & R \end{pmatrix}\hat{\eta}\right\} \mathbf{X} = 0, \tag{1.16}$$

where

$$\mathbf{X}_j^k(x) = \mathbf{X} = \{\tilde{L}_1 L_+, L_2 L_+, \tilde{L}_3 L_+, \tilde{L}_1 L_-, L_2 L_-, \tilde{L}_3 L_-\}^*,$$

$$L_\pm = (L_\pm)_j^k(x) = (a_1 L_1 + a_2^\pm L_2)p_j^k(x),$$

$$a_1 = m\kappa, a_2^\pm = -\frac{n}{a_\pm}, \quad \gamma = \beta\kappa^2, \quad n = -\frac{\lambda}{\beta},$$

$$a_{\pm} = \frac{M(\gamma \pm \sqrt{\gamma^2 - 4mn})}{2m}, \quad m = \beta d + \frac{\lambda M^2}{\beta}, \quad \kappa = M^{-1}$$

Condition (1.8) can be presented in the form:

$$\{\tilde{L}_1 - a_{\pm} L_2\}(L_{\pm})_j^k(0) = 0, \quad k, |j| = 0, 1, \dots. \tag{1.8'}$$

Let $m_1 = a_+/(1 + a_+^2)$ for the first system and $m_1 = a_-/(1 + a_-^2)$ for the second system of the form (1.15) which form the extended system (1.16). Then boundary conditions (1.8') are **dissipative** for system (1.16) since

$$\left(\begin{bmatrix} Q & O_3 \\ O_3 & Q \end{bmatrix} \mathbf{X}_j^k(0), \mathbf{X}_j^k(0) \right) = a_+(a_1 L_1 L_2 p_j^k(0) + a_2^+ L_2^2 p_j^k(0))^2 +$$

$$+ \frac{a_+}{1 + a_+^2}(a_1 L_1 \tilde{L}_3 p_j^k(0) + a_2^+ L_2 \tilde{L}_3 p_j^k(0))^2 + a_-(a_1 L_1 L_2 p_j^k(0) +$$

$$+ a_2^- L_2^2 p_j^k(0))^2 + \frac{a_-}{1 + a_-^2}(a_1 L_1 \tilde{L}_3 p_j^k(0) + a_2^- L_2 \tilde{L}_3 p_j^k(0))^2 > 0. \tag{1.17}$$

Remark 1.1. The results from above hold valid for the difference-differential **model II**.

It is convenient to use instead of (1.1) its corollary:

$$M\hat{\tau}\mathbf{Z} + B_1\hat{\xi}\mathbf{Z} + C_1\hat{\eta}\mathbf{Z} + = 0, \tag{1.1'}$$

where $\mathbf{Z}_j^k(x) = Z = T_0 A^{\frac{1}{2}} U$,

$$T_0 = \frac{1}{\sqrt{2}} \begin{bmatrix} 1 & 0 & -1 \\ 0 & -1 & 0 \\ 0 & -1 & 0 \\ 1 & 0 & 1 \end{bmatrix}, \quad B_1 = \begin{bmatrix} M & -1 \\ -1 & M \end{bmatrix} \times I_2,$$

$$C_1 = \begin{bmatrix} \omega M - 1 & 0 \\ 0 & \omega M + 1 \end{bmatrix} \times I_2, \quad I_2 = \begin{bmatrix} 1 & 0 \\ 0 & 1 \end{bmatrix},$$

$$\begin{smallmatrix}k\\j\end{smallmatrix}(x) = = \frac{M}{\sqrt{2}} \begin{bmatrix} 0 \\ -1 \\ 1 \\ 0 \end{bmatrix} \Omega, \quad \Omega_j^k(x) = \Omega = \hat{\eta}u - \hat{\xi}v.$$

We note that, by (1.1), the aggregate Ω (the difference-differential analog of the velocity vortex) satisfies the relation:

$$(\hat{T} + \hat{\xi})\Omega = 0, \quad x > 0, \quad k, |j| = 0, 1, \dots. \tag{1.18}$$

Multiplying (1.1′) scalarly by $2\tilde{\mathbf{Z}}$ and taking into account the relations

$$(2\tilde{\mathbf{Z}}, A\hat{\tau}\mathbf{Z}) = \hat{\tau}(\mathbf{Z}, A\mathbf{Z}) + (\alpha - \delta)\Delta(\hat{\tau}\mathbf{Z}, A\hat{\tau}\mathbf{Z}),$$

$$(2\tilde{\mathbf{Z}}, C\hat{\eta}\tilde{\mathbf{Z}}) = \hat{\eta}_0(\tilde{\mathbf{Z}}, C\tilde{\mathbf{Z}}) - \frac{\eta}{2h_y}(\bar{\eta}\tilde{\mathbf{Z}}, C\bar{\eta}\tilde{\mathbf{Z}}),$$

$$(2\tilde{\mathbf{Z}}, B\hat{\xi}\tilde{\mathbf{Z}}) = \hat{\xi}(\tilde{\mathbf{Z}}, B\tilde{\mathbf{Z}})$$

(A,B,C are symmetric matrices!), as the result we obtain

$$M\hat{\tau}(|\mathbf{Z}|^2) + \hat{\xi}(\tilde{\mathbf{Z}}, B_1\tilde{\mathbf{Z}}) + \hat{\eta}_0(\tilde{\mathbf{Z}}, C_1\tilde{\mathbf{Z}}) + M(\alpha - \delta)\Delta(|\hat{\tau}\mathbf{Z}|^2) -$$

$$- \frac{\eta}{2h_y}(\bar{\eta}\tilde{\mathbf{Z}}, C_1\bar{\eta}\tilde{\mathbf{Z}}) + 2(\tilde{\mathbf{Z}},) = 0, |\mathbf{Z}|^2 = (\mathbf{Z}, \mathbf{Z}).$$

By analogy, we multiply (1.18) by $2\tilde{\Omega}$ and, thus, we have

$$\hat{\tau}(\Omega)^2 + \xi(\tilde{\Omega})^2 + \omega\hat{\eta}_0(\tilde{\Omega})^2 + (\alpha - \delta)\Delta(\hat{\tau}\Omega)^2 - \omega\frac{\eta}{2h_y}(\bar{\eta}\tilde{\Omega})^2 = 0.$$

In a view of these relations, we easily derive that if $\alpha \geq \delta$, then

$$\hat{\tau}\{\|\mathbf{Z}^k\|^2 + \|\Omega^k\|^2\} - h_y \sum_{j=-\infty}^{\infty} \{(\tilde{\mathbf{Z}}_j^k(0), B_1\tilde{\mathbf{Z}}_j^k(0)) + (\tilde{\Omega}_j^k(0))^2\} \leq$$

$$\leq \tilde{N}_1\{\|\mathbf{Z}^k\|^2 + \|\mathbf{Z}^{k+1}\|^2 + \|\Omega^k\|^2\}, \qquad (1.19)$$

where

$$\|\mathbf{Z}^k\|^2 = Mh_y \sum_{j=-\infty}^{\infty} \int_0^{\infty} |\mathbf{Z}_j^k(x)|^2 dx,$$

$$\|\Omega^k\|^2 = h_y \sum_{j=-\infty}^{\infty} \int_0^{\infty} (\Omega_j^k(x))^2 dx.$$

While obtaining (1.19), we used the inequalities:

$$2|(\tilde{\mathbf{Z}},)| \leq 2|\tilde{\mathbf{Z}}||| \leq M|\tilde{\mathbf{Z}}|^2 + \kappa||^2 \leq$$

$$\leq 2\alpha^2 M|\varphi\mathbf{Z}|^2 + 2\delta^2 M|\mathbf{Z}|^2 + M(\Omega)^2,$$

moreover, $\tilde{N}_1 = \max\{2\alpha^2 M, M\}$. We also note that, by virtue of (1.2),

$$(\tilde{\mathbf{Z}}_j^k(0), B_1\tilde{\mathbf{Z}}_j^k(0)) = M(1 + M^2 d^2 - 2d)(\tilde{p}_j^k(0))^2 + M^3(\tilde{v}_j^k(0))^2, (\tilde{\Omega}_j^k(0))^2 =$$

$$= (\lambda + \kappa^2 - d)^2(\bar{\eta}\tilde{p}_j^k(0))^2.$$

Remark 1.2. For difference-differential **model II** the estimation analogous to (1.19) is as follows:

$$\hat{\tau}\{\|\mathbf{Z}^k\|^2 + \|\mathbf{\Omega}^k\|^2\} - \int\limits_{R}\{(\tilde{\mathbf{Z}}^k(0,y), B_1\tilde{\mathbf{Z}}^k(0,y)) + (\tilde{\mathbf{\Omega}}^k(0,y))^2\}dy \leq$$

$$\leq N_1\{\|\mathbf{Z}^k\|^2 + \|\mathbf{Z}^{k+1}\|^2 + \|\mathbf{\Omega}^k\|^2\},$$

where

$$\|\mathbf{Z}^k\|^2 = M\int\limits_{R_+^2} |\mathbf{Z}^k(x,y)|^2 dx dy,$$

$$\|\mathbf{\Omega}^k\|^2 = \int\limits_{R_+^2} |\mathbf{\Omega}^k(x,y)|^2 dx dy.$$

It follows from the third equation of (1.1) that

$$(\hat{T} + \hat{\xi})\hat{T}v = -\hat{T}\hat{\eta}p\kappa^2.$$

Hence

$$\hat{\tau}(\|\hat{T}v^k\|^2) - h_y \sum_{j=-\infty}^{\infty} (\tilde{\hat{T}}v_j^k(0))^2 \leq$$

$$\leq \tilde{N}_2\{\|\hat{T}\hat{\eta}p^k\|^2 + \|\hat{T}v^k\|^2 + \|\hat{T}v^{k+1}\|^2\}, \qquad (1.20)$$

where

$$\|\hat{T}v^k\|^2 = h_y \sum_{j=-\infty}^{\infty} \int_0^{\infty} |(\hat{T}v_j^k(x))^2 dx,$$

$$\|\hat{T}\hat{\eta}p^k\|^2 = h_y \sum_{j=-\infty}^{\infty} \int_0^{\infty} |(\hat{T}\hat{\eta}p_j^k(x))^2 dx,$$

$$\tilde{N}_2 = \max\{1, 2\alpha^2\}\kappa^2, \quad (\tilde{\hat{T}}v_j^k(0))^2 = \lambda^2((\tilde{\hat{\eta}}p_j^k(0))^2.$$

Remark 1.3. The analogous inequality for difference-differential **model II** is

$$\hat{\tau}(\|\hat{T}v^k\|^2) - \int\limits_{R^1} (\tilde{\hat{T}}\tilde{v}^k(0,y))^2 dy \leq$$

$$\leq \tilde{N}_2\{\|\hat{T}\hat{\eta}p^k\|^2 + \|\hat{T}v^k\|^2 + \|\hat{T}v^{k+1}\|^2\}, \qquad (1.20)$$

where

$$\|\hat{T}v^k\|^2 = \int\limits_{R_+^2} (\hat{T}v^k(x,y))^2 dx dy,$$

$$\|\hat{T}\hat{\eta}p^k\|^2 = \int\limits_{R_+^2} (\hat{T}\hat{\eta}p^k(x,y))^2 dxdy.$$

We now come to consideration of problem (1.16), (1.8'). Instead of (1.16) we will use its corollary

$$\left\{ \begin{pmatrix} D_1 & O_4 \\ O_4 & D_1 \end{pmatrix} \hat{T} - \begin{pmatrix} Q_1 & O_4 \\ O_4 & Q_1 \end{pmatrix} \hat{\xi} - \begin{pmatrix} R_1 & O_4 \\ O_4 & R_1 \end{pmatrix} \hat{\eta} \right\} \mathbf{W} = 0. \qquad (1.16')$$

Here

$$D_1 = \frac{M}{\beta^2} \cdot \begin{pmatrix} 1 & -M \\ -M & 1 \end{pmatrix} \otimes H, \quad Q_1 = \begin{pmatrix} 0 & -1 \\ -1 & 0 \end{pmatrix} \otimes H,$$

$$R_1 = \begin{pmatrix} -1 & 0 \\ 0 & 1 \end{pmatrix} \otimes H, \quad H = \begin{pmatrix} 1 & -m_1 \\ -m_1 & 1 \end{pmatrix} > 0 \ (m_1 < 1!),$$

$$\mathbf{W} = \mathbf{W}_j^k(x) = \begin{pmatrix} T_0 & O \\ O & T_0 \end{pmatrix} \cdot \mathbf{X}.$$

The matrix coefficients in (1.16') are block-diagonal, by this reason we will consider systems of the type:

$$\{D_1\hat{\tau} - Q_1\hat{\xi} + (\omega D_1 - R_1)\hat{\eta}\}\hat{\mathbf{W}} = 0,$$

where $\hat{\mathbf{W}}$ will be the corresponding restrictions of the vector $\mathbf{W}$. Multiplying the system scalarly by $2\tilde{\hat{\mathbf{W}}}$, we obtain

$$\hat{\tau}(\hat{\mathbf{W}}, D_1\hat{\mathbf{W}}) - \hat{\xi}(\tilde{\hat{\mathbf{W}}}, Q_1\tilde{\hat{\mathbf{W}}}) + \hat{\eta}_0(\tilde{\hat{\mathbf{W}}}, (\omega D_1 - R_1)\tilde{\hat{\mathbf{W}}})+$$

$$+(\alpha - \delta)\Delta(\hat{\tau}\hat{\mathbf{W}}, D_1\hat{\tau}\hat{\mathbf{W}}) - \frac{\eta}{2h_y}(\bar{\eta}\tilde{\hat{\mathbf{W}}}, (\omega D_1 - R_1)\bar{\eta}\tilde{\hat{\mathbf{W}}}) = 0,$$

At $\alpha > \delta$

$$\hat{\tau}\|\mathbf{W}^k\|_{D_1}^2 + h_y \sum_{j=-\infty}^{\infty} \left(\tilde{\mathbf{W}}_j^k(0), \begin{pmatrix} Q_1 & O_4 \\ O_4 & Q_1 \end{pmatrix} \tilde{\mathbf{W}}_j^k(0) \right) \leq 0,$$

where

$$\|\mathbf{W}^k\|_{D_1}^2 = h_y \sum_{j=-\infty}^{\infty} \int\limits_0^{\infty} \left(\begin{bmatrix} D_1 & O_4 \\ O_4 & D_1 \end{bmatrix} \cdot \mathbf{W}_j^k(x), \mathbf{W}_j^k(x) \right) dx =$$

$$= h_y \sum_{j=-\infty}^{\infty} \int\limits_0^{\infty} \left(\begin{bmatrix} D & O_3 \\ O_3 & D \end{bmatrix} \cdot \mathbf{X}_j^k(x), \mathbf{X}_j^k(x) \right) dx = \|\mathbf{X}^k\|_D^2.$$

Since, by virtue (1.17),

$$\left(\tilde{\mathbf{W}}_j^k(0), \begin{pmatrix} Q_1 & O_4 \\ O_4 & Q_1 \end{pmatrix} \tilde{\mathbf{W}}_j^k(0) \right) > 0,$$

finally, we have the estimation

$$\|\mathbf{X}^{k+1}\|_D^2 \le \|\mathbf{X}^k\|_D^2. \tag{1.21}$$

Remark 1.4. An analogous estimation can be inferred for **model II**:

$$\|\mathbf{X}^{k+1}\|_D^2 \le \|\mathbf{X}^k\|_D^2,$$

where

$$\|\mathbf{X}^k\|_D^2 == \int_{R_+^2} \left(\begin{bmatrix} D & O_3 \\ O_3 & D \end{bmatrix} \cdot \mathbf{X}^k(x,y), \mathbf{X}^k(x,y) \right) dx dy.$$

We have

$$L_1 L_2 p = \frac{L_1 L_+ - L_1 L_-}{a_2^+ - a_2^-}, \quad L_1^2 p = \frac{L_1 L_+ - a_2^+ L_1 L_2 p}{a_1},$$

$$L_2^2 p = \frac{L_2 L_+ - L_2 L_-}{a_2^+ - a_2^-}, \quad \tilde{L}_3^2 p = M^2 L_1^2 p - L_2^2 p,$$

$$L_2 \tilde{L}_3 p = \frac{\tilde{L}_3 L_+ - \tilde{L}_3 L_-}{a_2^+ - a_2^-}, \quad L_1 \tilde{L}_3 p = \frac{\tilde{L}_3 L_+ - a_2^+ L_2 \tilde{L}_3 p}{a_1},$$

and the aggregates

$$\hat{T}^2 p, \hat{T}\hat{\xi} p, \hat{T}\hat{\eta} p, \hat{\xi}^2 p, \hat{\xi}\hat{\eta} p, \hat{\eta}^2 p$$

can be estimated with the help of (1.21).

And, finally, we write down the obvious relations:

$$\hat{\tau} p - \hat{\xi} p + \omega \hat{\eta} p - T p + \hat{\xi} p = 0,$$

$$\hat{\tau}(Tp) - \hat{\xi}(Tp) + \omega \hat{\eta}(\hat{T}p) - \hat{T}^2 p + \hat{T}\hat{\xi} p = 0,$$

$$\hat{\tau}(\hat{\xi} p) - \hat{\xi}(\hat{\xi} p) + \omega \hat{\eta}(\hat{\xi} p) - \hat{T}\hat{\xi} p + \hat{\xi}^2 p = 0,$$

$$\hat{\tau}(\hat{\eta} p) - \hat{\xi}(\hat{\eta} p) + \omega \hat{\eta}(\hat{\eta} p) - \hat{T}\hat{\eta} p + \hat{\xi}\hat{\eta} p = 0,$$

$$\hat{\tau} v - \hat{\xi} v + \omega \hat{\eta} v - 2\hat{T} v - \frac{1}{M^2} \hat{\eta} p = 0.$$

We introduce into consideration the aggregate

$$\hat{J}_k = \varepsilon_1 \{ \|\mathbf{Z}^k\|^2 + \|\Omega^k\|^2 + \|\hat{T} v^k\|^2 \} + \|\mathbf{X}^k\|_D^2 +$$

$$+\|p^k\|^2 + \|\hat{T}p^k\|^2 + \|\hat{\xi}p^k\|^2 + \|\hat{\eta}p^k\|^2 + \|v^k\|^2, \qquad (1.22)$$

where $\varepsilon_1 > 0$ is a constant. It follows from (1.19)–(1.22) that

$$\hat{J}_{k+1} - \hat{J}_k \leq K_1\Delta(J_k + J_{k+1}) \qquad (1.23)$$

where

$$J_k = \|\mathbf{U}^k\|^2 + \|\Omega^k\|^2 + \|\hat{T}v^k\|^2 + \|\hat{T}p^k\|^2 + \|\hat{\xi}p^k\|^2 + \|\hat{\eta}p^k\|^2 +$$

$$+\|\hat{T}^2p^k\|^2 + \|\hat{T}\hat{\xi}p^k\|^2 + \|\hat{T}\hat{\eta}p^k\|^2 + \|\hat{\xi}^2p^k\|^2 + \|\hat{\xi}\hat{\eta}p^k\|^2 + \|\hat{\eta}^2p^k\|^2,$$

$$\|\mathbf{U}^k\|^2 = h_y \sum_{j=-\infty}^{\infty} \int_0^{\infty} |\mathbf{U}_j^k(x)|^2 dx, \quad K_1 > 0$$

is a constant. According to the lemma from [13], the estimation follows from (1.23)

$$J_k \leq \text{const} \cdot J_0, \quad k = 1, 2, \dots, \qquad (1.24)$$

and thus we establish well-posedness of difference-differential **model I**. The inequality $\alpha > \delta$ is the sufficient condition for its stability.

Remark 1.5. An estimation, analogous to (1.24), can be easily obtained for **model II**.

Remark 1.6. A more accurate estimation can be obtained (see [4])

$$G_k \leq \text{const} \cdot G_0, \quad k = 1, 2, \dots, \qquad (1.25)$$

where

$$G_k = \|\mathbf{U}^k\| + \|\hat{T}\mathbf{U}^k\|^2 + \|\hat{\xi}\mathbf{U}^k\|^2 + \|\hat{\eta}\mathbf{U}^k\|^2 + \|\hat{T}^2\mathbf{U}^k\|^2 +$$

$$+\|\hat{T}\hat{\xi}\mathbf{U}^k\|^2 + \|\hat{T}\hat{\eta}\mathbf{U}^k\|^2 + \|\hat{\xi}^2\mathbf{U}^k\|^2 + \|\hat{\xi}\hat{\eta}\mathbf{U}^k\|^2 + \|\hat{\eta}^2\mathbf{U}^k\|^2,$$

$$\|\hat{T}\mathbf{U}^k\|^2 = h_y \sum_{j=-\infty}^{\infty} \int_0^{\infty} |\hat{T}\mathbf{U}_j^k(x)|^2 dx$$

and so on.

2.2. METHOD OF LINES FOR GAS DYNAMICS EQUATIONS

In this section a difference-differential scheme to obtain approximate solutions to gas dynamics equations (0.20) is considered. Finally it reduces to a boundary value problem for a system of differential equations (in concrete cases it is solved by the orthogonal sweep method).

Theoretical justification of this calculation model on the example of **mixed problem II** is given in section 1. The case of two boundaries is investigated in [14].

This section is devoted mostly to description of the calculation algorithm. As an example we take the problem on supersonic flowing of perfect gas around a cone. This stationary solution of gas dynamics equations is found by the establishment method. An approximate solution is found with the help of the suggested calculation model. Results are compared with the results of some other authors.

Following [4], we formulate the mixed problem: in the domain

$$t > 0, \quad 0 < s < 1, \quad 0 < \varphi < \pi$$

the solution to system (0.20) in the spherical coordinate system

$$\frac{d}{dt}u - v^2 - w^2 = 0,$$

$$\frac{d}{dt}v + \frac{c^2}{\varepsilon}P_s + uv - w^2 \text{ctg}\Theta = 0,$$

$$\frac{d}{dt}w - \frac{c^2}{\varepsilon}a_2 P_s + \frac{c^2}{\sin\Theta}P_\varphi + uw + vw\text{ctg}\Theta = 0, \qquad (2.1)$$

$$\frac{d}{dt}P + \frac{1}{\varepsilon}v_s - \frac{a^2}{\varepsilon}w_s + \frac{1}{\sin\Theta}w_\varphi + 2u + w\text{ctg}\Theta = 0,$$

$$\frac{d}{dt}\Psi = 0$$

is sought which satisfies the boundary conditions at $s = 1$ (**on the head shock wave**):

$$u = u_\infty,$$

$$v = v_\infty + \frac{G(p)}{(1 + a_2^2(1))^{1/2}},$$

$$w = w_\infty - a_2(1)\frac{G(p)}{(1 + a_2^2(1))^{1/2}}, \qquad (2.2)$$

$$V = \frac{V_\infty(hp_\infty + p)}{(hp + p_\infty)},$$

$$\frac{\partial}{\partial t}\Theta_s + w_\infty a_2(1) = v_\infty + \frac{G_1(p)}{(1 + a_2^2(1))^{1/2}},$$

at $s = 0$ (**the flow does not pass through the boundary**):

$$v = 0, \qquad (2.3)$$

the **symmetry conditions** at $\varphi = 0, \pi$:

$$w = P_\varphi = u_\varphi = v_\varphi = \Psi_\varphi = 0, \quad \frac{\partial}{\partial\varphi}\Theta_s = 0 \qquad (2.4)$$

and the **initial data** at $t = 0$:

$$u = u_0(s,\varphi), \quad v = v_0(s,\varphi), \quad w = w_0(s,\varphi),$$

$$P = P_0(s,\varphi), \quad \Psi = \Psi_0(s,\varphi), \quad \Theta_s = \Theta_{s0}(\varphi). \qquad (2.5)$$

Here

$$\frac{d}{dt} = \frac{\partial}{\partial t} + \tilde{\Delta}\frac{\partial}{\partial s} + \frac{w}{\sin\Theta}\frac{\partial}{\partial\phi}, \tilde{\Delta} = \frac{a_1}{\varepsilon},$$

$$a_1 = v - s\frac{\partial}{\partial t}\Theta_s - wa_2, \quad a_2 = \frac{s}{\sin\Theta}\frac{\partial}{\partial\varphi}\Theta_s,$$

(R,Θ,φ) is the spherical coordinate system (in (2.1)$R = 1$); u, v, w are components of the velocity scaled by division by the maximal velocity W_{max};

$$P = \frac{1}{\gamma}\ln p, \quad \Psi = P + \ln V, \quad \gamma = 1.4,$$

p, ρ are the pressure and density scaled by division by p_T, ρ_T (the parametrs of bracing), $s = (\Theta - \Theta_b)/\varepsilon$, $\varepsilon = \Theta_s - \Theta_b$, Θ_b is the half-angle of the cone, $\Theta = \Theta_s(t,\varphi)$ is the equation of the shock wave on a sphere with the radius $R = 1, c^2 = ((\gamma - 1)\exp(\gamma - 1)P + \Psi)/2$ is the squared sound velocity,

$$u_\infty = U_\infty(\cos\alpha\cos\Theta_s - \sin\alpha\sin\Theta_s\cos\varphi),$$

$$v_\infty = -U_\infty(\sin\alpha\cos\Theta_s\cos\varphi + \cos\alpha\sin\Theta_s),$$

$$w_\infty = U_\infty\sin\alpha\sin\varphi,$$

α is the angle of attack (the velocity vector of the filling flow lies in the plane of symmetry),

$$U_\infty = \left(\frac{(\gamma - 1)M_\infty^2/2}{1 + (\gamma - 1)M_\infty^2/2}\right)^{1/2},$$

$M_\infty > 1$ is the Mach number of the filling flow (see fig.1),

$$G(p) = (\hbar(p - p_\infty)(V_\infty - V))^{1/2},$$

$$G_1(p) = \frac{V_\infty G(p)}{V_\infty - V},$$

$$V = \frac{1}{\rho}, \quad V_\infty = \frac{1}{\rho_\infty}, \quad p_\infty = \rho_\infty^\gamma,$$

$$\rho_\infty = \left(\frac{U_\infty^2}{(\gamma-1)M_\infty^2/2}\right)^{1/(\gamma-1)}$$

$$a_2(1) = \frac{1}{\sin\Theta_s}\frac{\partial}{\partial\varphi}\Theta_s, \quad h = \frac{\gamma+1}{\gamma-1}, \quad \hbar = \frac{\gamma-1}{2\gamma}.$$

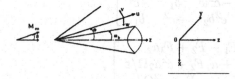

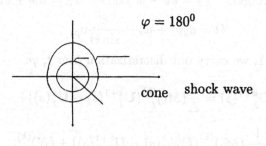

$$\varphi = 180^0$$

cone   shock wave

$$\varphi$$

$$\varphi = 0^0$$

Figure 1.

From the second and third relations in (2.2) at $s = 1$ we have

$$a_2(1) = \frac{w_\infty - w}{v - v_\infty} \qquad (2.6)$$

58

If $u, v, w, P, \Psi, \Theta_s$ do not depend on t, then problem (2.1)–(2.4) describes a stationary supersonic flowing around a circular cone. In accordance with the establishment method, solving numerically (2.1)–(2.5), we determine this solution.

The second, third and fourth equations of (2.1) are transformed into the system:

$$\mathbf{U}_s = M_1 \mathbf{U}_t + M_2 \mathbf{U}_\varphi + N, \tag{2.7}$$

where

$$\mathbf{U} = \begin{bmatrix} P \\ v \\ w \end{bmatrix}, \quad M_1 = -\frac{\varepsilon}{d} \begin{bmatrix} a_2 & -1 & a_2 \\ -c^2 & a_1 & 0 \\ c^2 a_2 & 0 & a_1 \end{bmatrix},$$

$$M_2 = -\frac{\varepsilon}{d \sin \Theta} \begin{bmatrix} b_1 & -w & b_2 \\ -c^2 w & b_1 & -c^2 \\ c^2 b_2 & -c^2 & b_1 \end{bmatrix},$$

$$N = -\frac{\varepsilon}{d} \begin{bmatrix} a_1 F_1 - F_2 + F_3 a_2 \\ -c^2 F_1 + a_1 F_2 + c^2 a_2 \Omega/\varepsilon \\ c^2 a_2 F_1 + a_1 F_3 + c^2 \Omega/\varepsilon \end{bmatrix},$$

$$d = a_1^2 - c^2 a_2^2 - c^2, \quad b_1 = a_1 w + c^2 a_2, \quad b_2 = a_1 + a_2 w,$$

$$F_1 = 2u + vctg\Theta, \quad F_2 = uv - w^2 ctg\Theta, \quad F_3 = uw + vw ctg\Theta,$$

$$\Omega = a_2 v_s + w_s - \frac{\varepsilon}{\sin \Theta} v_\varphi.$$

Following section 1, we carry out discretization over t, φ:

$$\xi \mathbf{U}_j^{k+1}(s) = \frac{1}{\Delta} (M_1)_j^{(k)} (\mathbf{U}_j^{k+1}(s) - \mathbf{U}_j^k(s)) +$$

$$+ \frac{1}{2h_\varphi} (M_2)_j^{(k)} (\mathbf{U}_{j+1}^{k+1}(s) - \mathbf{U}_{j-1}^{k+1}(s)) + (N)_j^{(k)},$$

$$j = 1, 2, ..., J - 1, k = 0, 1, ... \tag{2.8}$$

Here $\xi = \frac{\partial}{\partial s}$, $\mathbf{U}_j^k(s) = \mathbf{U}(k\Delta, s, (j-1)h_\varphi)$, Δ, h_φ are the discretization steps; $\pi = (J-1)h_\varphi$. We note also that the index k in parentheses means averaging of the sought functions which enter the entries of the matrices $M_{1,2}$, the vector N:

$$(\circ)^{(k)} = \frac{(\circ)^{k+1} + (\circ)^k}{2}.$$

By nonlinearity of the initial problem, calculations at $(k+1)$-th layer actually must be carried out by iterations. At $\varphi = 0, \pi, j = 1, J$ we have

$w_j^k = 0$. We present the second and fourth equations from (2.1) at $\varphi = 0, \pi$ as a system :

$$(\mathbf{W}_j)_s = (M_0)_j(\mathbf{W}_j)_t + (N_0)_j,$$

and after discretization over t:

$$\xi \mathbf{W}_j^{k+1}(s) = \frac{1}{\Delta}(M_0)_j^{(k)}(\mathbf{W}_j^{k+1}(s) - \mathbf{W}_j^k(s)) + (N_0)_j^{(k)},$$

$$j = 1, J, k = 0, 1, ... , \tag{2.9}$$

where

$$\mathbf{W} = \begin{bmatrix} P \\ v \end{bmatrix}, \quad \mathbf{W}_1^k(s) = \mathbf{W}(k\Delta, s, 0), \quad \mathbf{W}_J^k(s) = \mathbf{W}(k\Delta, s, \pi),$$

$$M_0 = \frac{\varepsilon}{c^2 - b^2} \begin{bmatrix} b & -1 \\ -c^2 & b \end{bmatrix}, \quad b = v - s\frac{\partial \Theta_s}{\partial t},$$

$$N_0 = \frac{\varepsilon}{c^2 - b^2} \begin{bmatrix} b(2u + vctg\Theta + \frac{1}{\sin\Theta}w_\varphi) - uv \\ buv - c^2(2u + vctg\Theta + \frac{1}{\sin\Theta}w_\varphi) \end{bmatrix}.$$

Consequently, for the vector $\mathbf{Y}$:

$$\mathbf{Y} = \begin{bmatrix} \mathbf{W}_1^{k+1} \\ \mathbf{U}_2^{k+1} \\ \cdots \\ \mathbf{U}_{J-1}^{k+1} \\ \mathbf{W}_J^{k+1} \end{bmatrix}$$

we obtain a system of ordinary differential equations

$$\xi \mathbf{Y} = \tilde{A}\mathbf{Y} + \tilde{F}. \tag{2.10}$$

Here $\tilde{A}$ is a square matrix of order $L = 3J - 2, \tilde{F}$ is a vector of the same dimension. Elements of the matrix $\tilde{A}$ and the vector $\tilde{F}$ can be easily found via (2.8), (2.9).

We write down boundary conditions at $s = 0, 1$ for system (2.10). Thus, by virtue of (2.3), we have:

$$v_j^k(0) = 0, \quad j = 1, ..., J, \quad k = 0, 1, ...,$$

that is

$$\tilde{C}\mathbf{Y}(0) = 0, \tag{2.11}$$

where $\tilde{C}$ is a matrix of order $J \times L$:

$$\tilde{C} = \begin{bmatrix} 0 & 1 & 0 & 0 & \cdots & 0 & 0 & 0 & 0 & 0 \\ 0 & 0 & 0 & 1 & \cdots & 0 & 0 & 0 & 0 & 0 \\ \cdots & \cdots & \cdots & \cdots & \cdots & \cdots & \cdots & \cdots & \cdots & \cdots \\ 0 & 0 & 0 & 0 & \cdots & 0 & 1 & 0 & 0 & 0 \\ 0 & 0 & 0 & 0 & \cdots & 0 & 0 & 0 & 0 & 1 \end{bmatrix}.$$

From (2.2) at $s = 1$ we come to

$$(\lambda_1)_j^k P_j^{k+1}(1) - v_j^{k+1}(1) = (\lambda_1)_j^k P_j^k(1) - v_j^k(1),$$

$$(\lambda_2)_j^k P_j^{k+1}(1) + w_j^{k+1}(1) = (\lambda_2)_j^k P_j^k(1) + w_j^k(1). \tag{2.12}$$

Here $j = 2, \cdots, J - 1$,

$$\lambda_1 = \frac{\gamma p G'(p)}{\sqrt{1 + a_2^2(1)}}, \quad \lambda_2 = a_2(1)\lambda_1,$$

$$G'(p) = \hbar \frac{(3 - \gamma)V_\infty + (\gamma + 1)V}{4G_1(p)}.$$

At $j = 1, J$ we have the following relation instead of (2.12):

$$(\lambda_0)_j^k P_j^{k+1}(1) - v_j^{k+1}(1) = (\lambda_0)_j^k P_j^k(1) - v_j^k(1), \tag{2.13}$$

where $\lambda_0 = \gamma p G'(p)$. In a view of (2.12), (2.13) we obtain

$$\tilde{B}\mathbf{Y}(1) = \tilde{\varphi}, \tag{2.14}$$

where $\tilde{B}$ is a matrix of order $(2J-2) \times L$, $\tilde{\varphi}$ is of order $2J-2$. Elements of the matrix $\tilde{B}$ and the vector $\tilde{\varphi}$ can be found with the help of relations (2.12), (2.13). We note also that another ways to obtain boundary conditions at $s = 1$ for system (2.10) are also possible.

Then, problem (2.10), (2.11), (2.14) is solved by the orthogonal sweep method [15].

The functions u and Ψ are found by the running count scheme.

$$\Psi_{ij}^{k+1} = \frac{\Psi_{ij}^k - d_1 \Psi_{i+1,j}^{k+1} + d_2 \Psi_{i,j-1}^{k+1}}{d}, \quad i = 2, ...I - 1, j = 2, ...J - 1;$$

$$\Psi_{i1}^{k+1} = \Psi_{I1}^{k+1}, \quad i = 1, ...I - 1;$$

$$\Psi_{1j}^{k+1} = \Psi_{I1}^{k+1}, \quad j = 2, ...J - 1;$$

$$\Psi_{iJ}^{k+1} = \Psi_{IJ}^{k+1}, \quad i = 1, ...I - 1;$$

$$u_{ij}^{k+1} = \frac{u_{ij}^k - d_1 u_{i+1,j}^{k+1} + d_2 u_{i,j-1}^{k+1} + \Delta[(v^2)_{(i),(j)}^{(k)} + (w^2)_{(i),(j)}^{(k)}]}{\tilde{d}},$$

$$i = 2, \cdots I - 1, \quad j = 2, ..., J - 1;$$

$$u_{ij}^{k+1} = \frac{u_{ij}^k - \tilde{d}_1 u_{i+1,j}^{k+1} + \Delta(v^2)_{(i),(j)}^{(k)}}{1 - \tilde{d}_1},$$

$$i = 1, \cdots, I - 1, \quad j = 1, J.$$

Here

$$\Psi_{ij}^k = \Psi(k\Delta, (i-1)h_s, (j-1)h_\varphi), \quad i = \overline{1, I},$$

$$(I - 1)h_s = 1, \quad d_1 = \frac{\Delta(\tilde{\Delta})_{(i)(j)}^{(k)}}{h_s}, \quad d_2 = \frac{\Delta(w/\sin\Theta)_{(i)(j)}^{(k)}}{h_\varphi},$$

$$\tilde{d} = 1 - d_1 + d_2; \quad \tilde{d}_1 = \Delta(b/\varepsilon)_{(i),j}, \quad j = 1, J;$$

$$\Psi_{Ij}^k = P_{Ij}^k + \ln\frac{V_\infty(hp_\infty + p_{Ij}^k)}{(hp_{Ij}^k + p_\infty)}, \quad j = 1, ..., J;$$

$$u_{Ij}^{k+1} = (u_\infty)_{(j)}^{(k)}, \quad j = 1, \cdots, J;$$

$$(\circ)_{(i)} = \frac{(\circ)_{j+1} + (\circ)_i}{2}, \quad (\circ)_{(j)} = \frac{(\circ)_j + (\circ)_{j-1}}{2}.$$

Then, using the last relation from (2.2), we obtain $\Theta_s(t, \varphi)$:

$$(\Theta_s)_j^{k+1} = (\Theta_s)_j^k + \Delta\{v_\infty - w_\infty a_2(1) + G_1(p)\sqrt{1 + a_2^2(1)}\}_{(j)}^{(k)}.$$

Remark 2.1. Initial data at $\alpha = 0$ were given by simple analytic formulae. At $\alpha \neq 0$ the known functions for less values of α serve as initial data.

Numerical calculations were carried out with the help of the algorithm from above. Since the initial data were assigned roughly on the first steps we used iterations, then we calculated without iterations up to establishment. Establishment was considered to be reached when

$$\sum_{j=1}^{J}\left\{|(\Theta_s)_j^{k+1} - (\Theta_s)_j^k| + \left|\left(\frac{\partial}{\partial t}\Theta_s\right)_j^{k+1} - \left(\frac{\partial}{\partial t}\Theta_s\right)_j^k\right|\right\} < 10^{-6}.$$

In calculations

$$M_\infty = 2, 7; \quad \Theta_b = 10°, 30°.$$

At $\alpha = 0$, $M_\infty = 7$, $\Theta_b = 10°$, $30°$ we have reached establishment for 15 time steps, at $\alpha = 2$, $M_\infty = 2$, $\Theta_b = 10°$, $30°$ for 30 time steps. The calculated values of parameters completely coincided with values in the table from [16].

With increasing of α a number of time steps also increased. Calculated parameters coincided with values in the table from [17] with high accuracy.

Remark 2.2. Results from [17] were used in this section.

Conclusion

Thus we have considered some problems of mathematical simulation (in particular, mathematical simulation in gas dynamics.)

Special attention was paid to discussion of two main, from our point of view, principles of mathematical simulation: the principle of simultaneous investigation of the mathematical and calculation models of a phenomenon and the principle of adequacy between the calculation model and the initial mathematical problem.

Speaking generally, the second principle follows the first one, but we distinguish it because of its great practical importance. The matter is that many researches support the opinion that the calculation model, since it has been formulated, can be investigated apart from the initial mathematical model. It is extremely serious fallacy. Then, while investigating stabiliity of difference schemes used in formulation of calculation models, researches often restrict themselves to investigation of the difference Cauchy problem only.

We assume that, even if a researcher does not strictly follow the principle of simultaneity nevertheless, he or she must follow the principle of adequacy, possibly not in the whole. It may be the following:

1) Usage of a priori information about the solution to the initial mathematical problem while constructing the calculation model.

2) Investigation of stability of the constructed calculation model with account of boundary conditions (at least at the linear level).

Corresponding examples are given in this paper.

Not going into the detailed analysis of the papers in which a priori information about the solution to the initial mathematical model is used, we would like to point at the papers [18,19] which conceptually are very close to the present lectures. The essence of these papers is to use analytic properties of solutions to corresponding problems while constructing calculation models to obtain the numerical solution to boundary value problems for Laplace equation. Infinite differentiability of solution, certain asymptotics in corner points of the boundary, certain growth of derivatives with growth of the order of differentiation can be considered as the mentioned properties. Account of such asymptotic properties allows to solve problems with higher accuracy without increasing of expenditures.

References

1. Courant, R., Friedrichs, K. and Levy, H.(1940) On difference equations of mathematical physics, *Advantages in Mathematics (Uspehi mat.nauk)* 8, 125–160.
2. Godunov, S.K. (1979) *Equations of mathematical physics*, Nauka, Moscow.
3. Ladyzhenskaja, O.A. (1973) *Boundary problems of mathematical physics*, Nauka, oscow.
4. Blokhin, A.M. (1986) *Energy integrals and their applications to problems of gas dynamics*, Nauka, Novosibirsk.
5. Volpert, A.I. and Hudyaev, S.I. (1972) Cauchy problem for composite systems of nonlinear differential equations, *Mathematical collection (Matematicheskij sbornik)* v. 87 (129), 504–528.
6. Blokhin, A.M. and Alajev R.D. (1984) On stability of modified difference scheme of McCormack for symmetric t-hyperbolic system, *Nonclassical equations of mathematical physics (Neklassicheskije uravnenija matematicheskoj fiziki)*, 24–42.
7. Blokhin, A.M., Alajev, R.D., and Druzhinin, I.Yu. (1986) Stability of explicit difference schemes for symmetric t-hyperbolic systems. *Proceedings of Sobolev's seminar. Novosibirsk. Institute of mathematics of Russian Academy of Sciences, Siberian Branch* N. 1, 26–39.
8. Blokhin, A.M. (1988) Application of difference analogs of dissipative energy integrals to investigations of stability of difference schemes, *Calculation problems in mathematical physics (Vychislitel'nyje problemy v zadachah mat. fiziki). Proceedings of Institute of mathematics of Russian Academy of Sciences, Siberian Branch, Novosibirsk* N. 11, 67–93.
9. Godunov, S.K., Ryaben'kij, V.S. (1962) *Introduction into theory of difference schemes*, Fizmatgiz, Moscow.
10. Blokhin, A.M., Druzhinin, I.Yu. (1988) Construction of difference examples of instability in problem on shock fitting, *Simulation in mechanics* 2 (19), 22–31.
11. Blokhin, A.M. (1990) Some topics from mathematical simulation of gas dynamics problems, *Numerical analysis and Mathematical modeling, Banach Center Publications* 24, 455–463.
12. Blokhin, A.M. and Druzhinin I.Yu. (1989) Influence of boundary conditions on stability of certain difference splitting scheme, *Embedding theorems and their applications to problems of mathematical physics*, 38–47.
13. Rihtmajer, R. and Morton, K. (1972) *Difference methods for solution of boundary-value problems*, Mir, Moscow.
14. Blokhin, A.M. and Zimmerman, V.R. (1992) Investigation of difference-differntial model of linear mixed problem on supersonic flowing aroud wedge, *Calculation problems in mathematical physics (Vychislitel'nyje problemy v matematicheskoj fizike). Proceedings of Institute of Mathematics of USSR Academy of Sciences, Siberian Branch. Novosibirsk*, 78–94.
15. Kuznetsov, S.V. (1985) Solution of boundary problems for ordinary differntial equations, *Calculations methods of linear algebra. Proceedings of Institute of mathematics of USSR Academy of Sciences, Siberian Branch* 6, 1985, 85–110.
16. Babenko, K.I, Voskresnskij, G.P., Lubimov, A.I., and Rusanov, V.V. (1964) *Spatial flowing around smooth solids by perfect gas*. Nauka, Moscow.
17. Blokhin, A.M., Pozdeev, A.A., and Zimmerman, V.R. (1992) Methods of lines for equations of gas dynamics: theoretical ground and calculation experiments, *Calculation problems in mathematical physics (Vychislitel'nyje problemy v zadachah matematicheskoj fiziki). Proceedings of Institute of mathematics of USSR Academy of Sciences, Siberian Branch* 22, 112-156.
18. Belyh, V.N. (1988) Calculation algorythms without saturation in nonstationary problems of hydrodynamics in perfect liquid with free boundaries, *Calculation problems in mathematical physics (Vychislitel'nyje problemy v zadachah matematicheskoj*

fiziki) Proceedings of Institute of mathematics of USSR Academy of Sciences, Siberian Branch. Novosibirsk **11** 3–67.

19. Vaskevich, V.L. (1989) On approoximate solution of Dirichle problem in composite spatial domains, *Numerical analysis (Chislennyj analiz). Proceedings of Institute of mathematics of USSR Academy of Sciences, Siberian Branch. Novosibirsk* /bf 15, 93–126.

ERROR ESTIMATES IN LINEAR SYSTEMS

C. BREZINSKI
Laboratoire d'Analyse Numérique et d'Optimisation
Université des Sciences et Technologies de Lille
59655-Villeneuve d'Ascq cedex, France
e-mail: `Claude.Brezinski@univ-lille1.fr`

Let us consider the system of p real linear equations in p unknowns $Ax = b$. When solving it by a direct method, an approximate solution is obtained and, when solving it by an iterative procedure, a sequence of approximate solutions is constructed. We will denote by x^* an approximation of $x = A^{-1}b$. Usually the quality of the approximate solution is judged by the norm of the residual vector $r = b - Ax^*$. We have

$$\frac{1}{\|A^{-1}\|} \leq \frac{\|r\|}{\|e\|} \leq \|A\| \tag{1}$$

where $e = x - x^*$ is the error. Thus, if $\|A\|$ and $\|A^{-1}\|$ are both close to 1, $\|r\|$ is a good estimate of $\|e\|$. However, this case seldom occurs and this is why we will now propose various procedures for estimating the error e. These estimates are valid for any matrices and for approximations of x obtained by any method, direct or iterative. They do not need estimates of the largest and smallest eigenvalues of the matrix and they provide lower and upper bounds for the error.

Although the error and the residual are related by $r = Ae$, it is not possible to compute the error from the residual. However, we will use this relation for obtaining estimates of the Euclidean norm of the error. More details can be found in [3].

Let us set $c_{2i-1} = (r, A^i r)$ and $c_{2i} = (A^i r, A^i r)$. We have $(e, e) = (A^{-1}r, A^{-1}r) = c_{-2}$. Thus, the problem of computing (e, e) consists of estimating c_{-2}. This question was already treated in a number of papers [6, 7, 10, 11, 12, 14] where bounds on c_{-2} were obtained by quadrature rules in the case where the matrix A is symmetric positive definite and the conjugate gradient method or a method minimizing a quadratic functional is used.

H. Bulgak and C. Zenger (eds.), Error Control and Adaptivity in Scientific Computing, 65–74.
© 1999 *Kluwer Academic Publishers. Printed in the Netherlands.*

In this paper, c_{-2} will be estimated from c_0, c_1 and c_2. Our approach holds for arbitrary matrices and methods. It is based on the extrapolation of the sequence (c_n), $n = 0, 1, \ldots$ at the point $n = -2$.

In the first Section, we will explain how to derive our estimates. The formulae obtained will be analyzed in the second Section where bounds for the norm of the error will also be given. The last Section is devoted to numerical examples. More complete results and other estimates are given in [3].

1. Derivation of the estimates

For deriving estimates of the norm of the error, let us consider the singular value decomposition (SVD) of the matrix A

$$A = U\Sigma V^T$$

with $UU^T = VV^T = I$ and $\Sigma = \text{diag}(\sigma_1, \ldots, \sigma_p)$ with $\sigma_1 \geq \sigma_2 \geq \cdots \geq \sigma_p > 0$. Let $u_1, \ldots, u_p$ and $v_1, \ldots, v_p$ denote the columns of the matrices U and V respectively. We have, for an arbitrary vector y,

$$Ay = \sum_{i=1}^{p} \sigma_i (v_i, y) u_i$$

$$A^{-1}y = \sum_{i=1}^{p} \sigma_i^{-1} (u_i, y) v_i.$$

It follows immediately

$$c_0 = (r, r) = (U^T r, U^T r) = \sum_{i=1}^{p} (u_i, r)^2 \tag{2}$$

$$= (V^T r, V^T r) = \sum_{i=1}^{p} (v_i, r)^2 \tag{3}$$

$$c_1 = (r, Ar) = \sum_{i=1}^{p} \sigma_i (u_i, r)(v_i, r) \tag{4}$$

$$c_2 = (Ar, Ar) = \sum_{i=1}^{p} \sigma_i^2 (v_i, r)^2 \tag{5}$$

$$c_{-1} = (r, A^{-1}r) = \sum_{i=1}^{p} \sigma_i^{-1} (u_i, r)(v_i, r) \tag{6}$$

$$c_{-2} = (A^{-1}r, A^{-1}r) = (e, e) = \sum_{i=1}^{p} \sigma_i^{-2} (u_i, r)^2. \tag{7}$$

So, the norm of the error can be computed by (7). However, this formula requires the knowledge of all the σ_i and all the u_i appearing in the sum.

Approximations of c_{-2} (respectively c_{-1}) could be obtained by using the a formula similar to (7) (resp. (6)) but with only one term in each sum. These terms will be estimated by considering that, c_0, c_1 and c_2 have the same form as (2-5) with only one term in each sum. Thus, we will look for α, β and σ satisfying the interpolation conditions

$$\left.\begin{array}{rcl} c_0 &=& \alpha^2 = \beta^2 \\ c_1 &=& \sigma\alpha\beta \\ c_2 &=& \sigma^2\beta^2. \end{array}\right\} \tag{8}$$

and then extrapolate for the value -2 (resp. -1) of the index. Thus, we see from (6) and (7) that c_{-1} and c_{-2} could be approximated by

$$c_{-1} = \sigma^{-1}\alpha\beta \quad \text{and} \quad c_{-2} = \sigma^{-2}\alpha^2.$$

Computing the unknowns α, β and σ from 3 of the interpolation conditions (8), we obtain several estimates of $c_{-2}^{1/2} = \|e\|$. Among these estimates, the most interesting one seems to be

$$e_3 = c_0/c_2^{1/2}.$$

The other estimates are

$$\begin{array}{rcl} e_1 &=& c_1^2/c_2^{3/2} \\ e_2 &=& c_0^{1/2}c_1/c_2 \\ e_4 &=& c_0^{3/2}/c_1 \\ e_5 &=& c_0^2 c_2^{1/2}/c_1^2. \end{array}$$

When A is symmetric positive definite, the quantity $c_{-1} = (r, A^{-1}r) = (e, Ae)$ is called the energy norm. From what precedes, we see that it can be estimated, for example, by c_0^2/c_1.

These formulae require the computation of Ar. However, in some methods such as Lanczos/Orthores, these products are already computed and, thus, the preceding estimates can be obtained for free. In other cases, they can only be computed from time to time.

It is also possible to construct approximations of c_{-2} by extrapolation formulae built by keeping a sum of terms of the form $\alpha^2\sigma^{-2}$ in (7) and the other formulae as well, see [3].

Remark 1
There are many iterative methods for solving linear systems where the iterates and the residuals are obtained recursively by formulae of the form

$$\begin{array}{rcl} x_{k+1} &=& x_k + \lambda_k z_k \\ r_{k+1} &=& r_k - \lambda_k A z_k \end{array}$$

where z_k is some vector and λ_k a parameter.

Instead of considering such methods, it is possible to compute the iterates x_k and the corresponding residuals r_k by an arbitrary iterative method and, then, to transform them into the new iterates y_k and the new corresponding residuals $\rho_k = b - Ay_k$ by formulae similar to the preceding ones, that is

$$
\begin{aligned}
y_k &= x_k + \lambda_k z_k \\
\rho_k &= r_k - \lambda_k A z_k.
\end{aligned}
$$

Under some assumptions (see [2]), the sequence $(\|\rho_k\|)$ converges faster than the sequence $(\|r_k\|)$. In such cases, y_k is usually a better approximation of the solution x than x_k. Thus, one can consider $y_k - x_k = \lambda_k z_k$ as a good approximation of the error $x - x_k$. Thus $|\lambda_k| \cdot \|z_k\|$ is a good estimate of $\|x - x_k\|$. On the other hand, as explained in [4], estimates of the error form the basis for constructing convergence acceleration methods which shows the connection between both topics.

Let us now analyze the estimates $e_1, \ldots, e_5$.

2. Justification of the estimates

We have

$$
(r, Ar)^2 \le (Ar, Ar)(r, r) \le \|A\|^2 (r, r)^2
$$

that is

$$
0 \le c_1^2 \le c_0 c_2 \tag{9}
$$

and it follows

Theorem 1

$$
e_1 \le e_2 \le e_3 \le e_4 \le e_5.
$$

We have (with the indexes in the sums always running from 1 to p, the dimension of the system)

$$
\begin{aligned}
\sigma_p^2 \sum (v_i, r)^2 &\le \sum \sigma_i^2 (v_i, r)^2 \le \sigma_1^2 \sum (v_i, r)^2 \\
\sigma_1^{-2} \sum (u_i, r)^2 &\le \sum \sigma_i^{-2} (u_i, r)^2 \le \sigma_p^{-2} \sum (u_i, r)^2
\end{aligned}
$$

that is, using the fact that $\kappa = \|A\| \cdot \|A^{-1}\| = \sigma_1/\sigma_p$,

$$
\kappa^{-2} c_0^2 \le c_2(e, e) \le \kappa^2 c_0^2. \tag{10}
$$

From the inequalities (9) and (10), we immediately obtain

Theorem 2

$$
\kappa^{-1} \le \frac{e_3}{\|e\|} \le \kappa. \tag{11}
$$

Since $1 \in [\kappa^{-1}, \kappa]$, e_3 is a good approximation of $\|e\|$ if A is well-conditioned. As we will see from the numerical examples, e_3 can also be a quite good estimate of $\|e\|$ even if κ is large. For the other estimates, similar intervals can be obtained but without the guarantee that 1 belongs to them.

From the inequalities $\|Ar\| \leq \|A\| \cdot \|r\|$ and $\|r\| = \|A^{-1}Ar\| \leq \|A^{-1}\| \cdot \|Ar\|$, we immediately obtain

$$\frac{\|r\|}{\|A\|} \leq e_3 \leq \|A^{-1}\| \cdot \|r\|.$$

So, e_3 and $\|e\|$ belong to the same interval (see (1)). Moreover, for a convergent iterative method, these inequalities show that the estimate e_3 also tends to zero.

If A is orthogonal, then $\kappa = 1$ and $\|e\| = e_3$.

3. Comparisons

Let us now compare the intervals given by (1) for $\|r\|/\|e\|$ with those given by (11) for $e_3/\|e\|$ in order to know if $\|r\|$ has a chance to be a good estimate of $\|e\|$. We remind that $1 \in [\kappa^{-1}, \kappa]$. For simplicity, we set $m_1 = 1/\|A^{-1}\|$ and $m_2 = \|A\|$. Several cases can arise

1. $m_2 \leq \kappa^{-1}$
   that is $\kappa \leq \|A\|^{-1}$. The two intervals are disjoint and $\|r\|$ is too small.
2. $m_1 \leq \kappa^{-1} \leq m_2 \leq \kappa$
   These conditions are equivalent to $\|A\| \leq 1 \leq \|A^{-1}\|$ and $\|A\|^{-1} \leq \kappa$. Thus $m_2 \leq 1$ and $\|r\|$ is too small.
3. $m_1 \leq \kappa^{-1} \leq \kappa \leq m_2$
   These conditions give $\|A\|$ and $\|A^{-1}\|$ both smaller or equal to 1, which is only possible if both norms are equal to 1.
4. $\kappa^{-1} \leq m_1 \leq \kappa \leq m_2$
   These inequalities are equivalent to $\|A^{-1}\| \leq 1 \leq \|A\|$ and $\|A^{-1}\|^{-1} \leq \kappa$. Since $1 \leq m_1$, $\|r\|$ is too large. This case is similar to the case 2 after replacing A by A^{-1}.
5. $\kappa \leq m_1$
   that is $\kappa \leq \|A^{-1}\|^{-1}$. The two intervals are disjoint and $\|r\|$ is too large. This case is the reverse of case 1 after replacing A by A^{-1}.
6. $\kappa^{-1} \leq m_1 \leq m_2 \leq \kappa$
   These inequalities lead to $\|A\|$ and $\|A^{-1}\|$ both greater or equal to 1. In this case $1 \in [m_1, m_2]$ also and $\|r\|$ and e_3 are good candidates for estimating $\|e\|$. However, as we will see in the numerical examples given in the next Section, $\|r\|$ can be quite a bad estimate of $\|e\|$ while e_3 is a good one.

All these cases will now be illustrated.

4. Numerical results

Let us now give some numerical results for illustrating the error estimate e_3 and compare it with $\|r\|$ as discussed in the previous Section. For each example, the solid line represents $\|e\|$, the dot–dashed line is $\|r\|$ and the dashed line corresponds to e_3.

The solution x was always chosen randomly and, then, b was computed by $b = Ax$. The methods tested are Lanczos/Orthomin, the BiCGSTAB of Van der Vorst [16], the CGS of Sonneveld [15], the method of Jacobi and that of Gastinel [8, 9] (which is always convergent but often slowly) and consists of the iterations

$$x_{n+1} = x_n + \frac{(r_n, r_n)}{(A^T r_n, A^T r_n)} \, A^T r_n.$$

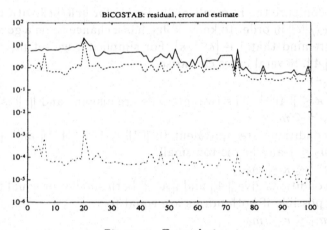

Figure 1. Example 1

All methods were initialized with $x_0 = 0$ and we took $y = r_0$ in Lanczos/Orthomin, in the BiCGSTAB and in the CGS. The matrices (except one) come out from the MATLAB matrix toolbox of Higham [13] and they all have dimension 100.

Example 1.

We consider the inverse of the matrix I+50*circul(100). Its condition number is 101.0408, $\|A\| = 4.0016 \cdot 10^{-4}$ and $\|A^{-1}\| = 2.5250 \cdot 10^5$. So, we are in the case 1 of Section 3.

The results obtained by the BiCGSTAB are given in the Figure 1.

71

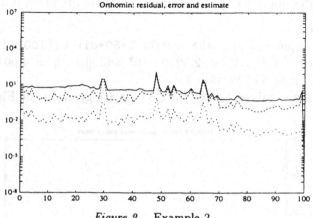

Figure 2. Example 2

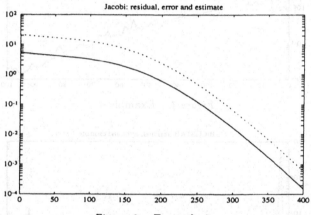

Figure 3. Example 3

Example 2.

In the second example, we took the inverse of the matrix `circul(100) +
parter(100)`. Its condition number is 107.7620, $\|A\| = 0.0213$ and $\|A^{-1}\| =
5.0501 \cdot 10^3$. So, we are in the case 2 of Section 3.

The results obtained by Lanczos/Orthomin are given in the Figure 2.

Example 3.

The matrix A is defined by

$$a_{ii} = 2, \quad i = 1,\ldots,100$$
$$a_{i,i+1} = 1, \quad i = 1,\ldots,99$$
$$a_{i+2,i} = 1, \quad i = 1,\ldots,98$$

all the other elements being zero. Its condition number is 2.9030, $\|A\| =
3.9977$ and $\|A^{-1}\| = 0.7262$. So, we are in the case 4 of Section 3.

The results obtained by the method of Jacobi are given in the Figure 3.

Example 4.

This example concerns the matrix I+50*circul(100). Its condition number is 101.0408, $\|A\| = 2.5250 \cdot 10^5$ and $\|A^{-1}\| = 4.0016 \cdot 10^{-4}$. So, we are in the case 5 of Section 3.

The results obtained by the BiCGSTAB are given in the Figure 4.

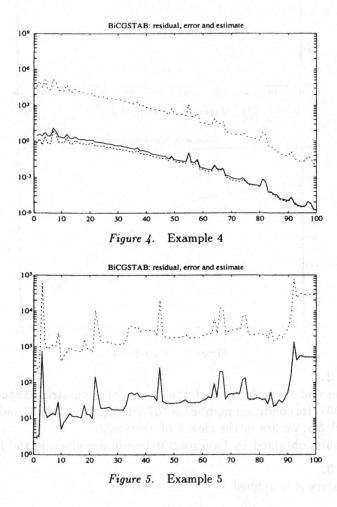

Figure 4. Example 4

Figure 5. Example 5

Example 5.

Let us now consider the matrix clement(100). Its condition number is $2.9408 \cdot 10^{15}$, $\|A\| = 99.9911$ and $\|A^{-1}\| = 2.941 \cdot 10^{13}$. So, we are in the case 6 of Section 3.

The results obtained by the BiCGSTAB are given in the Figure 5. As we can see, e_3 is a good approximation of $\|e\|$ while $\|r\|$ is not.

Example 6.

Let us now consider the matrix dingdong(100). Its condition number is 3.3162, $\|A\| = 1.5708$ and $\|A^{-1}\| = 2.1112$. So, we are again in the case 6 of Section 3.

The results obtained by the method of Gastinel are given in the Figure 6. Now, e_3 and $\|r\|$ are good approximations of $\|e\|$.

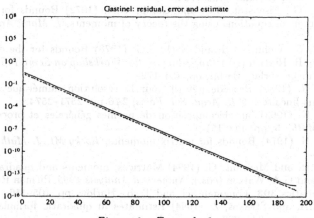

Figure 6. Example 6

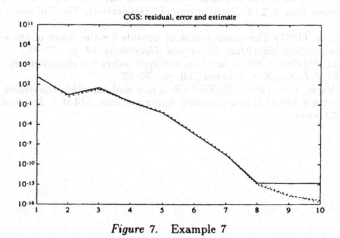

Figure 7. Example 7

Example 7.

For the matrix dingdong(100), the CGS gives the results of Figure 7. We have $\|A\| = 1.5708$ and $\|A^{-1}\| = 2.1112$ and, so, we are in the case 6 of Section 3. For this example, both $\|r\|$ and e_3 are good estimates of $\|e\|$.

74

References

1. Brezinski, C. (1988) A new approach to convergence acceleration methods, in A. Cuyt (ed.), *Nonlinear Numerical Methods and Rational Approximation*, Reidel, Dordrecht, pp. 373–405.
2. Brezinski, C. (1997) *Projection Methods for Systems of Equations*, North-Holland, Amsterdam.
3. Brezinski, C. Error estimates for the solution of linear systems, submitted.
4. Brezinski, C. Error estimates and convergence acceleration, this volume.
5. Brezinski, C. and Redivo Zaglia, M. (1991) *Extrapolation Methods. Theory and Practice*, North-Holland, Amsterdam.
6. Dahlquist, G., Eisenstat, S.C. and Golub, G.H. (1972) Bounds for the error in linear systems of equations using the theory of moments, *J. Math. Anal. Appl.*, **37**, pp. 151–166.
7. Dahlquist, G., Golub, G.H. and Nash, S.G. (1978) Bounds for the error in linear systems, in R. Hettich (ed.) *Proceedings of the Workshop on Semi-Infinite Programming*, Springer Verlag, Berlin, pp. 154–172.
8. Gastinel, N. (1958) Procédé itératif pour la résolution numérique d'un système d'équations linéaires, *C.R. Acad. Sci. Paris*, **246** pp. 2571–2574.
9. Gastinel, N. (1963) Sur-décomposition de normes générales et procédés itératifs, *Numer. Math.*, **5**, pp. 142–151.
10. Golub, G.H. (1974) Bounds for matrix moments, *Rocky Mt. J. Math.*, **4**, pp. 207–211.
11. Golub, G.H. and Meurant, G. (1994) Matrices, moments and quadrature, in D.F. Griffiths and G.A. Watson (eds.), *Numerical Analysis 1993*, Pitman Research Notes in Mathematics **303**, Longman Sci. and Tech., Harlow, pp. 105–156.
12. Golub, G.H. and Strakoš, Z. (1994) Estimates in quadratic formulas, *Numerical Algorithms*, **8**, pp. 241–268.
13. Higham, N.J. (1995) The test matrix toolbox for MATLAB (Version 3.0), *Numerical Analysis Report*, **276**, Departments of Mathematics, The University of Manchester.
14. Meurant, G. (1997) The computation of bounds for the norm of the error in the conjugate gradient algorithm, *Numerical Algorithms*, **16**, pp. 77–87.
15. Sonneveld, P.(1989) CGS, a fast Lanczos-type solver for nonsymmetric linear systems, *SIAM J. Sci. Stat. Comput.*, **10**, pp. 36–52.
16. van der Vorst, H.A. (1992) BiCGSTAB: a fast and smoothly converging variant of BiCG for the solution of nonsymmetric linear systems, *SIAM J. Sci. Stat. Comput.*, **13**, pp. 631–644.

ERROR ESTIMATES IN PADÉ APPROXIMATION

C. BREZINSKI
Laboratoire d'Analyse Numérique et d'Optimisation
Université des Sciences et Technologies de Lille
59655-Villeneuve d'Ascq cedex, France
e-mail: Claude.Brezinski@univ-lille1.fr

Let f be a formal power series

$$f(t) = c_0 + c_1 t + c_2 t^2 + \cdots.$$

We are looking for a rational function whose series expansion in ascending powers of the variable t coincides with that of f as far as possible.

Two main cases will be studied

- the denominator of the rational function is chosen arbitrarily and the numerator is computed in order to achieve the maximal order of approximation. Such approximants are called *Padé-type approximants*,
- both the numerator and the denominator are computed in order to obtain the maximal order of approximation. Such approximants are called *Padé approximants*.

All the intermediate cases can also be defined. The interested reader is referred to [4, 5, 6].

Padé approximants are very much used in many branches of applied sciences for obtaining approximations of functions only known by the first terms of their power series expansion. In practice, it is fundamental to be able to estimate the error of the approximants in order, in particular, to choose the degrees of the numerator and the denominator.

1. Padé–type approximants

Let c be the linear functional on the vector space of polynomials defined by

$$c(x^i) = \left\{ \begin{array}{ll} c_i, & i = 0, 1, \ldots \\ 0, & i < 0 \end{array} \right.$$

where the c_i's are the coefficients of the series f.

H. Bulgak and C. Zenger (eds.), Error Control and Adaptivity in Scientific Computing, 75–85.

The following result is a basic one

Lemma 1

$$f(t) = c\left(\frac{1}{1 - xt}\right).$$

Proof.

The result is obvious after expanding $1/(1 - xt)$ into a power series, applying c and using its linearity. ∎

Thus, the problem is now to compute an approximation of $c(1/(1 - xt))$. This is exactly a numerical quadrature problem. Indeed, a numerical quadrature formula allows to obtain an approximation of

$$c(g) = \int_a^b g(x)\omega(x)\,dx.$$

In our case, the linear functional is no longer represented by an integral with respect to a positive weight function ω, but c is completely defined by the knowledge of its moments c_i and the function g is $g(x) = 1/(1 - xt)$. A numerical quadrature formula consists of replacing the function g by its interpolation polynomial R_k at k points of $[a, b]$ and integrating it. This is exactly what we will do now. So, the first step is to give the expression of the interpolating polynomial of g. We have the

Lemma 2

Let $v_k(x) = (x - x_1)^{k_1} \cdots (x - x_n)^{k_n}$ with $k_1 + \cdots + k_n = k$ be an arbitrary polynomial of degree k.

The polynomial

$$R_k(x) = \frac{1}{1 - xt}\left(1 - \frac{v_k(x)}{v_k(t^{-1})}\right)$$

is the Hermite interpolation polynomial of $1/(1 - xt)$ at the zeros of v_k, that is

$$R_k^{(j)}(x_i) = \frac{d^j}{dx^j}(1 - xt)^{-1}|_{x=x_i}$$

for $i = 1, \ldots, n$ and $j = 0, \ldots, k_i - 1$.

Proof.

Let us first prove that R_k is a polynomial of degree $k - 1$ in x. We can write $v_k(x) = a_0 + a_1 x + \cdots + a_k x^k$ with $a_k \neq 0$. Thus

$$\frac{v_k(t^{-1}) - v_k(x)}{1 - xt} = a_1 t^{-1} + a_2 t^{-2}(1 + xt) + \cdots + a_k t^{-k}(1 + xt + \cdots + x^{k-1} t^{k-1}).$$

Let us now show that R_k satisfies the interpolation conditions of the lemma. We have

$$R_k^{(j)}(x) = \frac{d^j}{dx^j}(1 - xt)^{-1} - \frac{1}{v_k(t^{-1})}\frac{d^j}{dx^j}[v_k(x)(1 - xt)^{-1}].$$

Since $v_k^{(j)}(x_i) = 0$ for $i = 1, \ldots, n$ and $j = 0, \ldots, k_i - 1$, the second term in the right hand side of the preceding relation is zero for the same values of the indexes, which proves the result. ∎

An approximation of $f(t)$ can be obtained by "integrating" this interpolation polynomial, that is by computing $c(R_k(x))$. We have

$$c(R_k(x)) = \frac{1}{v_k(t^{-1})} \, c\left(\frac{v_k(t^{-1}) - v_k(x)}{1 - xt}\right).$$

Let us set

$$w_k(t) = c\left(\frac{v_k(x) - v_k(t)}{x - t}\right)$$

where c acts on x, t being a parameter. It is easy to see that w_k is a polynomial of degree $k - 1$ in t and that

$$c(R_k(x)) = \frac{\widetilde{w_k}(t)}{\widetilde{v_k}(t)}$$

where $\widetilde{v_k}(t) = t^k v_k(t^{-1})$ and $\widetilde{w_k}(t) = t^{k-1} w_k(t^{-1})$. Thus, $c(R_k(x))$ is a rational function with a numerator of degree at most $k - 1$ and a denominator of degree at most k. Such a rational function is called a *Padé–type approximant*, it is denoted by $(k - 1/k)_f(t)$, v_k is called the *generating polynomial* of the approximant and w_k the *polynomial associated* to v_k.

The main result is the following one

Property 1

$$(k - 1/k)_f(t) = f(t) + \mathcal{O}(t^k).$$

Proof.

We have

$$c(R_k(x)) = c\left(\frac{1}{1 - xt}\right) - \frac{t^k}{\widetilde{v_k}(t)} c\left(\frac{v_k(x)}{1 - xt}\right) = f(t) + \mathcal{O}(t^k). \blacksquare \qquad (1)$$

This result is similar to the property of an interpolation quadrature formula with k points to be exact on the vector space of polynomials of degree less than or equal to $k - 1$.

From $(k - 1/k)$, it is possible to construct Padé–type approximants with arbitrary degrees in the numerator and in the denominator but this case will not be considered in the sequel.

2. Padé approximants

Let us come back to the error formula (1). Lemma 1 leads us to the following expression for the error of $(k - 1/k)$

$$(k - 1/k)_f(t) = f(t) - \frac{t^k}{\widetilde{v_k}(t)} c\left(v_k(x)\left(1 + xt + \cdots + x^{k-1}t^{k-1} + \frac{x^k t^k}{1 - xt}\right)\right). \qquad (2)$$

If the polynomial v_k is chosen so that $c(v_k) = 0$, then the first term in (2) will be zero and the order of approximation will become $k+1$ (instead of k). If we impose, in addition, that $c(xv_k) = 0$, then the second term in (2) will be zero and the order of approximation will become $k+2$. Let us continue. v_k is an arbitrary polynomial of degree k but, in fact, it only depends on k arbitrary quantities since a rational function is defined apart from a multiplying factor. Moreover, v_k is monic. Thus, it is impossible to impose more than k conditions. So, let us choose v_k such that

$$c(x^i v_k(x)) = 0 \quad \text{for} \quad i = 0, \ldots, k-1. \tag{3}$$

If v_k satisfies these conditions, then it is entirely determined and the first k terms in (2) disappear and the order of approximation becomes $2k$. In that case, the rational function $c(R_k(x))$ is called a *Padé approximant* of f and it is denoted by $[k - 1/k]_f(t)$.

Such a choice for v_k is, in fact, equivalent to a Gaussian quadrature formula. Indeed, the conditions (3) are similar to those defining a family of orthogonal polynomials and the result about the order of approximation of $[k - 1/k]$ is nothing else than the property of a Gaussian quadrature formula with k points to be exact on the space of polynomials of degree at most $2k - 1$ (then, the error is $\mathcal{O}(t^{2k})$).

It is possible to construct Padé approximants with arbitrary degrees in the numerator and in the denominator.

A family of polynomials satisfying the conditions (3) is called the family of *formal orthogonal polynomials* (FOP) with respect to the linear functional c. These polynomials will be denoted P_k instead of v_k (resp. $\tilde{P}_k$ instead of $\tilde{v}_k$). Such FOP satisfy most of the algebraic properties of the usual orthogonal polynomials and, in particular, their three–term recurrence relationship (if they exist) which can be used for computing recursively the sequence of Padé approximants ($[k - 1/k]$) [1].

3. Error estimation

As we saw, the Padé approximant $[k - 1/k]$ can be understood as a formal Gaussian quadrature formula for the function $(1 - xt)^{-1}$. On the other hand, the error in a Gaussian quadrature formula can be estimated by using the so–called *Kronrod procedure*. So, we will see now how to use this procedure for estimating the error of $[k - 1/k]$, an idea introduced in [2].

For that purpose, we will add n new interpolation points to the k zeros of P_k. In other terms, we will construct the Padé–type approximant $(n + k - 1/n + k)$ with the generating polynomial $v = P_k V_n$, where V_n is a polynomial of degree n whose zeros are the n new interpolation points. We set

$$W_n(t) = c\left(P_k(x)\frac{V_n(x) - V_n(t)}{x - t}\right)$$

where c acts on x, t being a parameter. The polynomial w associated to v is given by

$$w(t) = W_n(t) + Q_k(t)V_n(t)$$

where Q_k is the polynomial associated to P_k.

$(V_n(x) - V_n(t))/(x - t)$ is a polynomial of degree $n - 1$ in x. Thus, by the orthogonality property of P_k with respect to every polynomial of degree strictly less than k, W_n is identically zero if $n \leq k$ and, in this case, $(n + k - 1/n + k) \equiv [k - 1/k]$. So, we have to take $n \geq k + 1$ and then W_n will have the degree $n - k - 1$. Thus, the smallest possible value for n is $n = k + 1$, and this is the choice that will be now considered.

In order to have a Padé–type approximant with the highest possible order of approximation, we will, as in Kronrod procedure, choose V_{k+1} such that

$$c(x^i P_k(x)V_{k+1}(x)) = 0 \quad \text{for} \quad i = 0, \ldots, k.$$

We will thus obtain an approximant $(2k/2k + 1)$ satisfying $f(t) - (2k/2k + 1)_f(t) = \mathcal{O}(t^{3k+2})$, which is exactly the result of Kronrod quadrature formula to be exact on the space of polynomials of degree at most $3k + 1$.

The difference $(2k/2k + 1) - [k - 1/k]$ is a good estimation of the error $f - [k - 1/k]$. Indeed, the following result holds

Theorem 1
Let $(2k/2k + 1)$ be the Padé–type approximant of f constructed with the generating polynomial $v = P_k V_{k+1}$ where V_{k+1} satisfies

$$c(x^i P_k(x)V_{k+1}(x)) = 0 \quad \text{for} \quad i = 0, \ldots, k.$$

Then

$$\frac{f(t) - [k - 1/k]_f(t)}{(2k/2k + 1)_f(t) - [k - 1/k]_f(t)} = 1 + t^{k+2}\frac{1}{c(x^k P_k)}c\left(\frac{x^{k+1}P_k V_{k+1}}{1 - xt}\right).$$

From the practical point of view, it is not necessary to know the zeros of V_{k+1} (nor those of P_k), contrarily to the usual case of definite integrals, and there is no limitation to the use of the procedure if the zeros are complex and/or multiple and they no longer have to belong to the interval of integration (since no interval arises in Padé–type approximation).

If we set

$$V_{k+1}(x) = a_0 + \cdots + a_k x^k + x^{k+1}$$
$$P_k(x) = b_0 + \cdots + b_k x^k,$$

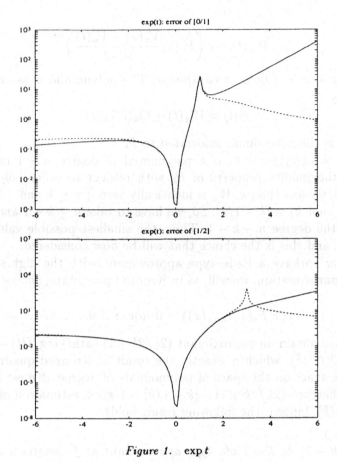

Figure 1. exp t

then the coefficients a_i are solution of the triangular system

$$
\begin{aligned}
c(v) &= a_k e_k + e_{k+1} = 0 \\
c(xv) &= a_{k-1} e_k + a_k e_{k+1} + e_{k+2} = 0 \\
&\vdots \qquad \vdots \\
c(x^k v) &= a_0 e_k + a_1 e_{k+1} + \cdots + a_k e_{2k} + e_{2k+1} = 0
\end{aligned}
$$

with

$$
e_i = c(x^i P_k) = b_0 c_i + \cdots + b_k c_{i+k}.
$$

This system is nonsingular if $e_k = c(x^k P_k) \neq 0$.

Let us give a numerical example concerning $f(t) = \exp t$. We have

$$[0/1] = \frac{1}{1-t} \qquad (2/3) = \frac{-11t^2 - 24t + 36}{(1-t)(7t^2 - 24t + 36)}$$

$$[1/2] = \frac{6+2t}{6-4t+t^2} \qquad (4/5) = \frac{379t^4/4 + 468t^3 - 630t^2 - 4200t + 15750}{(6-4t+t^2)(-62t^3 + 420t^2 - 1575t + 2625)}.$$

The results are displayed on Figures 1. The solid lines represent the error and the dashed lines the estimates for the error of $[0/1]$ and $[1/2]$, respectively. The ratios of the error by its estimation are shown on Figures 2. Obviously, the error is always zero at the point 0. So, this point has been suppressed in the following figures for avoiding infinity since we used a logarithmic scale.

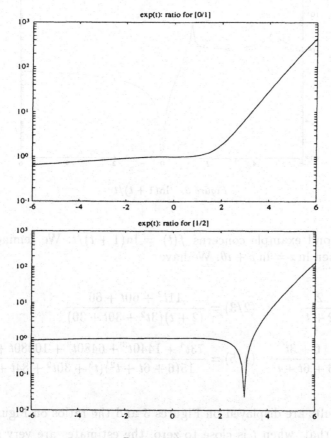

Figure 2. $\exp t$

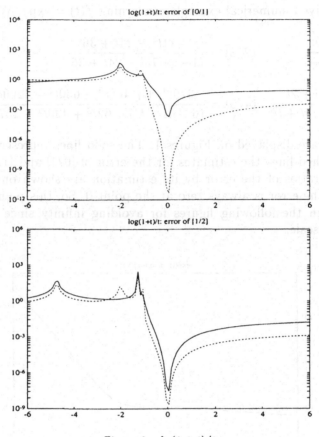

Figure 3. $\ln(1+t)/t$

The second example concerns $f(t) = \ln(1+t)/t$. We remind that, if $z = \rho e^{i\theta}$, then $\ln z = \ln \rho + i\theta$. We have

$$[0/1] = \frac{2}{2+t} \qquad (2/3) = \frac{11t^2 + 60t + 60}{(2+t)(3t^2 + 30t + 30)}$$

$$[1/2] = \frac{6+3t}{6+6t+t^2} \qquad (4/5) = \frac{73t^4 + 1440t^3 + 6480t^2 + 10080t + 5040}{15(6+6t+t^2)(t^3 + 30t^2 + 84t + 56)}.$$

The results are displayed on Figures 3 and the ratios on Figures 4.

We see that, when t is close to zero, the estimates are very good and even exact to several digits. Other numerical examples could be found in [2].

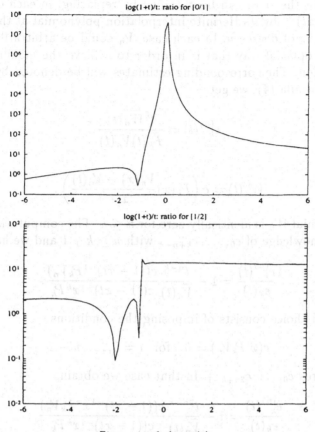

Figure 4. $\ln(1+t)/t$

4. Variants

Let us now describe cheaper (but less precise) variants of the procedure given above. More details can be found in [3].

Three theoretical expressions for the error of Padé approximants can be proved

$$e_k(t) = f(t) - [k-1/k]_f(t) = \frac{t^k}{\widetilde{P}_k(t)} \, c\left(\frac{P_k(t)}{1-xt}\right) \tag{4}$$

$$= \frac{t^{2k}}{\widetilde{P}_k(t)} \, c\left(\frac{x^k P_k(t)}{1-xt}\right) \tag{5}$$

$$= \frac{t^{2k}}{\widetilde{P}_k^2(t)} \, c\left(\frac{P_k^2(t)}{1-xt}\right). \tag{6}$$

Estimates for the error can be obtained by replacing, in each of these formulae, $(1 - xt)^{-1}$ by its Hermite interpolation polynomial at the zeros of a polynomial V_n of degree n. In each case, V_n could be arbitrarily chosen or chose in an optimal way that is in order to achieve the maximal order of approximation. The corresponding estimates will be denoted by $e_k^{(n)}(t)$.

Using formula (4), we get

$$e_k^{(n)}(t) = \frac{t^k \widetilde{W}_n(t)}{\widetilde{P}_k(t) \widetilde{V}_n(t)}$$

with

$$W_n(t) = c\left(P_k(x) \frac{V_n(x) - V_n(t)}{x - t} \right).$$

The polynomial W_n is identically zero for $n \leq k$. The computation of $e_k^{(n)}(t)$ needs the knowledge of $c_0, \ldots, c_{k+n-1}$ with $n \geq k + 1$ and we have

$$\frac{e_k^{(n)}(t)}{e_k(t)} = 1 - \frac{t^{n-k}}{\widetilde{V}_n(t)} \frac{c((1 - xt)^{-1} P_k V_n)}{c((1 - xt)^{-1} x^k P_k)}.$$

The optimal choice consists of imposing the conditions

$$c(x^i P_k V_n) = 0 \quad \text{for} \quad i = 0, \ldots, n - 1$$

which requires $c_0, \ldots, c_{2n+k-1}$. In that case we obtain

$$\frac{e_k^{(n)}(t)}{e_k(t)} = 1 - \frac{t^{2n-k}}{\widetilde{V}_n(t)} \frac{c((1 - xt)^{-1} x^n P_k V_n)}{c((1 - xt)^{-1} x^k P_k)}.$$

If $n = k + 1$ Kronrod's procedure is recovered.

Let us now use formula (5). We get

$$e_k^{(n)}(t) = \frac{t^{2k} \widetilde{W}_n(t)}{\widetilde{P}_k(t) \widetilde{V}_n(t)}$$

with

$$W_n(t) = c\left(x^k P_k(x) \frac{V_n(x) - V_n(t)}{x - t} \right).$$

The computation of $e_k^{(n)}(t)$ needs the knowledge of $c_0, \ldots, c_{2k+n-1}$ and we have

$$\frac{e_k^{(n)}(t)}{e_k(t)} = 1 - \frac{t^n}{\widetilde{V}_n(t)} \frac{c((1 - xt)^{-1} x^k P_k V_n)}{c((1 - xt)^{-1} x^k P_k)}.$$

The optimal choice consists of imposing the conditions

$$c(x^{i+k} P_k V_n) = 0 \quad \text{for} \quad i = 0, \ldots, n - 1$$

which requires $c_0, \ldots, c_{2n+2k-1}$. In that case we obtain

$$\frac{e_k^{(n)}(t)}{e_k(t)} = 1 - \frac{t^{2n}}{\widetilde{V}_n(t)} \frac{c((1-xt)^{-1}x^{n+k}P_k V_n)}{c((1-xt)^{-1}x^k P_k)}.$$

Finally, from formula (6), we get

$$e_k^{(n)}(t) = \frac{t^{2k}\widetilde{W}_n(t)}{\widetilde{P}_k^2(t)\widetilde{V}_n(t)}$$

with

$$W_n(t) = c\left(P_k^2(x)\frac{V_n(x) - V_n(t)}{x - t}\right).$$

The computation of $e_k^{(n)}(t)$ needs the knowledge of $c_0, \ldots, c_{2k+n-1}$ and we have

$$\frac{e_k^{(n)}(t)}{e_k(t)} = 1 - \frac{t^n}{\widetilde{V}_n(t)} \frac{c((1-xt)^{-1}P_k^2 V_n)}{c((1-xt)^{-1}P_k^2)}.$$

The optimal choice consists of imposing the conditions

$$c(x^i P_k^2 V_n) = 0 \quad \text{for} \quad i = 0, \ldots, n-1$$

which requires $c_0, \ldots, c_{2n+2k-1}$. In that case we obtain

$$\frac{e_k^{(n)}(t)}{e_k(t)} = 1 - \frac{t^{2n}}{\widetilde{V}_n(t)} \frac{c((1-xt)^{-1}x^n P_k^2 V_n)}{c((1-xt)^{-1}P_k^2)}.$$

All the procedures described in this paper can be extended to the various generalizations of Padé approximants which can be found in the literature [5, 6].

References

1. Brezinski, C. (1980) *Padé-Type Approximation and General Orthogonal Polynomials.* ISNM vol. 50, Birkhäuser, Basel.
2. Brezinski, C. (1988) Error estimate in Padé approximation, in M. Alfaro et al. (eds.), *Orthogonal Polynomials and their Applications*, LNM vol. 1328, Springer, pp. 1–19.
3. Brezinski, C. (1989) Procedures for estimating the error in Padé approximation, *Math. Comput.*, **53**, pp. 639–648.
4. Brezinski, C. (1991) Approximants de Padé, in J. Baranger (ed.), *Analyse Numérique*, Hermann, Paris, pp. 416–458.
5. Brezinski, C. and Van Iseghem, J. (1994) Padé approximations, in P.G. Ciarlet et J.L. Lions (eds.), *Handbook of Numerical Analysis*, vol. III, North-Holland, Amsterdam, pp. 47–222.
6. Brezinski, C. and Van Iseghem, J. (1995) A taste of Padé approximation, in A. Iserles (ed.), *Acta Numerica 1995*, Cambridge University Press, Cambridge, pp. 53–103.

ERROR ESTIMATES AND CONVERGENCE ACCELERATION

C. BREZINSKI
Laboratoire d'Analyse Numérique et d'Optimisation
Université des Sciences et Technologies de Lille
59655–Villeneuve d'Ascq cedex, France
e-mail: `Claude.Brezinski@univ-lille1.fr`

When a sequence (S_n) converges too slowly to its limit S, it can transformed into a new sequence (T_n) by a sequence transformation T. Under some assumptions, (T_n) converges to the same limit S and, under some additional assumptions, (T_n) converges faster than (S_n), that is, for a scalar sequence,

$$\lim_{n \to \infty} (T_n - S)/(S_n - S) = 0.$$

In that case T is said to accelerate the convergence of (S_n).

There are several ways of constructing sequence transformations. One of them is based on estimates of the error $S_n - S$. After having transformed (S_n) into the new sequence (T_n), a practical problem is to control the error $T_n - S$. This can be obtained by constructing intervals asymptotically containing the limit S. So, error estimates play a quite important role in the domain of convergence acceleration methods.

1. Construction of acceleration devices

Any scalar sequence transformation can always be written as

$$T_n = S_n + D_n, \quad n = 0, 1, \ldots$$

Thus, the sequence (T_n) converges faster than (S_n) if and only if

$$\lim_{n \to \infty} D_n/(S - S_n) = 0.$$

In that case, (D_n) is called a *perfect* estimation of the error. Thus, the construction of an acceleration device for a given sequence is equivalent to the knowledge of a perfect estimation of the error. If $D_n/(S - S_n)$ tends to

H. Bulgak and C. Zenger (eds.), Error Control and Adaptivity in Scientific Computing, 87–94.
© 1999 Kluwer Academic Publishers. Printed in the Netherlands.

$a \neq 1$ when n goes to infinity, (D_n) is called a *good* estimation of the error. In that case, let us set

$$D'_n = \frac{\Delta S_n}{\Delta D_n} D_n.$$

If $\exists \alpha < 1 < \beta$, $\exists N$ such that, $\forall n \geq N$, $(S_{n+1} - S)/(S_n - S) \notin [\alpha, \beta]$ (a sequence satisfying this assumption is called *nonlogarithmic*), then $\Delta S_n/\Delta D_n$ also tends to a and it follows that (D'_n) is a perfect estimation of the error. So, the transformation T defined by

$$T_n = S_n - \frac{\Delta S_n}{\Delta D_n} D_n, \quad n = 0, 1, \ldots$$

accelerates the convergence of (S_n). If $D_n = \Delta S_n$ we obtain

$$T_n = S_n - (\Delta S_n)^2/\Delta^2 S_n, \quad n = 0, 1, \ldots$$

which is Aitken Δ^2 process.

If $D_n/(S - S_n)$ has no limit, (D_n) is said to be a *bad* estimation of the error and it is, in general, impossible to accelerate the convergence of (S_n).

We will now explain how it is possible, from the classical convergence tests for sequences and series, to obtain perfect estimations of the error [3].

Before accelerating the convergence of a sequence, it is necessary to prove that it converges. This is usually done by using some convergence tests. We will see that these tests can also be used, if they are sufficiently sharp, for obtaining perfect estimations of the error. To be sharp enough, these tests must not only consist in an inequality (such as, for example, in d'Alembert test), but they must include an asymptotic information such as a comparison with a known sequence.

Let (S_n) be increasing and let (x_n) be a known increasing sequence with a known limit x. We set $R_n = S - S_n$, $r_n = x - x_n$ and $A_n = \Delta x_n/\Delta S_n = \Delta r_n/\Delta R_n$. We assume that $\exists A$ such that (A_n) converges to A. Under these assumptions, (S_n) converges and, $\forall n$,

$$r_n/A \ \leq R_n \ \leq r_n/A_n \quad \text{if } (A_n) \text{ is increasing}$$
$$r_n/A \ \geq R_n \ \geq r_n/A_n \quad \text{if } (A_n) \text{ is decreasing}.$$

Thus, in both cases, (r_n/A) and (r_n/A_n) are perfect estimations of the error and we obtain two sequences transformations for the acceleration of (S_n)

$$T_n \ = \ S_n - (x_n - x)/A, \quad n = 0, 1, \ldots$$
$$T_n \ = \ S_n - (\Delta S_n/\Delta x_n)(x_n - x), \quad n = 0, 1, \ldots$$

If $x_n = \Delta S_n$, the preceding test reduces to d'Alembert's and the second transformation is Aitken Δ^2 process. When $x_n = t^n$ with $0 < t < 1$, the test is Cauchy's. The choice $x_n = n\Delta S_n$ corresponds to the Raabe–Duhamel criterion and the second transformation is able to accelerate the convergence of some *logarithmic* sequences, that is sequences such that $(S_{n+1} - S)/(S_n - S)$ tends to 1 when n tends to infinity.

Let us now consider Kummer test. Let (a_n) be an auxiliary sequence such that

$$\lim_{n \to \infty} a_n \Delta S_{n-1} = 0.$$

We set

$$D_n = a_n \Delta S_{n-1} \Delta S_n / (a_n \Delta S_{n-1} - a_{n+1} \Delta S_n).$$

Under the assumptions that $\forall n, \Delta S_n > 0$ and $\exists \mu > 0$ such that

$$\lim_{n \to \infty} (a_n \Delta S_{n-1} - a_{n+1} \Delta S_n)/\Delta S_n = \mu,$$

it can be proved that (S_n) converges and, moreover, (D_n) is a perfect estimation of the error. This result can be generalized, see [3]. If $a_n = 1$, Kummer test is d'Alembert's and, if $a_n = n$, it is Raabe–Duhamel's.

Perfect estimations of the error can sometimes be obtained from integral tests for the convergence of series. If $S_n = \sum_{i=1}^{n} f(i)$ with, for all $x \geq 1$, $f(x) \geq 0$ and decreasing, then Cauchy's test says that, $\forall n$,

$$D_{n+1} \leq R_n = S - S_n \leq D_n = \int_n^\infty f(x)\, dx.$$

For example, if $f(x) = 1/x^2$, then $D_n = 1/n$ and, from the inequality $1/(n+1) < R_n < 1/n$, we see that $(1/n)$ is a perfect estimation of the error.

More results of the same type can be found in [3, 7, 8, 9].

Let us now assume that $\forall n, S_n = S + a_n D_n$ where (a_n) is an unknown sequence and (D_n) an error estimate. This case often arises in numerical analysis when, for example, it can be proved that $S_n - S = \mathcal{O}(D_n)$ or $S_n - S = o(D_n)$. According to the informations known on the sequences (a_n) and (D_n), various convergence acceleration processes can be constructed.

The first approach is due to Weniger [10] and it is based on the notion of annihilation difference operator, that is on an algebraic information on the sequence (a_n). Let $u = (u_n)$ be a sequence and L a linear difference operator on the space of sequences. We will denote by $L(u_n)$ the n–th term of the sequence $(L(u))$. L will be called an *annihilation operator* for u if, $\forall n (\geq N), L(u_n) = 0$. For example, the operator Δ is an annihilation operator for the constant sequence, $\forall n, u_{n+1} = u_n$.

If $S_n = S + a_n D_n$, then $S_n/D_n - S/D_n = a_n$. Applying L, we get $L(S_n/D_n) - SL(1/D_n) = L(a_n) = 0$. So, by construction, the sequence transformation T defined by

$$T_n = L(S_n/D_n)/L(1/D_n), \quad n = 0, 1, \ldots$$

will be such that, $\forall n, T_n = S$ if and only if, $\forall n, S_n = S + a_n D_n$. This set of sequences is called the *kernel* of the transformation T.

Let us give some examples. If $a_n = \alpha_0 + \alpha_1 n + \cdots + \alpha_{k-1} n^{k-1}$, the corresponding annihilation operator is Δ^k. So, we obtain a family of transformations depending on k

$$T_k^{(n)} = \Delta^k(S_n/D_n)/\Delta^k(1/D_n), \quad n = 0, 1, \ldots$$

For $D_n = \Delta S_n$ and $k = 1$, Aitken Δ^2 process is recovered while, for $k = 2$, we obtain the second column of the θ-algorithm. These transformations can be generalized to the case where a_n is a polynomial of degree $k - 1$ in x_n (an auxiliary sequence of numbers) by replacing the operator Δ^k by the divided differences operator of order k, δ_k, recursively defined by

$$\delta_{k+1}(u_n) = \frac{\delta_k(u_{n+1}) - \delta_k(u_n)}{u_{n+k+1} - u_n}$$

with $\delta_0(u_n) = u_n$. If $D_n = x_n$, we obtain Richardson extrapolation process. Shanks transformation, the E-algorithm, the θ-algorithm, the ρ-algorithm and other processes can be recovered in this framework. See [4] for more details.

Let us now see how to construct a sequence transformation when an asymptotic information on (a_n) is known. We assume that an approximation (b_n) of (a_n) is known and we define the sequence transformation by $T_n = S_n - b_n D_n, n = 0, 1, \ldots$. We have the

Theorem 1
A necessary and sufficient condition that (T_n) converges to S faster than (S_n) is that (b_n/a_n) tends to 1.

Let us now try to built a sequence (b_n) satisfying this condition. We consider the choice $b_n = \Delta S_n/\Delta D_n$. So, we recover the same transformation as above and we have

$$\frac{T_n - S}{S_n - S} = \frac{1 - a_{n+1}/a_n}{1 - D_n/D_{n+1}}$$

$$\frac{T_n - S}{S_{n+1} - S} = \frac{1 - a_n/a_{n+1}}{1 - /D_{n+1}/D_n}.$$

Two cases must be considered according to whether of not (D_{n+1}/D_n) tends to 1 (that is whether or not (D_n) is logarithmic according to the definition given above). We have the following result

Theorem 2
If (D_n) is nonlogarithmic and if $\exists \, 0 < m \leq M$ and $\exists \, N$ such that $\forall n \geq N$, $m \leq |D_{n+1}/D_n| \leq M$, then a necessary and sufficient condition that (T_n) converges to S faster than (S_n) and (S_{n+1}) is that (a_{n+1}/a_n) tends to 1.

For the logarithmic case, we have the

Theorem 3
If (D_n) is logarithmic, then a necessary and sufficient condition that (T_n) converges to S faster than (S_n) is that

$$\lim_{n\to\infty} \frac{1 - a_{n+1}/a_n}{1 - D_n/D_{n+1}} = 0.$$

A necessary and sufficient condition that (T_n) converges to S faster than (S_{n+1}) is that

$$\lim_{n\to\infty} \frac{1 - a_n/a_{n+1}}{1 - D_{n+1}/D_n} = 0.$$

Finally, we also have the

Theorem 4
If (D_{n+1}/D_n) tends to zero when n tends to infinity, then a necessary and sufficient condition that (T_n) converges to S faster than (S_{n+1}) is that (a_{n+1}/a_n) converges to 1.

Other results on this topics can be found in [6].

2. Error control

Let $T : (S_n) \longmapsto (T_n)$ be a sequence transformation. An important practical problem is to control the error $T_n - S$ in order to design a stopping criterion. As explained above, any sequence transformation can be written as $T_n = S_n + D_n$. Let us set $T_n(b) = S_n + (1 - b)D_n$. Obviously $T_n = T_n(0) = [T_n(b) + T_n(-b)]/2$. We will now study conditions for $T_n(b)$ and $T_n(-b)$ to have opposite signs. In that case, we will have $S \in I_n(b) = [\min(T_n(b), T_n(-b)), \max(T_n(b), T_n(-b))]$, which allows to control the error.

Let us define e_n by the relation $S_n - S = (-1 + e_n)D_n$. Of course, e_n is unknown but, as we will see, this is not a drawback. It is easy to check that $T_n - S = e_n D_n$. The condition $S \in I_n(b)$ is equivalent to $[T_n(b) - S][T_n(-b) - S] = (e_n^2 - b^2)D_n^2 < 0$. Thus, we have the

Theorem 5
A necessary and sufficient condition that $\forall b \neq 0$, $\exists N$ such that $\forall n \geq N$, $S \in I_n(b)$ is that $\forall n \geq N$, $|e_n| \leq |b|$.

92

Let us now assume that (T_n) converges to S faster than (S_n). This is true if and only if $(D_n/(S_n - S))$ converges to -1. We have $(T_n - S)/(S_n - S) = e_n D_n/(S_n - S)$. Thus, (T_n) converges faster than (S_n) if and only if (e_n) converges to zero. So, from the preceding theorem, we immediately obtain

Theorem 6

If (T_n) converges to S faster than (S_n) then, $\forall b \neq 0$, $\exists N$ such that $\forall n \geq N$, $S \in I_n(b)$.

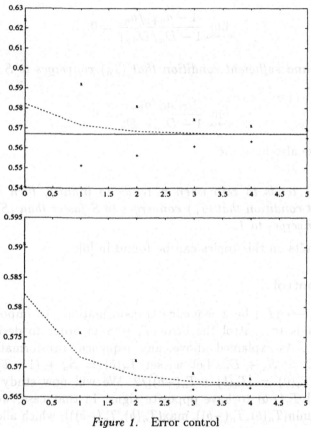

Figure 1. Error control

This result shows that it is quite easy to construct a sequence of intervals which asymptotically contain the limit. The main practical problem is that the index N is unknown. The length of $I_n(b)$ is proportional to $|bD_n|$ and, thus, in most cases, N is a decreasing function of $|b|$ which means that, for a large value of $|b|$, $I_n(b)$ is large but N is small. On the contrary, when $|b|$ is small, the interval $I_n(b)$ provides a good control on the error but only for large values of n. In fact, N tends to infinity when $|b|$ tends to zero.

Let us illustrate the preceding results. We consider the sequence $S_{n+1} = \exp(-S_n)$ with $S_0 = 1$. It converges to $S = \exp(-S) = 0.56714329\ldots$. When T is Aitken Δ^2 process, the convergence is accelerated and we get the results of Figure 1. The Figure on the top corresponds to $b = 0.1$ and the other one to $b = 0.02$. The exact result is indicated by a solid line and the dashed line represents T_n. the x and the + represent the intervals $I_n(b)$.

The main drawback of this procedure is that the sequences $(T_n(\pm b))$ do not converge faster than (S_n). So, we will now try to have simultaneously error control and acceleration. For that purpose, instead of a fixed value of the parameter b, we will use a sequence (b_n). Obviously, the result of Theorem 5 still holds with b replaced by b_n and it is easy to see that the sequences $(T_n(\pm b_n))$ both converge faster than (S_n) if and only if (b_n) tends to zero. The difficult point is that, for having, at the same time, error control and acceleration, we should have $\forall n, |e_n| \leq |b_n|$, a condition needing a quite precise knowledge of the asymptotic behavior of the sequence (e_n).

Let us assume that we have two transformations $T : (S_n) \longmapsto (T_n)$ and $V : (T_n) \longmapsto (V_n)$. We set $V_n(b) = V_n - b(V_n - T_n)$ and $J_n(b) = [\min(V_n(b), V_n(-b)), \max(V_n(b), V_n(-b))]$. We have the

Theorem 7
If $T_n - S = o(S_n - S)$ and $V_n - S = o(T_n - S)$ then, $\forall b \neq 0$, the sequences $(V_n(\pm b))$ both converge faster than the sequence (S_n) and $\exists N$ such that $\forall n \geq N$, $S \in J_n(b) \subseteq I_n(b)$.

Other results related to convergence acceleration and error control are given in [2] and [1].

References

1. Bellalij, M. (1990) A simultaneous process for convergence acceleration and error control, *J. Comput. Appl. Math.* **33**, pp. 217–231.
2. Brezinski, C. (1983) Error control in convergence acceleration processes, *IMA J. Numer. Anal.* **3**, pp. 65–80.
3. Brezinski, C. (1988) A new approach to convergence acceleration methods, in A. Cuyt (ed.), *Nonlinear Numerical Methods and Rational Approximation*, Reidel, Dordrecht, pp. 373–405.
4. Brezinski, C. and Matos, A.C. (1996) A derivation of extrapolation algorithms based on error estimates, *J. Comput. Appl. Math.*, **66**, pp. 5–26.
5. Brezinski, C. and Redivo Zaglia, M. (1991) *Extrapolation Methods. Theory and Practice.* North-Holland, Amsterdam.
6. Brezinski, C. and Redivo Zaglia, M. (1991) Construction of extrapolation processes, *Appl. Numer. Math.*, **8**, pp. 11–23
7. Matos, A.C. (1989) Acceleration methods for sequences such that $\Delta S_n = \sum_{i=1}^{\infty} a_i g_i(n)$, in W.F. Ames and C. Brezinski (eds.), *Numerical and Applied Mathematics, vol. 1.2*, Baltzer, Basel, pp. 447–451.
8. Matos, A.C. (1990) A convergence acceleration method based on a good estimation of the absolute error, *IMA J. Numer. Anal.*, **10**, pp. 243–251.

94

9. Matos, A.C. (1990) Extrapolation algorithms based on the asymptotic expansion of the inverse of the error; application to continued fractions, *J. Comput. Appl. Math.*, **32**, pp. 179–190.
10. Weniger, E.J. (1989) Nonlinear sequence transformations for the acceleration of convergence and the summation of divergent series, *Comput. Phys. Reports*, **10**, pp. 189–371.

PSEUDOEIGENVALUES, SPECTRAL PORTRAIT OF A MATRIX AND THEIR CONNECTIONS WITH DIFFERENT CRITERIA OF STABILITY

H. BULGAK
Research Center of Applied Mathematics, Selcuk University
Konya, Turkey

A new methodology for scientific computation aims at the design of problem solving environments with automatic result verification, providing full control over effects of the computational errors and the uncertainties in the data.

It is interesting to create the algorithms and programs which are stable to the influence of rounding-off errors and which make it possible to get the result with the evaluation of its precision or to detect the ill-conditions of the problem. Such algorithms and programs are called the algorithms and programs with guaranteed accuracy.

Sometimes the problems of linear algebra are found to be ill-conditioned when very small inescapable errors in the initial data can lead to considerable changes in the result and it causes difficulties in applications. This is the principal trouble in the problem of guaranteed accuracy. When constructing the method with guaranteed accuracy we must make sure that the problem is well posed and base carefully the algorithms to solve it, analysing minutely computer errors in the all intermediate steps of calculations which have an influence on the result.

For design an algorithm with guaranteed accuracy we must make clear the following seven items: 1) Standard Format, 2) Condition number, 3) The uncertainties in data, 4) A bound of well-conditions, 5) "Residual" problem, 6) An algorithm whose convergence must be connected with the condition number and the standard Format number, 7) Iterative refinement procedure.

The aim of this paper is to illustrate these principles on an example of examine following problems. The first, do all eigenvalues of matrix A lie in the interior of the left halfplane? This problem is well-known as Hurwitz problem. The second, do all eigenvalues of matrix A lie in the interior of the unique circle? In conclusion, we describe the algorithm with guaranteed accuracy for solving Lourier-Riccati matrix equation.

An invaluable help in the preparation of the paper were lent by Ayşe Bulgak, D.Eminov and V. Vaskevich. All Figures were prepared using the friendly dialogue system for drawing spectral portrait of the matrices. The friendly dialogue system which was created by D. Eminov together with the author. The present definition of a Format (see section 2) is suggested by Ayşe Bulgak. Several opportune remarks were made by G. Demidenko.

The author is very grateful to each of the contributors mentioned above.

H. Bulgak and C. Zenger (eds.), Error Control and Adaptivity in Scientific Computing, 95–124.
© *1999 Kluwer Academic Publishers. Printed in the Netherlands.*

1. Problem of Stability

The trivial solution of N order system

$$\frac{d}{dt} x(t) = Ax(t), \tag{1.1}$$

is stable (synonyms: system is stable, A is stable), if for any small positive number ε there exists the positive number $\delta = \delta(\varepsilon)$ such that for all $t > 0$ and $\| x(0) \| < \delta$ for solution of (1.1) the inequality holds $\| x(t) \| < \varepsilon$. Here $\| x \|$ is Euclidean vector norm .

The trivial solution of this system is called asymptotically stable (synonyms: system is asymptotically stable, A is asymptotically stable, A is Hurwitz-type matrix) if it is stable and $\| x(t) \| \to 0$ with t increasing to ∞.

The trivial solution of N order system

$$x(n+1) = Ax(n), \; n = 0, 1, 2, \ldots , \tag{1.2}$$

is discrete stable (synonyms: system is discrete stable, matrix A is discrete stable), if for any small positive number ε there exists the positive number $\delta = \delta(\varepsilon)$ such that for all $n > 0$ and $\| x(0) \| < \delta$ for solution of (1.2) the inequality $\| x(n) \| < \varepsilon$ is true.

The trivial solution of this system is called discrete asymptotically stable (synonyms: system is discrete asymptotically stable, matrix A is discrete asymptotically stable) if it is discrete stable and $\| x(t) \| \to 0$ with t increasing to ∞.

Above definitions are well-known. Checking stability or discrete stability of a given matrix A is known as stability problem. In this paper we give the algorithms with guaranteed accuracy for solving the stability problem. Further we assume that elements of A are real numbers.

2. Format

There are many different ways for real numbers representation. The normalised scientific notation is one of them. Let γ be a radix. An arbitrary real nonzero number z can be represented in normalised scientific notation as follows

$$z = s \, \gamma^{p(z)} \times m(z), \; 1/\gamma \le m(z) < 1.$$

This representation consists of three parts. A sign s is either + or -, a number $m(z)$ is in the set $[1/\gamma, 1)$, and a power $p(z)$ is an integer. The number $m(z)$ is called the mantissa of z and $p(z)$ is called the exponent of z. Normalised scientific notation is known as normalised floating point representation also.

Let γ, P_+ and k be natural numbers and let P_- be a negative integer number. Define a set **F** as follows

$$F = F(\gamma, P_-, P_+, k) = \{0\} \cup \{z, z = \pm \gamma^p (m_1 \gamma^{-1} + m_2 \gamma^{-2} + ... + m_k \gamma^{-k}),$$

$$P_- \le p \le P_+, \text{ p is an integer number,}$$

$$m_1, m_2, ..., m_k \text{ are integer numbers from 0 to } \gamma - 1, m_1 \ne 0\}.$$

We call the set $F = F(\gamma, P_-, P_+, k)$ as Format. There are similar definitions in [1], [2],[3]. The following statements are true.

1). $F \subset Q$, Q is the set of rational numbers.

2). $1 \in F$.

3). $F(\gamma, P_-, P_+, k) \subset F(\gamma, P_- - f, P_+ + h, k+m)$ for arbitrary natural numbers m, h, f. We call that $F(\gamma, P_-, P_+, k)$ is rough Format with respect to $F(\gamma, P_- - f, P_+ + h, k+m)$ and $F(\gamma, P_- - f, P_+ + h, k+m)$ is thin Format with respect to $F(\gamma, P_-, P_+, k)$.

4). If $P_- \to -\infty$, $P_+ \to \infty$ and $k \to \infty$ then $F = F(\gamma, P_-, P_+, k) \to R$.

5). Any Format has the greatest number ε_∞ and the least number $\varepsilon_{-\infty}$. These numbers are the functions of γ, P_-, P_+, k.

The formula $z = \pm m \times \gamma^p$ for elements of F allows us to write ε_∞ and $\varepsilon_{-\infty}$ as follows

$$\varepsilon_\infty = \gamma^{P_+} (1 - \gamma^{-k}); \quad \varepsilon_{-\infty} = -\gamma^{P_+} (1 - \gamma^{-k}).$$

6). There exists the least positive element $\varepsilon_0 = \gamma^{P_- -1}$ of F. The interval ($-\varepsilon_0, \varepsilon_0$) has 0 as the unique element from F.

7). The number $\varepsilon_1 = \gamma^{1-k}$ is an important characteristic of F. There are no elements of F in the interval ($1, 1 + \varepsilon_1$). The number $1 + \varepsilon_1$ belongs to F.

The values γ, ε_∞, ε_0, ε_1 are the number characteristics of F. The same Format can be defined by the numbers P_-, P_+, k and γ or by the numbers ε_∞, ε_0, ε_1, γ.

Define an operator

$$fl : [-\varepsilon_\infty, \varepsilon_\infty] \cap R \to F$$

which replaces a real number by an element of F. There are many different ways for definition of fl. We define $fl(z)$ for real z as the element of F which is the nearest to z. It is clear that $fl(z) = z$ for any z in F. Let $|z| \le \varepsilon_\infty$. There exist real $\alpha, \beta, |\alpha| \le \varepsilon_1, |\beta| \le \varepsilon_0$, $\alpha \beta = 0$ for which the following representations take place

$$fl(z) = z (1 + \alpha) + \beta, \quad | z - fl(z) | \le \varepsilon_1 | z | + \varepsilon_0.$$

Let a vector g and a matrix A be the real inside the Format's range, i.e.

$$\max_{i,j=1,2,...,N} | A_{ij} | \le \varepsilon_\infty \quad \text{and} \quad \max_{i=1,2,...,N} | g_i | \le \varepsilon_\infty.$$

Define the operator **fl** on the sets of matrices and vectors by the formulas $\mathbf{fl}(A)=\{\ \mathbf{fl}(A_{ij})\}$, $\mathbf{fl}(g) = \{\ \mathbf{fl}(g_i)\}$. The round-off errors arise if the elements of A and g are replaced by the numbers of **F**. We interpret these round-off errors as error matrix $\mathbf{fl}(A) - A$ and error vector $\mathbf{fl}(g) - g$ such that the inequalities

$$\| g - \mathbf{fl}(g) \| \le \varepsilon_1 \| g \| + \sqrt{N}\,\varepsilon_0 \quad \text{and} \quad \| A - \mathbf{fl}(A) \| \le \varepsilon_1 \sqrt{N} \| A \| + N\varepsilon_0$$

are true. Hence we must take into account of the so called "indefiniteness principle". In view of this principle we can guarantee only that instead of a matrix A we deal with a certain matrix A_0 closed to A.

Let **F** and **G** be two Formats and **G** is thin Format with respect to **F**. In this case the operator **fl** may be used for replacing the elements of **G** by the numbers of **F**. Hence for intermediate computing we can use more thin Format. But we must take into account of the round-off errors and return to the initial Format at last.

In any case before computing we must choose some Format which we call the standard Format. Standard Format depends on a problem.

It is important the following remark. Chosen Format we restrict the errors of computing both from above and from bellow.

3. Pseudoeigenvalues and Spectral Portrait of a Matrix

The problem of computation of eigenvalues is not the one which can be considered as correctly stated. In view of the "indefiniteness principle" we can guarantee only that instead of a matrix A we deal with a certain matrix A_0 closed to A. We know about this matrix only that the inequality $\| A - A_0 \| \le \varepsilon\|A\|$ is true with a small positive value ε. The ε characterises the number of digits in a cell of the computer or the uncertainties in data. Thereby, instead of eigenvalues of A we deal with the spots of its "ε-spectrum". Consider a set $\Sigma_\varepsilon(A)$ of complex numbers λ for every of each there exists a matrix $B(\lambda)$ satisfying the conditions as follows

$$\| B(\lambda) \| \le \varepsilon\| A \|, \quad \det (A + B(\lambda) - \lambda I) = 0 \text{ or } \| (A - \lambda I)^{-1} \| \ge 1/(\varepsilon\|A\|).$$

The elements of $\Sigma_\varepsilon(A)$ are known as the pseudoeigenvalues. The definition was introduced in [4]-[6] and discussed in [3],[7]-[10]. Further we call the set $\Sigma_\varepsilon(A)$ the "ε-spectrum" of A.

Let $S(0, z)$ be a circle with radius z and a centre in the original. It is easy to check that $\Sigma_\varepsilon(A) \subseteq S(0, (1+ \varepsilon) \|A\|)$. Hence for any $\varepsilon \le 0.01$ the "ε-spectrum" of A lies inside $S(0, 1.01 \|A\|)$. On figure 1 we give the picture of "0.01-spectrum" of the following matrix

$$A = \begin{pmatrix} -0.8 & 1 & 1 & 0 & 0 \\ 0 & -0.8 & 1 & 0 & 0 \\ 0 & 0 & -0.95 & 0 & 0 \\ 0 & 0 & 0 & 0.8 & 1 \\ 0 & 0 & 0 & 0 & 0.95 \end{pmatrix}.$$

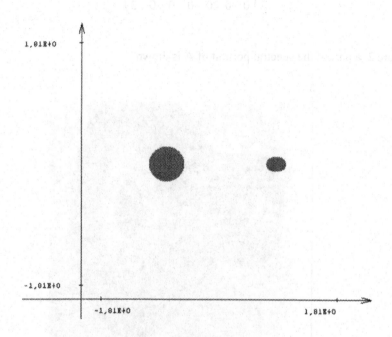

Figure 1.

Given numbers $\delta_1 > \delta_2 > \delta_3 > .... > \delta_k > 0$ we can write

$$S(\, 0, (1+\delta_1)\|A\|\,) \supset \Sigma_{\delta_1}(A) \supset \Sigma_{\delta_2}(A) \supset \Sigma_{\delta_3}(A) \supset ... \supset \Sigma_{\delta_k}(A).$$

Hence for all $\delta < 0.01$ we can draw the sets $\Sigma_\delta(A)$ enclosed in one to another and inside the circle $S(\, 0, 1.01\|A\|\,)$. We call the drawing the spectral portrait of A.

Give an example. Let $\delta_1 = 10^{-2}$, $\delta_2 = 10^{-3}$, $\delta_3 = 10^{-4}$, $\delta_4 = 10^{-5}$, $\delta_5 = 10^{-6}$, $\delta_6 = 10^{-7}$ and

$$A = \begin{pmatrix} 1 & 1 & 1 & 1 & 1 & 1 & 1 \\ 0 & 2 & 2 & 2 & 2 & 2 & 2 \\ 0 & 0 & 2 & 2 & 2 & 2 & 2 \\ 0 & 0 & 0 & 3 & 3 & 3 & 3 \\ 0 & 0 & 0 & 0 & 3 & 3 & 3 \\ 0 & 0 & 0 & 0 & 0 & 3 & 3 \\ 0 & 0 & 0 & 0 & 0 & 0 & 3 \end{pmatrix}.$$

On figure 2 a part of the spectral portrait of A is drawn.

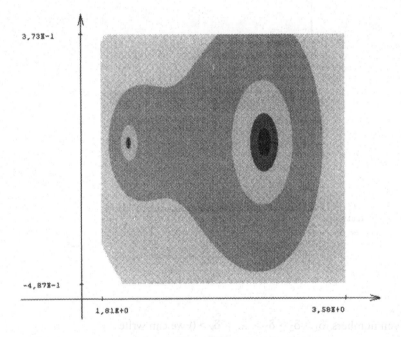

Figure 2.

Remark that it is sufficient to know the "ε-spectrum" of A to have complete information about dependence of its eigenvalues on a small perturbation of A.

For selfadjoint matrix $A = A^*$ each spot of its "ε-spectrum" has a diameter which does not exceed $\varepsilon\|A\|$. It means that the problem of eigenvalues of selfadjoint matrices is well conditioned. This is incorrect in the case of nonselfadjoint matrices. The above examples illustrate this. There exist other examples of such kind in [10].

The resume is as follows. To classify matrices with respect to quality of stability we need numerical characteristics which are based on spectrum of selfadjoint matrices but cannot based on the eigenvalues of nonselfadjoint matrices.

4. Factor of Stability, Quality of Stability and Stability Radius

There is well known Hurwitz problem formulated as follows. Do all eigenvalues of matrix A lie in the interior of the left halfplane? Also there is mechanical formulation of the problem: Is the trivial solution of the system (1.1) asymptotically stable?

We can answer for the questions by virtue of Lyapunov theorem formulated as follows. *The N order system* (1.1) *is asymptotically stable if all eigenvalues* $\lambda_j(A)$ *of A lie in the interior of the left halfplane, i.e. if there exists* $\delta > 0$ *such that the inequalities holds*

$$\text{Real } \lambda_j(A) \le -\delta \|A\| < 0, \quad j = 1, 2, \dots N.$$

For every solution of the asymptotically stable system (1.1) the following inequality

$$\|x(t)\| \le \theta' \, e^{-\delta'\|A\|t} \, \|x(0)\|, \, t > 0$$

is true. Here δ' is any positive number less than δ, $\theta' \ge 1$ and θ' does not depend on the initial vector $x(0)$ but θ' depends on δ'. It is clear that $\delta' \|A\|$ is less then $|\max \text{Real } \lambda_j(A)|$. The quantity δ' is known as the factor of stability of (1.1), see [11].

But we couldn't use the factor of stability as a practical measure of stability for (1.1). To illustrate this we consider an example now. Take two 25×25 matrices

$$A = \begin{pmatrix} -1 & & & & \\ 10 & -1 & & O & \\ & \ddots & \ddots & & \\ & O & 10 & -1 & \\ & & & 10 & -1 \end{pmatrix} \text{ and } B = \begin{pmatrix} -1 & & & \\ & -1 & & O \\ & & \ddots & \\ & O & & -1 \\ & & & & -1 \end{pmatrix}.$$

Factors of stability of A and B are equal to one another. Let dx/dt=Ax and dy/dt=By. It is clear that $y(t) = e^{-t}y(0)$ is monotonically decreasing to 0 for t→∞. The eigenvalues of symmetric matrix B are stable with respect to small perturbations of its elements. Hence matrix B is a "good" asymptotically stable.

If we choose $x_j(0) = 0, j = 2, 3, \dots, 25$ and $x_1(0) = \xi$ then for the solutions of (1.1) the equality holds

$$x_j(t) = \xi \, \frac{10^{j-1}}{(j-1)!} \, t^{j-1} e^{-t}.$$

Hence function $\|x(t)\|$ has a large "hump" on the set $t > 0$ before decreasing to 0. Moreover, the matrix

$$A_\omega = A + \begin{pmatrix} 0 & 0 & \cdots & 0 & \omega \\ & & & & 0 \\ & & O & & \vdots \\ & & & & 0 \\ & & & & 0 \end{pmatrix}$$

is closed to A and the estimation holds $\|A_\omega - A\| \le \omega$ with $\omega = -10\times 8^{-25}$. The number $\lambda = 0.25$ is the positive eigenvalue of A_ω. Thus small perturbation of A leads to non asymptotically stable matrix A_ω. Hence A is not "good" asymptotically stable.

Consider another example of matrix which is not "good" asymptotically stable. Take two-parameter pencil of 20×20 matrices

$$A_\omega(L) = \begin{pmatrix} -20 & 2.5L & & & & \\ 0 & -19 & 3.5L & & O & \\ \vdots & & \ddots & \ddots & & \\ \vdots & & & \ddots & \ddots & \\ 0 & O & & & -2 & 20.5L \\ \omega & 0 & 0 & \cdots & 0 & -1 \end{pmatrix}.$$

The matrix $A_\omega(L)$ has the following characteristic equation

$$\det (A_\omega(L) - \lambda I) = \prod_{j=1}^{20} (-j - \lambda) - \frac{2}{3} \omega L^{19} \prod_{j=1}^{20} (-j - 0.5) = 0.$$

Let ω_0 be equal to $1.5 L^{-19}$. For $L = 10$ the number ω_0 equals $1.5_{10}-19$. The matrix $A_0(L)$ is the Hurwitz-type one and $A_{\omega_0}(L)$ is non-Hurwitz-type matrix. The number $\lambda_1(A_{\omega_0}(L)) = 0.5$ is an eigenvalue of $A_{\omega_0}(L)$. The elements of $A_{\omega_0}(L)$ and $A_0(L)$ do not differ more than by the value $|\omega_0| = 1.5 |L|^{-19}$. This value can be made arbitrary small by an appropriate choice of L. Hence very small perturbations may transform the Hurwitz-type matrix to unstable one.

The examples of such kind as above exist in view of the instability of the eigenvalues of nonselfadjoint matrices. Hence we cannot use the factor of stability as a condition number for the stability problem.

In practice we need some numerical characteristics which detect the Hurwitz-type of a matrix and all matrices closed to it. These characteristics must be similar to the condition number for a linear algebraic equation $Ax = f$.

Define the characteristic $\kappa(A)$ of A on the solutions of (1.1) by the formula

$$\kappa(A) = 2 \| A \| \underset{T>0, \|x(0)\|=1}{Sup} \int_0^T \| x(t) \|^2 \, dt \tag{4.1}$$

with $\|A\| = \underset{\|x\|=1}{max} \| A x \|$ as the spectral norm of A. If A is not asymptotically stable then we set $\kappa(A) = \infty$.

As we move an asymptotically stable matrix A to a nonzero matrix with at least one purely imaginary eigenvalue the value $\kappa(A)$ increases unboundedly. The value $\kappa(A)$ characterises the property of stability of (1.1). We call $\kappa(A)$ the quality of stability of A [12],[13].

Lyapunov proved that for any asymptotically stable matrix A there exists the solution H of Lyapunov matrix equation $A^*H + HA + I = 0$ and H is a symmetric positive definite matrix. If there exists H positive definite solution of Lyapunov matrix equation $A^*H + HA + I = 0$ then $\kappa(A) = 2\|A\| \|H\|$. In the other case $\kappa(A)=\infty$.

Lemma 4.1. *For an asymptotically stable A an inequality holds*

$$\|e^{tA}\| \leq \sqrt{\kappa(A)} \; e^{-t\frac{\|A\|}{\kappa(A)}}, t > 0. \tag{4.2}$$

Proof. For the solutions of (1.1) the following relationships holds

$$\frac{d}{dt}(Hx(t), x(t)) = (H\frac{d}{dt}x, x) + (Hx, \frac{d}{dt}x) = - (x, x) \leq - \frac{1}{\lambda_N (H)} (Hx(t), x(t)).$$

Here $(x, y) = x_1y_1 + x_2y_2 + ... + x_Ny_N$. Hence for arbitrary solution of (1.1) the following inequality

$$\lambda_1(H) (x(t), x(t)) \leq \lambda_N(H) (x(0), x(0)) \; e^{-\frac{t}{\lambda_N (H)}}$$

is true. Here $\lambda_N(H)$ and $\lambda_1(H)$ are the greatest and the least eigenvalues of H. Hence we have the inequality [24]

$$\| x(t) \| \leq \sqrt{\frac{\lambda_N(H)}{\lambda_1(H)}} \; e^{-\frac{t}{2\lambda_N (H)}} \; \| x(0) \|. \tag{4.3}$$

By virtue of $\lambda_1(H) \geq 0.5/\|A\|$ and using $x(t) = e^{tA}x(0)$ we conclude that (4.2) holds.

Lemma 4.2.[14] *For an asymptotically stable* A *we have* $\|A\| \|A^{-1}\| \le \kappa^{3/2}(A)$.

The (4.1-3) state different connections between the quality of stability and solutions of (1.1).

The following theorem is true [14].

Theorem 4.1. *Let* A *be an asymptotically stable matrix (* viz. $\kappa(A) < \infty$ *) then* $A + B$ *also is an asymptotically stable matrix for an arbitrary perturb matrix* B *such that* $\dfrac{\|B\|}{\|A\|} \le \dfrac{1}{15\kappa(A)}$ *and* $|\kappa(A + B) - \kappa(A)| \le 2.15\,\kappa^2(A)\,\dfrac{\|B\|}{\|A\|}$ *is true.*

This theorem guarantees that the distance from A to the set of unstable matrices is greater then $\|A\|/(15\kappa(A))$. It is interesting to find this distance. This question was discussed in [15] by C. Van Loan who defined the parameter

$$\theta(A) = \|A\| \max_{Re z \le 0} \| (A + zI)^{-1} \|$$

and proved that $1/\theta(A)$ is the distance. For a stable A we have

$$\theta(A) = \|A\| \max_{-\infty < z < \infty} \| (A + izI)^{-1} \| .$$

We can use $\theta(A)$ as a measure of sensitivity to a small perturbation for an asymptotically stable matrix. This remark naturally follows from analysis of the spectral portrait of A. Let $\varepsilon = 1/\theta(A) = r(A)$. The number $r(A)$ is known as the stability radius of A [16]. If the right halfplane and $\Sigma_\varepsilon(A)$ have no overlap and $\|B\|/\|A\| < \varepsilon$ then $A + B$ is an asymptotically stable matrix.

In the case of an arbitrary matrix A we can compute approximations to $\Sigma_\varepsilon(A)$ and $\theta(A)$ only. For this reason we propose to use the quality of stability $\kappa(A)$ instead of $\theta(A)$. The properties of $\kappa(A)$ and $\theta(A)$ are fully analogous. But $\kappa(A)$ admits an explicit representation in terms of quantities which may be calculated by standard procedures of linear algebra. To calculate $\theta(A)$ exactly it would be necessary to compute the maximum value of function $\|A\| \| (A + izI)^{-1} \|$. We can not be done this by algebraic operations only. So it is more convenient to use $\kappa(A)$ for the analysis of stability of concrete systems.

Theorem 4.2. *The quantity of stability* $\kappa(A)$ *and stability radius* $r(A) = \theta^{-1}(A)$ *are connected by the following inequalities*

$$\frac{1}{15}\theta(A) \le \kappa(A) \le 2\theta^2(A), \qquad \theta^{2/3}(A) \le \kappa(A). \qquad (4.4)$$

To prove (4.4) we need in the following lemma [14].

Lemma 4.3. *Let* A *be an asymptotically stable matrix (viz.* $\kappa(A) < \infty$ *). Then the inequalities*

$$\sigma_1 (A + izI) \geq \frac{1}{14.2} \frac{\|A\|}{\kappa(A)}$$

are true for any real z . Here $\sigma_1(X)$ *is the minimal singular value of X.*

Proof of theorem 4.2. In view of the equality $\| (A + izI)^{-1} \| = 1/\sigma_1(A + izI)$ and the definition of $\theta(A)$ the inequality holds

$$\frac{1}{14.2} \theta(A) \leq \kappa(A).$$

Further, for the asymptotically stable matrix A the integral representation

$$(A + izI)^{-1} = - \int_0^\infty e^{t(A+izI)} \, dt$$

is true. This together with (4.2) allow us to write

$$\| (A + izI)^{-1} \| \leq \int_0^\infty \sqrt{\kappa(A)} e^{-t\frac{\|A\|}{\kappa(A)}} \, dt = \frac{\kappa^{3/2}(A)}{\|A\|} \quad \text{and} \quad \theta(A) \leq \kappa^{3/2}(A) .$$

Further there is the well known integral representation for the positive definite solution H of Lyapunov matrix equation. The representation is as follows

$$H = \frac{1}{2\pi} \int_{-\infty}^\infty (A^* + izI)^{-1} (A - izI)^{-1} \, dz.$$

For $z > \|A\|$ we have

$$\| (A - izI)^{-1} \| = 1/\sigma_1(A - izI) \leq \frac{1}{z - \|A\|}, \quad \| (A^* + izI)^{-1} \| \leq \frac{1}{z - \|A\|}. \quad (4.5)$$

For $T > 0$ we set

$$R_T = \frac{1}{2\pi} [\int_T^\infty (A^* + izI)^{-1} (A - izI)^{-1} \, dz + \int_{-\infty}^{-T} (A^* + izI)^{-1} (A - izI)^{-1} dz].$$

By virtue of (4.5) for $T > \|A\|$ we can write the evaluation

$$\| R_T \| \leq \frac{1}{\pi} \int\limits_{T}^{\infty} \frac{1}{(z - \|A\|)^2} \, dz = \frac{1}{\pi(T - \|A\|)}.$$

We obtain the evaluation of $\|R_T\|$ based on the H integral representation and taken into account of the definitions R_T, $\kappa(A)$ and the following chain of inequalities

$$\frac{\kappa(A)}{2\|A\|} = \|H\| = \| \frac{1}{2\pi} \int\limits_{-T}^{T} (A^* + izI)^{-1} (A - izI)^{-1} \, dz + R_T \|$$

$$\leq \frac{1}{2\pi} \int\limits_{-T}^{T} \| (A - izI)^{-1} \|^2 \, dz + \frac{1}{\pi(T - \|A\|)} \leq \frac{\theta^2(A)}{\pi\|A\|} [\frac{T}{\|A\|} + (\frac{T}{\|A\|} - 1)^{-1}].$$

Hence

$$\kappa(A) \leq \frac{2}{\pi} \theta^2(A) [\frac{T}{\|A\|} + (\frac{T}{\|A\|} - 1)^{-1}].$$

Setting $T = 2\|A\|$ we obtain the estimate $\kappa(A) \leq 2\theta^2(A)$. The theorem 4.2 is proved.

5. Spectral Radius, Quality of Discrete Time Stability and Discrete Stability Radius

Consider the stability problem of discrete time systems (1.2). The definitions, theorems and examples of this section and section 4 are similar. But they are not identical.

There is a well-known Lyapunov theorem: *The N order discrete time system* (1.2) *is discrete asymptotically stable if all eigenvalues* $\lambda_j(A)$ *lie in the interior of the unique circle, i. e. if there exists positive* $\delta < 1$ *such that the inequalities hold*

$$| \lambda_j (A) | \leq \delta, \ j = 1, 2, ..., N.$$

If A is discrete asymptotically stable then every solution of (1.2) satisfies the following inequality

$$\| x(n) \| \leq \theta' \ (\delta')^n \| x(0) \|, \ n > 0.$$

Here δ' is an arbitrary positive number less than δ, $\theta' \geq 1$ and θ' does not depend on $x(0)$, but θ' depends on δ'. It is clear that $\delta' < \rho(A) = \max | \lambda_j(A) |$. The quantity $\rho(A)$ is known as the spectral radius of A.

But we couldn't use the spectral radius as a practical measure of stability for (1.2). To illustrate this we consider an example now. Take two 25×25 matrices

$$A = \begin{pmatrix} 0.5 & 10 & & & \\ & 0.5 & 10 & & O \\ & & \ddots & \ddots & \\ & O & & 0.5 & 10 \\ & & & & 0.5 \end{pmatrix} \quad \text{and} \quad B = \begin{pmatrix} 0.5 & & & \\ & 0.5 & & O \\ & & \ddots & \\ & O & & 0.5 \\ & & & & 0.5 \end{pmatrix}.$$

The spectral radii of A and B are equal to one another. Let $x(n+1) = Ax(n)$, $n=0,1,2, \ldots$ and $y(n+1) = By(n)$, $n = 0, 1, 2, \ldots$. It is clear that $y(t) = (0.5)^n y(0)$ is monotonically decreasing to 0 for $n \rightarrow \infty$. The eigenvalues of symmetric matrix B are stable with respect to small perturbations of its elements. Hence the matrix B is a "good" discrete asymptotically stable.

If we choose $x_j(0) = 0$, $j = 1, 2,\ldots, 24$ and $x_{25}(0) = \xi$ then the equality holds

$$x_j(n) = \xi \frac{10^{25-j}}{(25-j)!} n^{25-j} \frac{1}{2^n} .$$

Hence the function $\|x(n)\|$ has a large "hump" on the set $n > 0$ before decreasing to 0. Moreover, the matrix

$$A_\omega = A + \begin{pmatrix} 0 & & & & \\ 0 & & & & \\ \vdots & & O & & \\ 0 & & & & \\ \omega & 0 & \cdots & 0 & 0 \end{pmatrix}$$

is closed to A and the estimation holds $\|A_\omega - A\| \le \omega$ with $\omega = -10^{-24}$. The number $\lambda = -1.5$ is an eigenvalue of A_ω which has the modulus greater then 1. Thus a small perturbation of A leads to non discrete asymptotically stable matrix A_ω. Hence A is not "good" discrete asymptotically stable.

Define the characteristic $\omega(A)$ on the solutions of (1.2) by the formula

$$\omega(A) = \underset{T>0, \|x(0)\|=1}{\mathrm{Sup}} \sum_{n=0}^{T} \|x(n)\|^2 . \qquad (5.1)$$

If A is not discrete asymptotically stable then we set $\omega(A) = \infty$. As we move discrete asymptotic stable matrix A to a nonzero matrix with at least one eigenvalue on the unique circle the value $\omega(A)$ increases unboundedly.

The value $\omega(A)$ characterises the property of discrete stability of (1.2). We call $\omega(A)$ the discrete quality of stability of A. If there exists positive definite solution H of Lyapunov matrix equation $A^*HA - H + I = 0$ then $\omega(A) = \|H\|$. In the other case $\omega(A) = \infty$.

Lyapunov proved that for a discrete asymptotically stable matrix A the solution H of Lyapunov matrix equation $A^*HA - H + I = 0$ exists and H is a symmetric positive definite matrix.

Lemma 5.1. *For a discrete asymptotically stable A the inequalities hold*

$$\| A^n \| \leq \sqrt{\omega(A)} \; e^{-\frac{n}{2\omega(A)}}, n = 0, 1, 2, \dots \,. \tag{5.2}$$

Proof. For the solutions of (1.2) the following equation holds

$$(Hx(n+1), x(n+1)) = (Hx(n), x(n)) - (x(n), x(n)).$$

Hence for arbitrary solution of (1.2) the following inequality

$$\lambda_1(H) \,(x(n), x(n)) \leq \lambda_N(H) \,(x(0), x(0)) \; e^{-\frac{n}{\lambda_N(H)}}$$

is true. Hence we have the inequality

$$\| x(n) \| \leq \sqrt{\frac{\lambda_N(H)}{\lambda_1(H)}} \; e^{-\frac{n}{2\lambda_N(H)}} \; \| x(0) \| . \tag{5.3}$$

Hence and by virtue of $\lambda_1(H) \geq 1$ and $x(n) = A^n x(0)$ we conclude that (5.2) hold.

Lemma 5.2. *For a discrete asymptotically stable A we have* $\|A\|^2 \leq \omega(A)$ *and* $\omega(A) \geq 1$. *Moreover, if* $\| A \| < 1$ *then* $\omega(A) \leq 1/ (1 - \| A \|^2)$.

Proof. Since $H = I + A^*HA$ then $\lambda_{\min}(H) \geq 1$ and

$$\omega(A) = \| H \| = 1 + \max_{\|y\|=1} (A^*HAy, y) \geq 1 + \lambda_{\min}(H) \max_{\|y\|=1} (A^*Ay, y) \geq 1 + \| A \|^2.$$

The (5.1-3) state different connections between the discrete quality of stability and solutions of (1.2). The following theorem is true.

Theorem 5.1. *Let* A *be a discrete asymptotically stable matrix (viz.* $\omega(A) < \infty$ *) then for an arbitrary perturb matrix* B *such that* $\quad \|B\| \leq \dfrac{1}{20\omega^{3/2}(A)} \quad$ *the matrix* A + B *also is a discrete asymptotically stable matrix and*

$$| \omega(A + B) - \omega(A) | \leq 5 \, \omega^{5/2}(A)\|B\|.$$

The theorem 5.1 may be proved by applying the following lemma.

Lemma 5.3. *Let* A *be a discrete asymptotically stable matrix (viz.* $\omega(A) < \infty$ *) then for a real z the inequality holds*

$$\sigma_1 (A + e^{iz}I) \geq \frac{1}{6\pi\omega(A)} \; .$$

The lemma which is similar to lemma 5.3 is proven in [3].

It is interesting to find the distance from discrete asymptotically stable A to the set of discrete unstable matrices. Theorem 5.1 guarantees that the distance is greater then $1/(20\omega^{3/2}(A))$.

A concept of the distance between a discrete asymptotically stable matrix A and the set of unstable matrices is natural in view of the ε-pseudospectrum definition. As a measure of sensitivity to a small perturbation for a discrete asymptotically stable matrix we can use a number

$$r(A) = \min_{|\lambda_j (A+B)|\geq 1} \| B \|$$

which is called the stability radius [16]. This remark naturally follows from analysis of the spectral portrait of A. Let $\varepsilon = r(A)$. If the unique circle and the "ε-spectrum" of A have no overlap then A is a discrete asymptotic stable together with all matrices A+B with $\|B\|/\|A\| < \varepsilon$.

Let

$$A(a) = \begin{pmatrix} -0.8 & a & a & a & a \\ 0 & -0.8 & a & a & a \\ 0 & 0 & -0.95 & a & a \\ 0 & 0 & 0 & 0.8 & a \\ 0 & 0 & 0 & 0 & 0.95 \end{pmatrix}.$$

We give the parts of the spectral portraits of A(5) and A(20) inside the unique circle on the figures 3 and 4 respectively. The radii of stability of the matrices are the same one but the sensitivity of the spectrum of A(5) are better then of A(20) one.

110

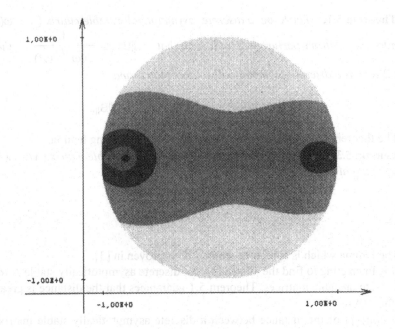

Figure 3.

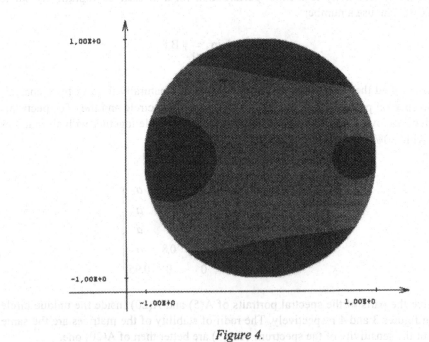

Figure 4.

In general case, the pseudospectrum $\Sigma_\varepsilon(A)$ cannot be easily computed. But we can compute approximations of $\Sigma_\varepsilon(A)$. This remark also concerns with r(A).

The stability radius r(A) has representation

$$r^{-1}(A) = \max_{0 \le \varphi < 2\pi, \xi \ge 1} \| (A + \xi e^{i\varphi} I)^{-1} \|.$$

To calculate r(A) exactly by this formula, it would be necessary to define the maximum of the function $\| (A + \xi e^{i\varphi} I)^{-1} \|$. We can not compute the maximum by using the algebraic operations only. For this reason, it is more convenient to use $\omega(A)$ for the analysis of the discrete asymptotic stability.

The discrete quantity of stability $\omega(A)$ and the stability radius r(A) are connected by the inequalities

$$\frac{1}{6\pi} r^{-1}(A) \le \omega(A) \le r^{-2}(A).$$

There is the proof of this inequalities in [3]. It is similar to the proof of theorem 4.2.

6. Uncertainties in Data, "Residual" Problem and Practical Discrete Asymptotic Stability

Chosen the $\omega(A)$ as a condition number we design algorithm with guaranteed accuracy for examining the discrete asymptotic stability of a given matrix A.

The quality of discrete stability $\omega(A)$ may be used also as a measure of stability to perturbations for Lyapunov equation

$$A^*XA - X + C = 0, C = C^*. \tag{6.1}$$

We call the equation

$$(A + \delta A)^* Y (A + \delta A) - Y + C + \delta C = 0 \tag{6.2}$$

a perturbation of (6.1). The solution of (6.2) is called the perturbed solution and denoted by $X + \delta X$. Now we are ready to bound the difference between the solutions of (6.1) and (6.2).

Theorem 6.1. *Let A be a discrete asymptotically stable matrix (viz. $\omega(A) < \infty$) and X = X* be the solution of (6.1). Perturb A and C by δA and $\delta C = (\delta C)^*$ respectively. If $\|\delta A\| \le \dfrac{1}{6\pi\omega(A)}$ then A + δA is a discrete asymptotically stable. Moreover, if $\omega(A)[2 \|A\| \|\delta A\| + \|\delta A\|^2] < 0.5$ then for the unique solution of (6.2) Y=X+δX the estimation holds $\|\delta X\| \le 2\omega(A)\|\delta C\| + 2\omega^2(A) \|C\| [2\|A\| + \|\delta A\|] \|\delta A\|.$*

The theorem 6.1 is proven in [3] and together with the theorem 6.1 it answers to the question for the uncertainties in data.

For working out algorithm with guaranteed accuracy it is important to know the error of the solution to Lyapunov matrix equation. The following "residual" theorem is true.

Theorem 6.2. *If* $X=X^*$ *is a positive definite matrix and* $\Delta = \|A^*XA - X + I\| < 1$ *then the matrix A is a discrete asymptotically stable and the inequalities hold*

$$\| H - X \| \leq \Delta \| H \| \quad and \quad | \omega(A) - \|X\| | / \omega(A) \leq \Delta.$$

Here H *is the positive definite solution of* (6.1) *with* C=I.

The proof of theorem 6.2 is in [3].

Let $\omega^* > 1$. We call A a practical discrete asymptotic ω^*-stable or discrete ω^*-stable matrix if $\omega(A) \leq \omega^*$.

Lemma 6.1. *Let A be a practical discrete asymptotic* ω^**-stable. Then for every solution* x(n) *of* (1.2) *and any integer* T>0 *the inequalities hold*

$$\sum_{n=0}^{T} \|x(n)\|^2 \, dt \; < \; \omega^* \, \|x(0)\|^2 \; and \; \|x(T)\| < \sqrt{\omega^*} \, e^{-T/(2\omega^*)} \, \|x(0)\|.$$

As a rule we know A with some indefiniteness only. The indefiniteness is measured by a perturb matrix δA, $\|\delta A\| < \varepsilon$. Here ε is a given small positive number. Let ω^* be equal to $(20/\varepsilon)^{2/3}$. By theorem 6.1 we have the following conclusion. If A is discrete ω^*-stable and $\| \delta A \| < 1/(20\omega^*)^{2/3}$ then $A + \delta A$ is discrete $1.25\omega^*$-stable matrix.

Let A be a practical discrete asymptotic stable matrix and C be an arbitrary matrix. The number ω^* and chosen small positive number $\rho < 1$ allow us to choose standard Format for computing (6.1) such that for any $C = C^*$ we can compute matrix Y such that the inequality holds

$$\| A^*YA - Y + C \| < \rho \, \|C\|$$

with some positive ρ which is less then 1 and does not depend on C. To this end it is sufficient to apply iterative refinement procedure for compute H with standard Format precision.

7. Computing a Discrete Lyapunov Matrix Equation with Guaranteed Accuracy

For solving positive definite solution $X = H(C)$ of Lyapunov matrix equation (6.1) with $C = C^* > 0$ we take the representation

$$H(C) = \sum_{n=0}^{\infty} A^n{}^*CA^n.$$

Only for a discrete asymptotically stable A the matrix $H(C)$ exists and the sequence

$$H_j(C) = \sum_{n=0}^{2^j-1} A^n{}^*CA^n , j = 0, 1, 2, \ldots \tag{7.1}$$

is convergent to $H(C)$. We can compute $H_j(C)$ by the formulas [17]:

$$H_0 = C, B_0 = A,$$

$$\tag{7.2}$$

$$H_k(C) = H_{k-1}(C) + [B_{k-1}]^* H_{k-1}(C) B_{k-1} \text{ and } B_k = [B_{k-1}]^2 , k = 1, 2, \ldots, j.$$

In view of the inequalities

$$\| H(C) - H_j(C) \| \leq \omega(A) \, e^{-2^j / \omega(A)} \, \|H(C)\|, j = 0, 1, 2, \ldots ,$$

we have estimation of the speed of convergence $H_j(C)$ to $H(C)$.

Further, for a given positive $\rho < 1$ we denote by $M = M(\omega(A), \rho)$ an integer which satisfies to the inequality

$$\omega(A) \, e^{-2^M / \omega(A)} \leq \frac{\rho}{2}$$

and is the minimal integer with this property. For $j \geq M$ we have

$$\| H(C) - H_j(C) \| / \| H(C)\| \leq \frac{\rho}{2} .$$

Naturally this estimation does not take into account of the round-off errors. The round-off errors of (7.2) on the class of symmetric positive definite matrix H have substantially been analysed in [3]. It was discovered that the convergence of (7.2) depends on the quality of stability $\omega(A)$ and on Format used.

Let A be a practical discrete asymptotically stable, i.e. $\omega(A) \leq \omega^*$ for some $\omega^* > 1$. Take $\rho < 1$ and $M = M(\omega^*, \rho)$. Denote by D_j and X_j, $j = 0, 1, \ldots, M$, the matrices which are computed by (7.2) with the round-off errors. Hence we can write

$$X_0 = C, \quad D_0 = A,$$

$$X_j = X_{j-1} + [D_{j-1}]^* X_{j-1} D_{j-1} + \Phi_j \quad \text{and} \quad D_j = [D_{j-1}]^2 + \Psi_j, \ j = 1, 2, \ldots, M.$$

(7.3)

Here Φ_j, Ψ_j are the matrices simulated by the round-off errors and determined by the computation accuracy applied for the elementary operations.

Theorem 7.1. *Let ω^* and $\rho < 1$ be positive real numbers and $M = M(\omega^*, \rho)$. Choose the positive α and r_0 such that*

$$2^{M+1} r_0 < 1 \quad and \quad 24 \, \omega^{*2} \sqrt{\omega^*} \, r_0 + 8\alpha \, \omega^{*2} \leq \rho.$$

Assume that the matrices D_j and X_j are determined by squaring algorithm (7.3) and

$$\| \Psi_m \| \leq r_0 / [2\omega^{*2}] \, \| D_{m-1} \|^2, \quad \| \Phi_m \| \leq \alpha \, \omega^* \| C \|, \ m = 1, 2, \ldots, M. \tag{7.4}$$

In this case for the matrices defined by (7.3) and (7.2) the inequalities hold

$$\| X_m - H_m(C) \| \leq \frac{\rho}{2} \| C \|,$$

$$\| A^* X_M A - X_M + C \| \leq [\frac{\rho}{2}(\|A\|^2 + 1) + \omega^* e^{-\frac{2^M}{\omega^*}}] \|C\|, \tag{7.5}$$

$$\| D_m - A^{2^m} \| \leq \frac{2^m r_0}{1 - 2^m r_0} e^{-2^{m-1}/\omega^*} \quad and \quad \lambda_{min}(X_M) \geq \lambda_{min}(C) - \rho\|C\|.$$

The theorem 7.1 is proven in [3].

For computing we can use some Format **G** which is thin with respect to standard Format **F** and provides (7.4).

By virtue of the theorem we can construct an algorithm with guaranteed accuracy for computing of positive definite solution of (6.1) together with computing $\omega(A)$. Recall that the answer is either state that A is practically discrete unstable ($\omega(A) > \omega^*$) or H(C) with standard Format precision. The scheme of the algorithm is as follows. As a data of the process we take ω^* and A.

Step 1. Compute $\|A\|$ and check an inequality $\|A\|^2 > \omega^*$. If it is true then we stop process with the answer $\omega(A) > \omega^*$.

Step 2. Computing the approximations to $\omega(A)$ and to the solution of (6.1) with $C=I$.

Step 2.1. Choose a positive number $\rho = 1/(2 \|A\|^2 + 3)$ and take $M = M(\omega^*, \rho)$.

Step 2.2. Compute B_M and H_M by recurrent formulas.

Step 2.2.1. Take $B_0 = A$; $H_0 = I$.

Step 2.2.2. For $k = 1, 2, ..., M$ compute

$$H_k = H_{k-1} + [B_{k-1}]^* H_{k-1} B_{k-1} \text{ and } B_k = [B_{k-1}]^2 .$$

Check the inequalities $\| B_k \| > 2\sqrt{\omega^*}$ and $\| H_k \| > 2\omega^*$. If one of them is true then we stop process with the answer $\omega(A) > \omega^*$.

Remark. Here we must using the parameters α and r_0 of theorem 7.1 to choose Format for computing the elementary operations.

Step 2.3. Check the inequalities

$$\| B_M \| \le \rho, \| A^* H_M A - H_M + I \| \le (\|A\|^2 + 3/2) \rho \text{ and } \lambda_{min}(H_M) \ge 1/2 - 3\rho/2.$$

If one of them is false then we stop process with the answer $\omega(A) > \omega^*$. Otherwise applying residual theorem we conclude that $\varpi = \|H_M\|$ and H_M approximate $\omega(A)$ and H with a relative error not greater then ($\|A\|^2 + 3/2$) ρ and A is a discrete asymptotically stable.

Step 3. Using the formulas of step 2 as a standard iterative refinement procedure with computing residual matrix and accumulation of the approximations to the solution $H = H(I)$ inside Format which is more thin with respect to standard one. By this way we compute H and $\omega(A)$ with Format precision.

8. Computing of Exponential Function of Asymptotically Stable Matrix

We need compute the matrix e^{tA} to find numerical value of the quality of stability $\kappa(A)$. For this reason we are turning to the problem of computing of an exponential matrix function now.

Let A be an arbitrary matrix and $\|tA\| < 0.5$. It is easy inside one Format **F** to compute e^{tA} by the formulas

$$A_0 = B_0 = I , \ B_k = 1/k \ tAB_{k-1} , \ A_k = A_{k-1} + B_k , k = 1, 2, ..., m. \quad (8.1)$$

The formulas (8.1) based on Taylor series representation of exponential function. The matrix A_m approximates e^{tA} with the norm of remainder bounded by as follows

$$\| e^{tA} - A_m \| \le [(t\|A\|)^{m+1} /(m+1)!] e^{\|tA\|} .$$

Analysing (8.1) in view of the round-off errors it is easy prove that there exists number m for which computed matrix $(A_m)_{comp}$ connected with matrix in question by the inequality

$$\| e^{tA} - (A_m)_{comp} \| \leq 100N\epsilon_1.$$

For computing (8.1) we can choose more thin Format and to find e^{tA} with standard Format precision $\| e^{tA} - (A_m)_{comp} \| \leq \epsilon_1$.

Sometimes, if $t\|A\|$ is sufficiently small it should be enough to approximate e^{tA} by $I+tA$. Using similar approximations A.J. Povzner and B.V. Pavlov in [18] successfully compute e^{TA} with great T. To this end they are starting with computing e^{tA} for tA of the small norm and further they are applying the idea of doubling argument.

Chosen $\tau = 2^{-M}T$ such that $0.25 \leq \tau \| A \| \leq 0.5$ they are computing the matrices given by the formulas

$$B_0 = e^{\tau A} \quad \text{and} \quad B_j = [B_{j-1}]^2, \quad j = 1, 2, ..., M. \tag{8.2}$$

It is clear that

$$B_j = \exp \{ 2^{j-1} \tau A \} \quad \text{and} \quad B_M = \exp \{ 2^M \tau A \} = e^{TA}.$$

Each squaring gives us the doubling argument t of e^{tA}. Since increasing it in 1024 times takes only 10 matrix multiplication ($1024 = 2^{10}$) the algorithm (8.2) has a very high speed of convergence to e^{TA}. It is known as the scaling and squaring algorithm.

How can we check that the computed approximation to e^{TA} has enough precision if the matrix e^{TA} is unknown?

To answer this question it is possible to use the ideas of backward error analysis. There are some necessary conditions for application of the backward error analysis. We formulate the conditions as follows.

1. Computing the solution of $Ax = f$ we can interpret the arising round-off errors as the errors in the data. By the other words if y is computed approximation of x, that there exists a perturb matrix E and a vector g such that $(A + E)y = f + g$. Moreover, for some small positive number ϵ the inequalities hold $\|E\| \leq \epsilon\|A\|$, $\|g\| \leq \epsilon \|f\|$.

2. We must know the condition number $\mu(A) = \|A\| \|A^{-1}\|$ for applying it in the estimation $\|x-y\| / \|y\| \leq \mu(A)\epsilon / [1 - \mu(A)\epsilon]$.

3. We need calculate the condition number $\mu(A)$ by some stable procedure.

C. Moler and C. Van Loan tried to use the backward error analysis for the computation of e^{TA}. The value

$$\nu(A,T) = \max_{|E|=1} \| \int_0^T e^{(T-s)A} E\, e^{sA}\, ds\, [\|A\|/\| e^{TA} \|] \|$$

was suggested as a condition number in [19]. In the same paper they proved the approximate equality

$$\| e^{T(A+E)} - e^{TA} \| / \| e^{TA} \| \approx v(A,T) \, \|E\|/\|A\|.$$

This formula shows that with the increasing $v(A,T)$ the stability e^{TA} to a small perturbation of A decreases.

We explain the introduction of $v(A,T)$ by the following reason. The matrix function $Y(t) = e^{t(A+E)}$ is the unique solution of the initial problem

$$\frac{d}{dt} Y(t) = AY(t) + E\, Y(t), \quad Y(0) = I.$$

Hence the equality holds

$$e^{t(A+E)} - e^{tA} = \int_0^t e^{(t-s)A} E\, e^{sA}\, ds$$

and we have

$$\| e^{t(A+E)} - e^{tA} \| / \| e^{tA} \| \leq \| \int_0^t e^{(t-s)A} E\, e^{sA}\, ds \| / \| e^{tA} \|.$$

In the general case we don't know any algorithm for computing e^{TA} ($T\|A\| \gg 1$) whose result may be interpreted as the exponential function with some perturbed matrix A. We think that $v(A,t)$ has no application in practice by this reason.

Finally we see that the backward analysis didn't find application in the computation of matrix exponential function problem yet.

There are works where analysis of the round-off errors influence in the calculation of e^{tA} by the scaling and squaring algorithm are elaborated. R.C. Ward suggested in [20] to compute a posteriori estimation of the accuracy computation of e^{TA}. The idea of computing such estimation is simple.

Let $(B_j)_{comp}$, $j = 0, 1, \ldots, M$, be the matrices which were found by the computer realisation of (8.2). We assume that the matrices $(B_j)_{comp}$ and matrix e^{TA} are connected by the equalities

$$(B_0)_{comp} = e^{TA} + \Psi_0 \text{ and } (B_j)_{comp} = [(B_{j-1})_{comp}]^2 + \Psi_j, \quad j = 1, 2, \ldots, M. \qquad (8.3)$$

The vectors Ψ_j, $j = 0, 1, 2, \ldots, M$, simulate the influence of the round-off errors arising in the computer realisation of corresponding operations. Let

$$G_j = (B_j)_{comp} - e^{2^j \tau A}, \quad j = 0, 1, \ldots, M.$$

We can easily verify an inequality

$$\|G_j\| = 2 \| e^{2^{j-1}\tau A} \| \|G_{j-1}\| + \|G_{j-1}\|^2 + \| (B_j)_{comp} - [(B_{j-1})_{comp}]^2 \|.$$

Since $\| e^{2^{j-1}\tau A} \|$ is unknown we can using an inequality

$$\| e^{2^{j-1}\tau A} \| \le \| (B_{j-1})_{comp} \| + \| (B_{j-1})_{comp} - e^{2^{j-1}\tau A} \| = \| (B_{j-1})_{comp} \| + \| G_{j-1} \|$$

to obtain an estimation

$$\|G_j\| \le 2 \| (B_{j-1})_{comp} \| \|G_{j-1}\| + 3\|G_{j-1}\|^2 + \| (B_j)_{comp} - [(B_{j-1})_{comp}]^2 \|. \tag{8.4}$$

The values $\|G_0\|$ and $\| (B_j)_{comp} - [(B_{j-1})_{comp}]^2 \|$ are close to the Format constant ε_1. Thus we can use (8.4) as the recurrent formulae for computing a posteriori estimations of the e^{TA} computation accuracy.

Sometimes Ward's estimations are no sufficient. Let

$$A = \begin{pmatrix} -20.15 & 239.3 & 0 & 0 & 0 \\ 0 & -20.15 & 239.3 & 0 & 0 \\ 0 & 0 & -20.15 & 239.3 & 0 \\ 0 & 0 & 0 & -20.15 & 239.3 \\ 0 & 0 & 0 & 0 & -20.15 \end{pmatrix}.$$

The equalities hold $\| A \| = 292.4$ and $\| e^A \| = 0.2417$. Chosen $\tau = 2^{-9}$ we have

$$\| e^{\tau A} \| = 3.651; \qquad \| e^{2\tau A} \| = 3.925; \qquad \| e^{2^2\tau A} \| = 5.254; \qquad \| e^{2^3\tau A} \| = 14.86;$$

$$\| e^{2^4\tau A} \| = 90.94; \quad \| e^{2^5\tau A} \| = 635.1; \quad \| e^{2^6\tau A} \| = 27335;$$

$$\| e^{2^7\tau A} \| = 3478 \quad \text{and} \quad \| e^{2^8\tau A} \| = 359.6.$$

Chosen $\varepsilon_1 = 0.1_{10}-14$ we can write $2^9 \prod_{j=1}^{8} \| e^{2^j\tau A} \| \varepsilon_1 \approx 1000.$

There are the interesting questions pertaining to the stability of the scaling and squaring approach. The difficulties arise during the squaring process if $\|e^{tA}\|$ grows before it decays. Then $\|e^{tA}\|$ has a "hump" on the set $t > 0$.

If the "hump" of $\|e^{tA}\|$ includes the points $2\tau, 4\tau, ..., 2^M\tau$, i.e. $\| \exp[2^j\tau A] \| > 1$, $j = 1, 2, ..., M$ then the inequality $\varepsilon_1 \prod_{j=1}^{m} \| e^{2^j\tau A} \| >> \| e^A \|$ may be true. The Ward's estimations are no sufficient in this case.

The "hump" of the $\|e^{tA}\|$ function may be sufficiently large [21]. The quantity of stability $\kappa(A)$ may be interpreted as the estimation of the function $\|e^{tA}\|$ "hump". There is a fast survey of the works dedicated to the problem of computation of e^{tA} in [19].

Let A be a practical asymptotically stable, $\kappa(A) < \kappa^*$ and $T\|A\| >> 1$. There exist positive value τ and integer M such that $0.25 \leq \tau \|A\| \leq 0.5$ and $2^M \tau = T$. Consider the matrices B_j given by (8.3) with Ψ_j as the error matrices determined by the computation accuracy applied for the elementary operations.

Theorem 8.1. *Given a positive* ρ *choose a positive* r_0 *such that*

$$1 - 2^M r_0 > 0 \ and \ \frac{2^M r_0}{1 - 2^M r_0} e^{-2^M \tau \frac{\|A\|}{\kappa^*}} < \rho.$$

Assume that the matrices B_j *are determined by scaling and squaring algorithm* (8.3) *and*

$$\|\Psi_0\| \leq \frac{r_0}{2(\kappa^*)^2} \quad and \quad \|\Psi_m\| \leq \frac{r_0}{4\kappa^*} \|B_{m-1}\|^2, m = 1, 2, \dots, M. \qquad (8.5)$$

Then the inequalities hold

$$\|B_{m-1} - e^{2^m \tau A}\| \leq \frac{2^m r_0}{1 - 2^m r_0} e^{-2^m \tau \frac{\|A\|}{\kappa^*}} \quad and \quad \|B_M - e^{TA}\| \leq \rho.$$

The theorem 8.1 is proven in [22]. Now we can use some Format **G** which is thin with respect to standard Format **F** and provides (8.5). Hence we can compute e^{tA} for $t \leq T$ with guaranteed accuracy.

9. Computing Lyapunov Matrix Equation with Guaranteed Accuracy

For solving positive definite solution H(C) of Lyapunov matrix equation

$$A^* H(C) + H(C)A + C = 0, \quad C = C^* > 0$$

we take the integral representation

$$H(C) = \int_0^\infty e^{tA^*} C \, e^{tA} \, dt.$$

This integral exists only for an asymptotically stable A. For an asymptotically stable A and any positive number h the sequence

$$H_j(C) = \int_0^{2^j h} e^{tA^*} C \, e^{tA} \, dt \, , j = 0, 1, 2, \ldots \tag{9.1}$$

is convergent to H(C). E.J. Davison and F.T. Man suggested in [17] the following formulas for $H_j(C)$ computation :

$$B_0 = e^{hA} \, ; \ H_0(C) = \int_0^h e^{tA^*} C \, e^{tA} \, dt;$$

$$\tag{9.2}$$

$$H_k(C) = H_{k-1}(C) + [B_{k-1}]^* \, H_{k-1}(C) \, B_{k-1} \text{ and } B_k = [B_{k-1}]^2 \, , k = 1, 2, \ldots, j.$$

They pointed out that convergence of $H_k(C)$ determined by (9.2) depends on h. As a rule the authors start with $h = 0.001\sqrt{N/trA^2}$. If the process (9.2) does not converge they decrease h and restart (9.2) until the convergence improves. From our point of view this way leads unjustified losses of time. In [17] it was discovered that the convergence of (9.2) does not depend on h and it is sufficient to take $h = 0.5/\|A\|$. In principle the convergence of (9.2) depends on the quality of stability $\kappa(A)$ and on Format used.

Return to (9.1). In view of the inequalities

$$\| H(C) - H_j(C) \| \leq \kappa(A) \, e^{-2^{j+1} h \frac{|A|}{\kappa(A)}} \, \|H(C)\|, j = 0, 1, 2, \ldots,$$

we have estimation of the speed of convergence $H_j(C)$ to $H(C)$.

Further, for given positive $\rho < 1$ we denote by $M = M(\kappa(A), \rho, h)$ an integer which satisfies to the inequality

$$\kappa(A) \, e^{-2^{M+1} h \frac{|A|}{\kappa(A)}} \leq \frac{\rho}{2}$$

and is the minimal integer with this property. For $j \geq M$ we have

$$\| H(C) - H_j(C) \| / \| H(C)\| \leq \frac{\rho}{2} \, .$$

Naturally this estimation didn't take into account of the round-off errors. For a symmetric positive definite matrix H the round-off errors of (9.2) have substantially been analysed in [23].

10. Lourier-Riccatti Matrix Equation

We describe the main step of the algorithm with guaranteed accuracy for solving Lourier-Riccati matrix equation

$$A^*\Lambda + \Lambda A + 2\,\|A\|\,[\,C^*(CFC^*)^{-1}C - \Lambda B(B^*GB)^{-1}B^*\Lambda\,] = 0\,. \qquad (10.1)$$

This algorithm was suggested and founded in [25]. It bases on the computation the two-point boundary value problem. It includes a series of orthogonal transformations of the matrix made up of the Lourier-Riccati equation coefficients. As a result we obtain the system of equations whose condition is not worse than the condition of the Lourier-Riccati matrix equation. As the measure of condition we use four numerical characteristics suggested by S.K.Godunov [26] which are the norms of the solution of the four matrix equations: the initial and dual equations and two Lyapunov matrix equations. The algorithm is as fast as the known algorithm of the sign-function type but is more stable.

The equation (10.1) arises from so-called linear-quadratic optimal control problem: minimises the value

$$J(x(0), u) = 2\,\|A\| \int_0^\infty [(C^*(CFC)^{-1}C\,x(t), x(t)) + (B(B^*GB)^{-1}B^*u(t), u(t))]\,dt$$

in the solutions of system

$$\frac{d}{dt}x = Ax + Bu.$$

Here the matrices A, B, C, F and G with a few natural assumptions related to stabilizability and detectability, the solution to this problem is well-known to be in feed-back form

$$u(t) = -2\,\|A\|\,(B^*GB)^{-1}B^*\,\Lambda\,x(t),$$

where Λ is the unique symmetric nonnegative definite solution of (10.1).

For definition a condition number S.K.Godunov suggested [26] to complete the equation (10.1) with a dual Lourier-Riccati matrix equation

$$MA^* + AM + 2\,\|A\|\,[\,B(B^*GB)^{-1}B^* - M\,C^*(CFC^*)^{-1}CM\,] = 0,$$

and with two Lyapunov matrix equations

$$HK^* + KH + 2\,\|A\|\,G^{-1} = 0 \text{ and } EL + L^*E + 2\|A\|\,\|\,C^*(CFC^*)^{-1}C\,\|\,I = 0.$$

Here

$$K = A - 2 \|A\| \ B(B*GB)^{-1}B*\Lambda \quad \text{and} \quad L = A - 2 \|A\| \ M \ C*(CFC*)^{-1}C$$

must be the asymptotically stable matrices, and to use quantities $\|\Lambda\|$, $\|M\|$, $\|H\|$, $\|E\|$ as a condition number.

For computational Λ approximations it is convenient to use a two-point boundary value problem: $0 \leq t \leq T$,

$$\frac{d}{dt}\begin{pmatrix} X(t) \\ Y(t) \end{pmatrix} = W \begin{pmatrix} X(t) \\ Y(t) \end{pmatrix}, \quad \Im \begin{pmatrix} X(0) \\ Y(0) \end{pmatrix} = I, \ \Re \begin{pmatrix} X(T) \\ Y(T) \end{pmatrix} = 0, \quad (10.2)$$

where

$$W = \begin{pmatrix} A & 2\|A\|B(B*GB)^{-1}B* \\ 2\|A\|C*(CFC*)^{-1}C & -A* \end{pmatrix}, \quad \Im = [I_{2N}, 0], \ \Re = [0, I_{2N}].$$

It is known that $Y(0,T)$ converges to the solution of the Lourier-Riccati matrix equation with the increase of T. The speed of convergence depends on the condition of this equation. The proposed algorithm is iterative. On all its next steps the interval of the boundary value problem (10.2) is doubling.

Give the main steps of the algorithm.

1) Compute $\tau = 0.5/\|W\|$, $A_0 = e^{-\tau W}$ and suppose $B_0 = I$.

2) We carry out a successive recurrent determination of the matrix pairs A_1, B_1; A_2, B_2 ; ...; A_m ,B_m ; ... by means of the procedure which is reduced at every step of the recursion to the choice of the orthogonal transformation P_j, such as

$$P_j = \begin{pmatrix} B_{j-1} & A_{j-1} & 0 \\ A_{j-1} & 0 & B_{j-1} \end{pmatrix} = \begin{pmatrix} \times & \cdots & \cdots & \cdots \\ & \ddots & \cdots & \cdots \\ 0 & & \times & \cdots \end{pmatrix} = \begin{pmatrix} \cdots & \cdots & \cdots \\ O & A_j & B_j \end{pmatrix}.$$

After l steps of the algorithm we have equations

$$A_{11}^{(l)} X(T) + B_{12}^{(l)} Y(0) + B_{11}^{(l)} = 0 ;$$

$$A_{21}^{(l)} X(T) + B_{22}^{(l)} Y(0) + B_{21}^{(l)} = 0 ,$$

where $A_{ij}^{(l)}$, $B_{ij}^{(l)}$ - the N×N matrix elements of the matrices A_l, B_l.

3) Choice of the orthogonal transformation Q, such that

$$Q = \begin{pmatrix} A_{11}^{(l)} & B_{12}^{(l)} & B_{11}^{(l)} \\ A_{21}^{(l)} & B_{22}^{(l)} & B_{21}^{(l)} \end{pmatrix} = \begin{pmatrix} \times & \cdots & \cdots & \cdots \\ & \ddots & \cdots & \cdots \\ 0 & & \times & \cdots \end{pmatrix} = \begin{pmatrix} \cdots & \cdots & \cdots \\ O & R_l & F_l \end{pmatrix}.$$

where R_l is an upper triangular.

It turns out that $R_l^{-1} F_l \to \Lambda$ as $l \to \infty$. The velocity of the convergence of the approximations to Λ is estimated by the condition number [24]

$$\| R_l^{-1} F_l - \Lambda \| \leq 2 \|\Lambda\| [1 + \|M\| \|\Lambda\|] \kappa(K^*) e^{-2^{l+1} \tau \frac{|\kappa|}{\kappa(K^*)}}.$$

It is clear that $\|H\|/\lambda_{max}(G) \leq \kappa(K^*) 2\|A\|/\|K\| \leq \|H\|/\lambda_{min}(G)$.

So, we can see that if $\|\Lambda\|, \|M\|, \|H\|, \|E\|$ are less than some boundary value λ^*, m^*, h^*, e^*, it is possible to determine the number of steps so that the residual on matrix $R_l^{-1} F_l$ will be less than the given small δ. Notice, that the freedom of choice λ^*, m^*, h^*, e^* and δ^* is limited in particular by the accuracy of elementary matrix operations.

As δ it is convenient to choose the value which guarantees that we are in the region of convergence of iterative refinement procedure. This procedure allows us to compute Λ with all true digits.

If after l steps we do not get into the convergence region of iterative refinement procedure it means that our problem is ill-conditioned, i.e. at least one of the inequalities $\|\Lambda\| \geq \lambda^*$, $\|M\| \geq m^*$, $\|H\| \geq h^*$, $\|E\| \geq e^*$ is true. In this case the detection of the ill-conditions is the result of the algorithm.

References

1. Kulisch, U.W. and Miranker, W.L. (1986) The arithmetic of the digital computer: A New Approach, *SIAM Review* **28** , 1 - 40.
2. Godunov, S.K., Antonov, A.G., Kiriluk, O.P. and Kostin V.I. (1993) *Guaranteed Accuracy in Numerical Linear Algebra*, Kluwer Academic Publishers, Dordrecht.
3. Akın, Ö. and Bulgak, H. (1998) *Linear Difference Equations and Stability Theory*, Selcuk University, Research Centre of Applied Mathematics, Konya (in Turkish).
4. Godunov, S.K. and Ryaben'kii, V.S. (1964) *Theory of Difference Schemes: An Introduction*, North-Holland, Amsterdam.
5. Trefethen, L.N. (1990) Approximation theory and numerical linear algebra, in J.C. Mason and M.G. Cox (eds.), *Proceedings of the Second International Conference on Algorithms for Approximation*, Chapman and Hall, London, pp. 336-359.

6. Kostin, V.I. and Razzakov, Sh.I. (1985) On the convergence of the orthogonal-power method for calculating the spectrum, *Numerical methods of linear algebra*, Nauka, Novosibirsk, pp. 55-84 (in Russian).

7. Trefethen, L.N. (1992) Pseudospectra of matrices, in D.F. Griffiths and G.A.Watson (eds.), *Proceedings of the 14 Dundee Biennial Conference on Numerical Analysis*, Longman Sci. Tech. Publ., Harlow, pp. 234-266.

8. Kostin, V.I. (1991) On definition of matrices' spectra, *High Perfomance Computing II*.

9. Chaitin-Chatelin, F. and Fraysse, V. (1996) *Lectures on Finite Precision Computations*, SIAM, Philadelphiya.

10. Godunov, S.K. (1997) *Modern Aspects of the Linear Algebra*, Nauchnaya Kniga, Novosibirsk (in Russian).

11. Wonham, W.M. (1979) *Linear multivariable control: a geometric approach*, Springer -Verlag, New-York.

12. Bulgakov, A.Ya. (1980) An effectively calculable parameter for the stability quality of systems of linear differential equations with constant coefficients, *Siberian Math J.* **21**, 339-347.

13. Godunov, S.K. (1990) The Problem of Guaranteed Precision in Numerical Methods of Linear Algebra, *Amer. Math. Soc. Transl. (2)* **147**, pp. 65-73.

14. Bulgakov, A.Ya. (1991) The basis of guaranteed accuracy in the problem of separation of invariant subspaces for nonselfadjoint matrices, *Siberian Advances in Mathematics*, **1**, 64-108 and **2**, 1-56.

15. Van Loan, C.F. (1984) How near is a stable matrix to unstable matrix? *Linear algebra and its role in systems theory*, Brunswick, Maine, 465-478.

16. Hinrichsen, D. and Pritchard, A.J. (1986) Stability radius for structured perturbations and the algebraic Riccati equations, *Systems Control Letters* **8**, 105-113.

17. Davison, E.J. and Man, F.T. (1968) The numerical solution of A*Q+QA=-C, *IEEE Trans. Aut. Control* **AC-13**, 448-449.

18. Povzner, A.J. and Pavlov, B.V. (1973) On one method of the numerical integration of systems of ordinary differential equations, *Zh. Vychislit. Mat. Mat. Fiz.* **13**, 256-259 (in Russian).

19. Moler, C.B. and Van Loan, C.F. (1978) Nineteen dubious ways to compute the exponential of a matrix, *SIAM Review* **20** , 801 - 836.

20. Ward, R.C. (1977) Numerical computation of the matrix exponential with accuracy estimate, *SIAM J. Numerical Anal.* **14**, 600-610.

21. Golub, G.H. and Van Loan, C.F. (1989) *Matrix Computations*, John Hopkins University Press, Baltimore.

22. Bulgakov, A.Ya. (1985) Computation of the exponential function of an asymptotically stable matrix, *Numerical methods of linear algebra*, Nauka, Novosibirsk, pp. 4-17 (in Russian).

23. Bulgakov, A.Ya. and Godunov, S.K. (1985) Calculation of positive definite solutions of Lyapunov's equation, *Numerical methods of linear algebra*, Nauka, Novosibirsk, pp. 17-38 (in Russian).

24. Daletskii, Yu. L. and Krein, M.G. (1970) *Stability of solutions to differential equations in Banach space*, Nauka, Moscow (in Russian).

25. Bulgakov, A.Ya. (1995) The guaranteed accuracy for solving Riccati matrix equation, *Siberian Advances in Mathematics* **5**, 1-49.

26. Godunov, S.K. (1992) Norms of the solutions of Lourier -Riccati matrix equations as criteria of the quality of stabilizability and detectability, *Siberian Advances in Mathematics* **2**, 135-157.

ERROR CONTROL FOR ADAPTIVE SPARSE GRIDS

HANS-JOACHIM BUNGARTZ, CHRISTOPH ZENGER
Institut für Informatik, Technische Universität München
D - 80290 München, Germany

Abstract. The divide-and-conquer paradigm as well as other closely related principles of algorithmic design like recursion or hierarchy have turned out to be advantageous for a wide spectrum of numerical topics, especially in situations where adaptively refined grids have to be taken into account. Starting from a really classic example, Archimedes' solution to the problem of integrating a parabola segment, we discuss the impact of such design patterns on numerical quadrature, interpolation, and the numerical solution of elliptic PDEs.

1. Introduction

In the early days of numerical programming, recursive algorithms and hierarchical data structures were possibly of some theoretical interest, but they were not used in practical programs. The most important programming language in this context, FORTRAN 4 and later FORTRAN 77, did not even allow recursive program structures. It was not before FORTRAN 90, which adopted many features from other languages like C or PASCAL, that recursive programming and hierarchical data structures were supported. Nevertheless, many important innovative ideas in numerical programming are most adequately described by recursive constructs. Prominent examples are the Fast Fourier Transform, multilevel techniques, and domain decomposition methods. Moreover, design principles like modularity or top-down design, for example, are more easily implemented in recursive programs working on hierarchical data structures.

In this paper, we want to show that adaptivity and error control have close and very natural relations to recursion and hierarchy. To this end, we introduce the concept of hierarchical bases, the origins of which can be traced back to Archimedes' decomposition of a parabola segment by

H. Bulgak and C. Zenger (eds.), Error Control and Adaptivity in Scientific Computing, 125–157.

an infinite sequence of triangles. In this context, the stopping criterion, which is a constitutive part of every recursive program, can be realized in a very natural way via an error estimator based upon the hierarchical decomposition, i. e. on the coefficients in the hierarchical representation. Even the dependence of an algorithm on the dimension of the underlying space can be treated by a linear recursion following the ideas of Cavalieri.

Section 2 is dedicated to the presentation of the idea of Archimedes. In Section 3, this concept is generalized to higher dimensional spaces. Here, we leave well-known territory: Though the one-dimensional case leads to nothing else but the classic adaptive trapezoidal rule, its extension to higher dimension ends up in a rather modern quadrature rule which was published first by Russian mathematicians in the sixties of this century for the non-adaptive case and which seems to be new in an adaptive context. In order to lay the foundations of the more heuristic approach of Archimedes from the point of view of approximation theory, Sects. 4 and 5 deal with the underlying function spaces. We discuss their most important properties and derive the grid patterns resulting from the application of error bounds. Furthermore, a generalization of the piecewise linear case to higher order approximation is given. Finally, in Section 6, we apply the concept developed so far to the adaptive solution of elliptic partial differential equations. Again, we present only the basic principles and restrict ourselves to the Laplacian on the unit cube. With some conclusions, we close the discussion.

2. Archimedes Quadrature

Archimedes was the first who found a formula for the area of a parabola segment. He decomposed the segment into a union of an infinite number of triangles with geometrically declining area. By summing up the resulting convergent series, he was able to compute the area exactly. As an illustration of this old principle, the first few triangles are shown in Fig. 1.

We extend this idea to the calculation of the definite integral of some function f in a recursive way. To this end, we first define the functionals

$$F(x \mapsto f(x), a, b) := \int_a^b f(x) \, dx, \tag{1}$$

representing the area to be computed, and

$$S(x \mapsto f(x), a, b) = F(x \mapsto f(x), a, b) - (b - a) \cdot \frac{f(a) + f(b)}{2}, \tag{2}$$

representing the area of the segment S illustrated in the left part of Fig. 2. Next, we establish a recursive functional relationship of the segment

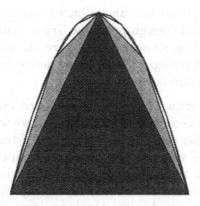

Figure 1. Decomposing the parabola segment into triangles

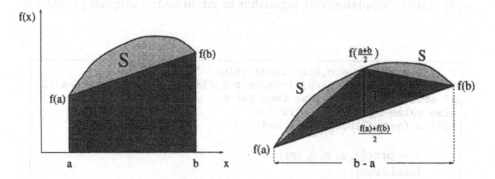

Figure 2. Segment S (left) and recursive relation (right)

functional S:

$$S(x \mapsto f(x), a, b) = T(x \mapsto f(x), a, b)$$
$$+ \; S\left(x \mapsto f(x), a, \frac{a+b}{2}\right)$$
$$+ \; S\left(x \mapsto f(x), \frac{a+b}{2}, b\right), \tag{3}$$

where

$$T(x \mapsto f(x), a, b) := \frac{b-a}{2} \cdot \left(f(\frac{a+b}{2}) - \frac{f(a)+f(b)}{2}\right). \tag{4}$$

Recursion (3) follows directly from the additivity of integration on subintervals. The functional T represents the area of the triangle in the right part of Fig. 2.

The reduction done so far can be interpreted as the application of a very general algorithmic design principle: Try to reduce the complexity of a problem by computing its easy parts – here the computation of the area of a trapezoid and of triangles – and try to derive functional relationships between the different unknown functionals.

The recursive formula (3) cannot be exploited directly to compute the integral, because the recursion never stops, and we would have to evaluate more and more smaller and smaller segments. Fortunately, in a numerical context, we need the value of the integral only approximately. Thus, we may stop as soon as the computed area of the triangles is smaller than some prescribed tolerance ε. The underlying algorithm is presented as a small program implemented in the Maple programming language, see Fig. 3. The notation should be understandable also for those who are not familiar with Maple. Applying this algorithm to Archimedes' original problem of

```
>   s:= proc(f,x,a,b,eps) local value;
value:=(subs(x=(a+b)/2,f)-(subs(x=a,f)+subs(x=b,f))/2)*(b-a)/2;
if abs(evalf(value))<eps then value
else value+s(f,x,a,(a+b)/2,eps)
+s(f,x,(a+b)/2,b,eps) fi; end;
```

$$
\begin{aligned}
s := &\mathbf{proc}(f,\, x,\, a,\, b,\, eps) \\
&\mathbf{local}\ value; \\
&value := (\text{subs}(x = (a+b)/2,\, f) \\
&\qquad\qquad - (\text{subs}(x = a,\, f) + \text{subs}(x = b,\, f))/2) \times (b-a)/2; \\
&\mathbf{if}\ abs(evalf(value)) < eps\ \mathbf{then}\ value \\
&\mathbf{else}\ value + s(f,\, x,\, a,\, (a+b)/2,\, eps) \\
&\qquad\qquad + s(f,\, x,\, (a+b)/2,\, b,\, eps) \\
&\mathbf{fi} \\
&\mathbf{end}
\end{aligned}
$$

```
>   ar:= proc(f,x,a,b,eps)
(subs(x=a,f)+subs(x=b,f))*(b-a)/2 +s(f,x,a,b,eps); end;
```

$$
\begin{aligned}
ar := &\mathbf{proc}(f,\, x,\, a,\, b,\, eps) \\
&(\text{subs}(x = a,\, f) + \text{subs}(x = b,\, f)) \times (b-a)/2 + s(f,\, x,\, a,\, b,\, eps) \\
&\mathbf{end}
\end{aligned}
$$

Figure 3. One-dimensional Archimedes quadrature

the parabola, we have to evaluate the following Maple statements:

```
> f:= x*(1-x);
```
$$f := x\,(1 - x)$$
```
> a:=ar(f,x,0,1,0.01);
```
$$a := \frac{21}{128}$$
```
> evalf(a-1/6);
```
$$-.002604166667$$
```
> a:= ar(f,x,0,1,0.00001);
```
$$a := \frac{1365}{8192}$$
```
> evalf(a-1/6);
```
$$-.00004069010417$$

As we can see, both computed results are indeed approximations of the exact value $\frac{1}{6}$, and the error caused by the truncation of the series gets smaller if we reduce ε. Nevertheless, the algorithm of Fig. 3 has some serious flaws. First, it is not very robust. The area of a triangle, which is used as an error indicator to decide whether to continue the evaluation or not, may be very small even if the area of the overall segment is not small at all. This can happen if the local interval contains a turning point of f. Of course, this inherent lack of robustness can never be overcome totally in adaptive algorithms. It can only be reduced, by checking the area of the triangles on two subsequent levels of refinement, for example.

Moreover, there is no direct relationship between the error control variable ε and the actual error of our computation. As a remedy for this problem, we estimate the error by replacing the function on the smallest intervals, i.e. on the level where the recursion stops, by parabolic arcs whose area can be computed exactly. If the area of the triangle is t, the area under the parabolic arc is $\frac{4}{3}t$. Thus, the sum of all those differences $\frac{1}{3}t$ provides an estimate of the overall error. Note that this algorithm means nothing else but neglecting derivatives of f of order higher than 2 on the finest level, which, asymptotically, yields good estimates if f is sufficiently smooth. The above ideas are implemented in our second program shown in Fig. 4. It returns a two-element vector of the computed integral and the estimated error. For those who want to experiment on this code in order to get a deeper insight into the underlying algorithm, the program also collects the set p of points where f is evaluated.

For the remainder of this section, we apply the program of Fig. 4 to several examples. For the case of a parabola, the error estimate is exact (as expected due to its definition), and the points of the resulting grid are equidistant (see Fig. 5). For the exponential function $f(x) = e^{5x}$, Fig. 6 shows that the grid is very much refined on the right part of the interval. Furthermore, the error estimate is close to the actual error and even better if ε gets smaller. In order to show the robustness of our program, Fig. 7 gives the results for the oscillating function $f(x) := \sin(20x)$, which has

```
>   s:= proc(f,x,a,b,eps,st,p) local value,sti;
value:=(subs(x=(a+b)/2,f)-(subs(x=a,f)+subs(x=b,f))/2)*(b-a)/2;
sti:=abs(evalf(value))<4*eps;
p:=eval(p) union {[(a+b)/2,evalf(subs(x=(a+b)/2,f))]};
if st and sti then [value,value/3];
else
[value,0]+s(f,x,a,(a+b)/2,eps,sti,p)+s(f,x,(a+b)/2,b,eps,sti,p)
fi; end;
```

$$s := \mathbf{proc}(f, x, a, b, eps, st, p)$$
$$\mathbf{local}\ value,\ sti;$$
$$value := (\mathrm{subs}(x = (a + b)/2, f)$$
$$- (\mathrm{subs}(x = a, f) + \mathrm{subs}(x = b, f))/2) \times (b - a)/2;$$
$$sti := \mathrm{abs}(\mathrm{evalf}(value)) < 4 \times eps;$$
$$p := \mathrm{eval}(p)\ \mathrm{union}\ \{[(a + b)/2,\ \mathrm{evalf}(\mathrm{subs}(x = (a + b)/2, f))]\};$$
$$\mathbf{if}\ st\ \mathbf{and}\ sti\ \mathbf{then}\ [value,\ 1/3 \times value]$$
$$\mathbf{else}\ [value,\ 0] + \mathrm{s}(f, x, a, (a + b)/2, eps, sti, p)$$
$$+ \mathrm{s}(f, x, (a + b)/2, b, eps, sti, p)$$
$$\mathbf{fi}$$
$$\mathbf{end}$$

```
>   ar:= proc(f,x,a,b,eps,st,p)
[(subs(x=a,f)+subs(x=b,f))*(b-a)/2,0] +s(f,x,a,b,eps,st,p); end;
```

$$ar := \mathbf{proc}(f, x, a, b, eps, st, p)$$
$$[(\mathrm{subs}(x = a, f) + \mathrm{subs}(x = b, f)) \times (b - a)/2,\ 0] + \mathrm{s}(f, x, a, b, eps, st, p)$$
$$\mathbf{end}$$

Figure 4. One-dimensional quadrature with error estimation

several turning points in the unit interval.

The program developed so far is very elementary. It is just an adaptive variant of the trapezoidal rule and could be improved in many ways, for example with the help of extrapolation or higher order interpolation. The purpose of this section was to show that a straightforward recursive approach immediately yields adaptivity and error control as an inherent and crucial part of the algorithmic design. In the next section, we shall see how this simple principle leads to rather new algorithms, if it is generalized to higher dimensional integration.

3. Archimedes-Cavalieri Quadrature

Now, we proceed to multi-dimensional quadrature and start with the two-dimensional case. Since we have functions of several variables, we have to

```
>  p:={};
>  f:= x*x;
```
$$p := \{\}$$
$$f := x^2$$
```
>  a:=ar(f,x,0,1,0.001,false,'p');
```
$$a := [\frac{171}{512}, \frac{-1}{1536}]$$
```
>  a[1]-1/3;
```
$$\frac{1}{1536}$$
```
>  PLOT(POINTS(p[],SYMBOL(BOX)),AXESSTYLE(BOX),VIEW(0..1,0..1));
```

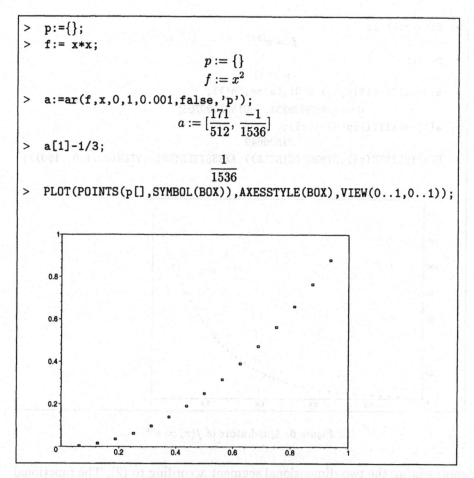

Figure 5. Quadrature of $f(x) := x^2$

use a notation that is a little bit more involved. As before, we first define the functionals

$$F_2((x_1, x_2) \mapsto f(x_1, x_2), (a_1, a_2), (b_1, b_2))$$
$$:= \int_{a_1}^{b_1} \int_{a_2}^{b_2} f(x_1, x_2) \, dx_2 \, dx_1, \qquad (5)$$

representing the integral to be computed, and

$$S_2((x_1, x_2) \mapsto f(x_1, x_2), (a_1, a_2), (b_1, b_2))$$
$$:= F_2((x_1, x_2) \mapsto f(x_1, x_2), (a_1, a_2), (b_1, b_2))$$
$$- F_1\left(x_1 \mapsto (b_2 - a_2) \cdot \frac{f(x_1, a_2) + f(x_1, b_2)}{2}, a_1, b_1\right), \qquad (6)$$

132

```
> f:=exp(5*x);
                                f := e^(5 x)
> p:={};
                                p := {}
> a:=evalf(ar(f,x,0,1,0.01,false,'p'));
                   a := [29.49799831, -.01535499706]
> a[1]-evalf((exp(5)-1)/5);
                           .01536649
> PLOT(POINTS(p[],SYMBOL(CIRCLE)),AXESSTYLE(BOX),VIEW(0..1,0..150));
```

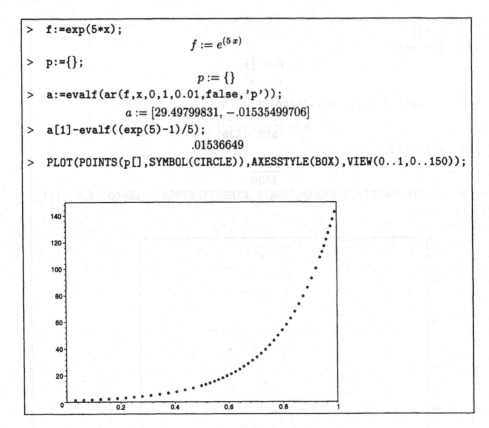

Figure 6. Quadrature of $f(x) := e^{5x}$

representing the two-dimensional segment according to (2). The functional F_1 in (6) is just the (one-dimensional) functional F from (1). Note that this generalization to 2 D is just what is known as Cavalieri's principle of iterated integration.

In an analogous way to the one-dimensional case in (2), we get a recursive functional relationship for the segment functional S_2:

$$
S_2((x_1, x_2) \mapsto f(x_1, x_2), (a_1, a_2), (b_1, b_2))
$$
$$
:= F_1 \left(x_1 \mapsto \frac{b_2 - a_2}{2} \cdot \left(f(x_1, \frac{a_2 + b_2}{2}) - \frac{f(x_1, a_2) + f(x_1, b_2)}{2} \right), a_1, b_1 \right)
$$
$$
+ S_2 \left((x_1, x_2) \mapsto f(x_1, x_2), (a_1, a_2), (b_1, \frac{a_2 + b_2}{2}) \right)
$$
$$
+ S_2 \left((x_1, x_2) \mapsto f(x_1, x_2), (a_1, \frac{a_2 + b_2}{2}), (b_1, b_2) \right) .
$$
$$
(7)
$$

Formula (7) can easily be generalized to the multi-dimensional case by

```
>  p:={};
>  f:= sin(20*x);
```
$$p := \{\}$$
$$f := \sin(20\,x)$$
```
>  a:=evalf(ar(f,x,0,1,0.01,false,'p'));
```
$$a := [.01971000680, .01129243442]$$
```
>  evalf((1-cos(20))/20- a[1]);
```
$$.00988589011$$
```
>  PLOT(POINTS(p[],SYMBOL(BOX)),AXESSTYLE(BOX),VIEW(0..1,-1..1));
```

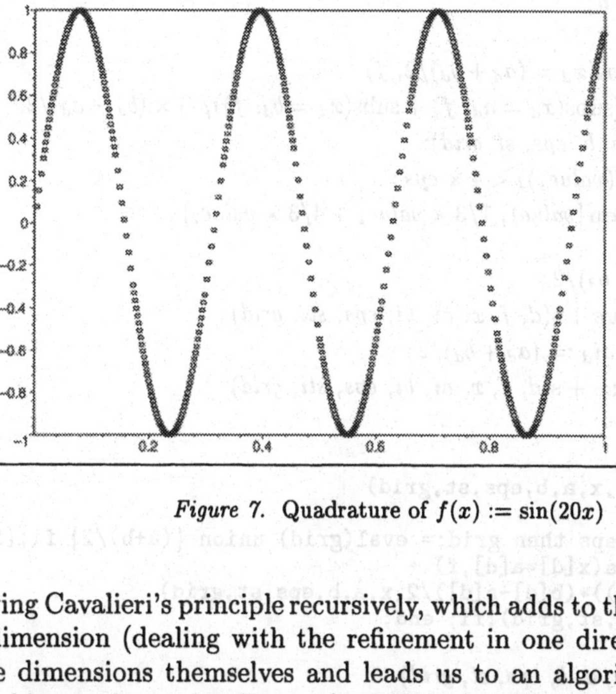

Figure 7. Quadrature of $f(x) := \sin(20x)$

applying Cavalieri's principle recursively, which adds to the recursion within one dimension (dealing with the refinement in one direction) a recursion of the dimensions themselves and leads us to an algorithm for a general and arbitrary dimensionality n. An implementation of this algorithm, once again realized as a Maple program, can be seen in Fig. 8.

Fig. 8 already shows the more elaborate version, including the trick to improve robustness of the algorithm and including an error estimator. Moreover, we added some graphical interface in order to be able to plot the grid, i.e. the points where the function is evaluated. With the above explanations, the essential features of the program should be easy to understand.

The index d denotes the dimensionality of the integral. In the one-dimensional case, the S-functional vanishes if f is linear. In the general d-dimensional case, the S-functionals vanish if f is d-linear, i.e. linear with

```
>  s:= proc(d,f,x,a,b,eps,st,grid)
local value,sti,ai,bi;ai:=a;bi:=b;
value:=ar(d-1,(subs(x[d]=(a[d]+b[d])/2,f)-(subs(x[d]=a[d],f)+
subs(x[d]=b[d],f))/2)*(b[d]-a[d])/2,x,a,b,eps,st,grid);
sti:=abs(evalf(value[1]))<4*eps;
if st and sti then [value[1],value[1]/3+value[2]*4/3];
else bi[d]:=(a[d]+b[d])/2;value:=value+s(d,f,x,ai,bi,eps,sti,grid);
bi[d]:=b[d]; ai[d]:=(a[d]+b[d])/2;
value:=value+s(d,f,x,ai,bi,eps,sti,grid)
fi; end;
```

$s := \mathbf{proc}(d, f, x, a, b, eps, st, grid)$
$\quad \mathbf{local}\ value,\ sti,\ ai,\ bi;$
$\qquad ai := a;\quad bi := b;$
$\qquad value := \mathrm{ar}(d - 1,$
$\qquad\qquad\qquad (\mathrm{subs}(x_d = (a_d + b_d)/2,\ f)$
$\qquad\qquad\qquad\quad - (\mathrm{subs}(x_d = a_d,\ f) + \mathrm{subs}(x_d = b_d,\ f))/2) \times (b_d - a_d)/2,$
$\qquad\qquad\qquad x,\ a,\ b,\ eps,\ st,\ grid);$
$\qquad sti := \mathrm{abs}(\mathrm{evalf}(value_1)) < 4 \times eps\,;$
$\qquad \mathbf{if}\ st\ \mathbf{and}\ sti\ \mathbf{then}\ [value_1,\ 1/3 \times value_1 + 4/3 \times value_2]$
$\qquad \mathbf{else}$
$\qquad\qquad bi_d := (a_d + b_d)/2\,;$
$\qquad\qquad value := value + \mathrm{s}(d,\ f,\ x,\ ai,\ bi,\ eps,\ sti,\ grid)\,;$
$\qquad\qquad bi_d := b_d;\quad ai_d := (a_d + b_d)/2\,;$
$\qquad\qquad value := value + \mathrm{s}(d,\ f,\ x,\ ai,\ bi,\ eps,\ sti,\ grid)$
$\qquad \mathbf{fi}$
$\quad \mathbf{end}$

```
>  ar:= proc(d,f,x,a,b,eps,st,grid)
if d=0 then
if abs(eval(f))>eps then grid:= eval(grid) union {(a+b)/2} fi;[f,0]
else ar(d-1,(subs(x[d]=a[d],f) +
subs(x[d]=b[d],f))*(b[d]-a[d])/2,x,a,b,eps,st,grid)
+s(d,f,x,a,b,eps,st,grid);fi; end;
```

$ar := \mathbf{proc}(d,\ f,\ x,\ a,\ b,\ eps,\ st,\ grid)$
$\quad \mathbf{if}\ d = 0\ \mathbf{then}$
$\qquad \mathbf{if}\ eps < \mathrm{abs}(\mathrm{eval}(f))\ \mathbf{then}\ grid := \mathrm{eval}(grid)\ \mathrm{union}\ \{(a + b)/2\}\ \mathbf{fi};\ [f,\ 0]$
$\qquad \mathbf{else}\ \mathrm{ar}(d - 1,\ (\mathrm{subs}(x_d = a_d,\ f) + \mathrm{subs}(x_d = b_d,\ f)) \times (b_d - a_d)/2,\ x,\ a,\ b,$
$\qquad\qquad eps,\ st,\ grid) + \mathrm{s}(d,\ f,\ x,\ a,\ b,\ eps,\ st,\ grid)$
$\qquad \mathbf{fi}$
$\quad \mathbf{end}$

Figure 8. Multi-dimensional quadrature

respect to each coordinate direction. In addition to that, S_d vanishes if f is a polynomial of degree $p \le 2d - 1$. Consequently, in this case, we do not need any function evaluations in the interior of the unit cube. Therefore, the simplest polynomial of interest is

$$f(x_1, \ldots, x_d) := x_1^2 \cdot x_2^2 \cdot \ldots \cdot x_d^2, \tag{8}$$

which is of order $2d$. For the numerical results, see Fig. 9.

```
>  f:= (x[1]*x[2])^2;
                              f := x₁² x₂²
>  grid:={}:
>  a:=(ar(2,f,x,[0,0],[1,1],0.000001,false,'grid'));
                              14561    23
                        a := [------, -------]
                             131072  1179648
>  [evalf(1/9-a[1]),evalf(a[2])];
                     [.00001949734158, .00001949734158]
>  PLOT(POINTS(grid[],SYMBOL(CIRCLE)),AXESSTYLE(BOX),VIEW(0..1,0..1));
```

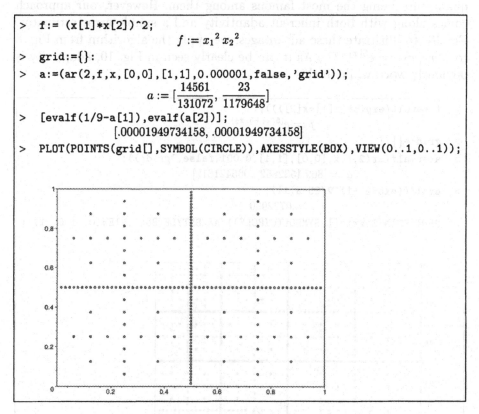

Figure 9. Quadrature of $f(x_1, x_2) := x_1^2 \cdot x_2^2$

For the example of (8) shown in Fig. 9, we observe that the resulting grids are no longer equidistant (as in the one-dimensional case), but a rather complicated sparse subset of an equidistant grid which is, however, of some regular structure. The approximation spaces related to these *sparse grids* will be discussed in the next two sections. The non-adaptive numerical integration on sparse grids dates back to Smolyak [14] (see the discussion in Sect. 5), more recent results can be found in [11]. For more information concerning the adaptive case, see [4, 5].

Sparse grids are much more economical than standard equidistant (or *full*) grids. If we integrate sufficiently smooth functions with the help of

136

a multi-dimensional trapezoidal rule, we need $O(\varepsilon^{-d/2})$ function evaluations to obtain an error of $O(\varepsilon)$. In contrast to that, with sparse grids, one needs only $O(\varepsilon^{-1/2-\delta})$ for any positive δ. Thus, the order of the number of function evaluations, put in relation with the resulting error, is essentially independent of the dimensionality d. From literature, some other quadrature schemes having such a property are known, with Monte Carlo quadrature being the most famous among them. However, our approach comes along with both inherent adaptivity and a simple error estimator. Finally, to illustrate these advantages, we apply the algorithm from Fig. 8 to $f(x_1, x_2) := e^{5(x_1+x_2)}$. As it can be clearly seen in Fig. 10, adaptivity is obviously worthwhile in this case.

```
>   f:=evalf(exp(5*(x[1]+x[2])));
                  f := e^(5. x_1 + 5. x_2)
>   grid:={}:
>   a:=evalf(ar(2,f,x,[0,0],[1,1],0.001,false,'grid'));
                  a := [869.1532862, .065421612]
>   evalf((exp(5)-1)^2/25-a[1]);
                  .0722930
>   PLOT(POINTS(grid[],SYMBOL(CIRCLE)),AXESSTYLE(BOX),VIEW(0..1,0..1));
```

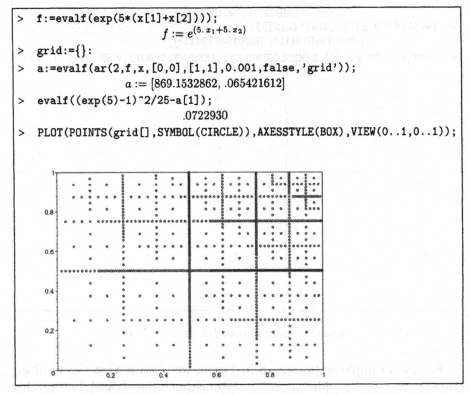

Figure 10. Quadrature of $f(x_1, x_2) := e^{5(x_1+x_2)}$

4. Hierarchical Tensor Product Spaces

The approach of Archimedes to the approximate calculation of one-dimensional integrals and its straightforward generalization to $d > 1$ led us to sparse patterns of grid points of a quite regular structure. For a formal derivation of those sparse grids, we recall the use of hierarchical bases of tensor-product-type for the approximation of functions and define a suitable hierarchical subspace decomposition of the underlying function spaces. Starting from that, an optimization process allows the definition of approximation spaces that are optimal with respect to a norm-depending cost-benefit setting. For a detailed analysis and discussion of the following, including derivations and proofs, we refer to [6, 8].

4.1. SUBSPACE DECOMPOSITION

On the d-dimensional unit cube $\bar{\Omega} := [0, 1]^d$, we consider multivariate functions $u : \bar{\Omega} \to \mathbb{R}$ with (in some sense) bounded weak mixed derivatives

$$D^\alpha u := \frac{\partial^{|\alpha|_1} u}{\partial x_1^{\alpha_1} \dots \partial x_d^{\alpha_d}} \tag{9}$$

up to $|\alpha|_1 = 2d$. Here and in the following, we use a multi-index notation $\alpha \in \mathbb{N}_0^d$ with standard component-wise arithmetic operations and relations, with constants like $0 := (0, \dots, 0)$, and with the discrete norms $|\alpha|_1 := \sum_{j=1}^d \alpha_j$ and $|\alpha|_\infty := \max_{1 \le j \le d} \alpha_j$. For $q \in \{\infty, 2\}$, we study the spaces

$$X^q(\Omega) := \left\{ u : \mathbb{R}^d \supset \bar{\Omega} \to \mathbb{R}, \ D^\alpha u \in L_q(\Omega), \ |\alpha|_\infty \le 2 \right\}, \tag{10}$$

together with their homogeneous counterparts $X_0^q(\Omega)$ of functions vanishing on the boundary (which, for reasons of simplicity, are of primary interest, here), and we introduce the respective semi-norms

$$|u|_\infty := \left\| D^2 u \right\|_\infty, \qquad |u|_2 := \left\| D^2 u \right\|_2. \tag{11}$$

Now, with $\mathbf{l} = (l_1, \dots, l_d) \in \mathbb{N}^d$ indicating the level in a multivariate sense, we consider the family $\{\Omega_\mathbf{l}, \mathbf{l} \in \mathbb{N}^d\}$ of d-dimensional rectangular and equidistant grids on $\bar{\Omega}$ with mesh size $\mathbf{h}_\mathbf{l} := (h_{l_1}, \dots, h_{l_d}) := 2^{-\mathbf{l}}$, i.e. with, in general, different mesh widths in the different coordinate directions. The grid points of $\Omega_\mathbf{l}$ are just the points

$$\mathbf{x}_{\mathbf{l},\mathbf{i}} := (x_{l_1,i_1}, \dots, x_{l_d,i_d}) := \mathbf{i} \cdot \mathbf{h}_\mathbf{l}, \qquad 0 \le \mathbf{i} \le 2^\mathbf{l}. \tag{12}$$

In $\Omega_\mathbf{l}$'s inner grid points, we define piecewise d-linear basis functions

$$\phi_{\mathbf{l},\mathbf{i}}(\mathbf{x}) := \prod_{j=1}^d \phi_{l_j,i_j}(x_j) \tag{13}$$

as products of one-dimensional piecewise linear functions $\phi_{l_j,i_j}(x_j)$ with support $[x_{l_j,i_j} - h_{l_j}, x_{l_j,i_j} + h_{l_j}] = [(i_j - 1)h_{l_j}, (i_j + 1)h_{l_j}]$, which can all be

138

obtained from the standard hat function by dilation and translation. For a fixed l, the $\phi_{l,i}$ span the space

$$V_l := \text{span}\left\{\phi_{l,i} : 1 \leq i \leq 2^l - 1\right\} \qquad (14)$$

of piecewise d-linear functions with respect to the interior of Ω_l. Obviously, the $\phi_{l,i}$ form a basis of V_l, its standard *nodal point basis*. In contrast to that, the hierarchical increments

$$W_k := \text{span}\left\{\phi_{k,i} : i \in I_k\right\} \qquad (15)$$

with $I_k := \left\{i \in \mathbb{N}^d : 1 \leq i \leq 2^k - 1,\ i_j \text{ odd for all } 1 \leq j \leq d\right\}$ lead to the standard piecewise d-linear *hierarchical basis* of V_l due to

$$V_l = \bigoplus_{k \leq l} W_k, \qquad (16)$$

which generalizes the well-known one-dimensional basis shown in Fig. 11 to the d-dimensional case by means of our tensor product approach.

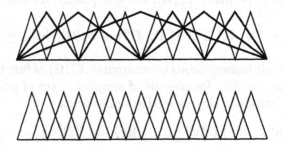

Figure 11. Piecewise linear hierarchical basis vs. nodal point basis

Now, we can define

$$V := \sum_{l_1=1}^{\infty} \cdots \sum_{l_d=1}^{\infty} W_{(l_1,\dots,l_d)} = \bigoplus_{l \in \mathbb{N}^d} W_l \qquad (17)$$

with its natural hierarchical basis $\left\{\phi_{l,i} : i \in I_l, l \in \mathbb{N}^d\right\}$. Except for completion with respect to the H^1-norm, V is just the underlying Sobolev space $H_0^1(\Omega)$, i.e. $\bar{V} = H_0^1(\bar{\Omega})$. Note that any $u \in H_0^1(\bar{\Omega})$ and, consequently, any $u \in X_0^q(\Omega)$ can be uniquely split by

$$u(x) = \sum_l u_l(x), \quad u_l(x) = \sum_{i \in I_l} v_{l,i} \cdot \phi_{l,i}(x) \in W_l, \qquad (18)$$

where the $v_{l,i} \in \mathbb{R}$ are the coefficient values of the hierarchical basis representation of u (also called *hierarchical surplus*).

Concerning the subspaces W_l, the crucial point is how important W_l is for the interpolation of some given $u \in X_0^q(\Omega)$ and what computational and storage cost come along with it. The following lemma provides answers to these two questions.

Lemma 1 *The dimension of the subspace W_l, i. e. the number of degrees of freedom or basis functions related to W_l, is given by*

$$|W_l| = 2^{|l-1|_1}. \tag{19}$$

Furthermore, let $u \in X_0^q(\Omega)$, $q \in \{2, \infty\}$, be split according to (18). With respect to the L_∞-, the L_2-, and the energy norm $\|u\|_E := \left(\int_\Omega \nabla u \cdot \nabla u \, dx\right)^{1/2}$, the following estimates for the components $u_l \in W_l$ hold:

$$
\begin{aligned}
\|u_l\|_\infty &\leq 2^{-d} \cdot 2^{-2 \cdot |l|_1} \cdot |u|_\infty, \\
\|u_l\|_2 &\leq 3^{-d} \cdot 2^{-2 \cdot |l|_1} \cdot |u|_2, \\
\|u_l\|_E &\leq \frac{1}{2 \cdot 12^{(d-1)/2}} \cdot 2^{-2 \cdot |l|_1} \cdot \left(\sum_{j=1}^{d} 2^{2 \cdot l_j}\right)^{1/2} \cdot |u|_\infty.
\end{aligned}
\tag{20}
$$

4.2. FINITE DIMENSIONAL APPROXIMATION SPACES

Now, the information gathered above concerning the hierarchical subspaces W_l will be used to construct finite dimensional approximation spaces for $X_0^q(\Omega)$. The idea is to select a finite subset of subspaces W_l characterized by their multi-indices l and to restrict the summation in (18) to this subset of active levels. The simplest and natural choice is

$$V_n^{(\infty)} := \bigoplus_{|l|_\infty \leq n} W_l, \tag{21}$$

which is just the usual space of piecewise d-linear functions on the rectangular grid of equidistant mesh size $h_n = 2^{-n}$ in each coordinate direction. If we arrange the subspaces W_l in a (potentially infinite) scheme of subspaces as indicated in Fig. 12 for the 2 D case, $V_n^{(\infty)}$ corresponds to a square sector of subspaces, see Fig. 13. Obviously, the number of inner grid points in the underlying grid is

$$\left|V_n^{(\infty)}\right| = (2^n - 1)^d = O(2^{d \cdot n}) = O(h_n^{-d}). \tag{22}$$

140

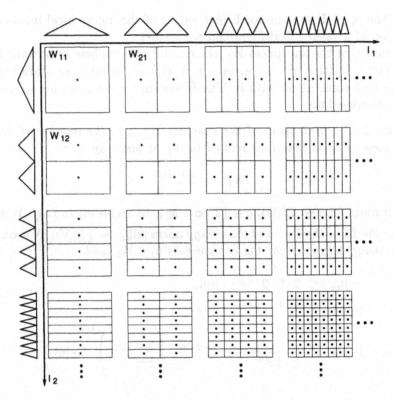

Figure 12. Scheme of subspaces for $d = 2$: Each square represents one subspace W_l with its associated grid points. The supports of the corresponding basis functions have the same mesh size h_l and cover the domain Ω.

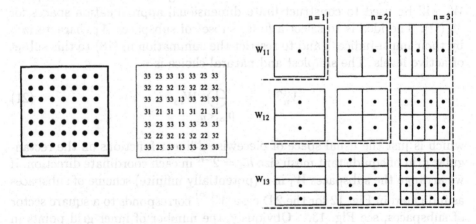

Figure 13. The (full) grid of $V_3^{(\infty)}$, $d = 2$, and the assignment of grid points to subspaces

For the error $u - u_n^{(\infty)}$ of the interpolant $u_n^{(\infty)} \in V_n^{(\infty)}$ of some given $u \in X_0^q(\Omega)$ with respect to the different norms we are interested in, the following lemma summarizes the respective results. Note that we get the same order of accuracy as in standard approximation theory, although our regularity assumptions differ from those normally used there.

Lemma 2 *For* $u \in X_0^q(\Omega)$, *the following estimates for the different norms of* $u - u_n^{(\infty)}$ *hold:*

$$\|u - u_n^{(\infty)}\|_\infty \leq \frac{d}{6^d} \cdot 2^{-2n} \cdot |u|_\infty \qquad = O(h_n^2),$$

$$\|u - u_n^{(\infty)}\|_2 \leq \frac{d}{9^d} \cdot 2^{-2n} \cdot |u|_2 \qquad = O(h_n^2), \qquad (23)$$

$$\|u - u_n^{(\infty)}\|_E \leq \frac{d^{3/2}}{2 \cdot 3^{(d-1)/2} \cdot 6^{d-1}} \cdot 2^{-n} \cdot |u|_\infty = O(h_n).$$

If we look at (22) and (23), the crucial drawback of $V_n^{(\infty)}$, the so-called *curse of dimensionality*, can be seen clearly. With increasing d, the number of degrees of freedom that are necessary to achieve an accuracy of $O(h)$ or $O(h^2)$, resp., grows exponentially. In order to tackle this problem, we look for optimal discrete approximation spaces $V^{(\mathrm{opt})}$ that are solutions of

$$\max_{u \in X_0^q: \, |u|=1} \|u - u_{V^{(\mathrm{opt})}}\| = \min_{U \subset V: \, |U|=w} \max_{u \in X_0^q: \, |u|=1} \|u - u_U\| \qquad (24)$$

for some prescribed cost or work count w. I. e., the aim is to profit from invested work as much as possible. Of course, any potential solution $V^{(\mathrm{opt})}$ of (24) has to be expected to depend on the norm and on the semi-norm used to measure the error of u's interpolant $u_U \in U$ or its smoothness, resp. According to our hierarchical setting, only spaces U of the type $U := \bigoplus_{l \in I} W_l$ for an arbitrary finite index set $\mathbf{I} \subset \mathbb{N}^d$ are candidates. Such a proceeding can be seen as an a priori optimization, since the resulting spaces and grids depend on the problem classes, but not on the given problem, which is somewhat unusual in the PDE context, but a quite common approach in numerical quadrature (think of the Gauß quadrature rules or of the so-called *low-discrepancy-formulas*; see [12], e. g.).

We begin with a reformulation of (24). Defining $c(\mathbf{l}) := |W_l|$ as the (local) cost of W_l and $b(\mathbf{l}) \in \mathbb{N}^d$ as a measure for W_l's (local) benefit, that is W_l's contribution to the overall interpolant (the squared and suitably scaled upper bounds from (20), for example), we can restrict our search for an optimal grid $\mathbf{I} \subset \mathbb{N}^d$ to all $\mathbf{I} \subset \mathbf{I}^{(\mathrm{max})} := \{1, \ldots, N\}^d$ for a sufficiently large N without loss of generality. Next, based on the corresponding local

142

quantities, the global cost and benefit are suitably defined:

$$C(\mathbf{I}) := \sum_{\mathbf{l} \in \mathbf{I}^{(\max)}} x(\mathbf{l}) \cdot c(\mathbf{l}), \qquad B(\mathbf{I}) := \sum_{\mathbf{l} \in \mathbf{I}^{(\max)}} x(\mathbf{l}) \cdot b(\mathbf{l}), \qquad (25)$$

where the binary $x(\mathbf{l})$ equals 1 if and only if $\mathbf{l} \in \mathbf{I}$. This leads us to the desired reformulation of (24):

$$\max_{\mathbf{I} \subset \mathbf{I}^{(\max)}} B(\mathbf{I}) \qquad \text{with} \qquad \sum_{\mathbf{l} \in \mathbf{I}^{(\max)}} C(\mathbf{I}) = w, \qquad (26)$$

which is just a *binary knapsack problem* that can be written as

$$\max_{\mathbf{x}} \mathbf{b}^T \mathbf{x} \qquad \text{with} \qquad \mathbf{c}^T \mathbf{x} = w, \qquad (27)$$

where $\mathbf{b} \in \mathbb{N}^M$, $\mathbf{c} \in \mathbb{N}^M$, $\mathbf{x} \in \{0,1\}^M$, and, without loss of generality, $w \in \mathbb{N}$. The *rational* variant of (27), i. e. $\mathbf{x} \in ([0,1] \cap \mathbb{Q})^M$, is easy to solve:

1. rearrange the order such that $b_1/c_1 \geq b_2/c_2 \geq \ldots \geq b_M/c_M$,

2. let $r := \max \left\{ j : \sum_{i=1}^{j} c_i \leq w \right\}$, $\quad s := w - \sum_{i=1}^{r} c_i$,

3. $x_i := 1$ for $i \leq r$, $\quad x_{r+1} := \frac{s}{c_{r+1}}$, $\quad x_i := 0$ for $i \geq r + 2$

results in an optimal solution vector $\mathbf{x} \in ([0,1] \cap \mathbb{Q})^M$. Since our fictitious work count w can be arbitrarily chosen, we can force x to be binary and, thus, an optimal solution of the binary problem.

Consequently, the *global* optimization problem (24) or (27), resp., can be reduced to the discussion of the *local* cost-benefit ratios b_i/c_i or $b(\mathbf{l})/c(\mathbf{l})$ of the underlying subspaces $W_{\mathbf{l}}$. Those subspaces with the best cost-benefit ratios are taken into account first, and the smaller these ratios become, the more negligible the underlying subspaces turn out to be. Thus, for the construction of optimal approximation spaces in the next section, these local cost-benefit ratios have to be studied in detail. As already mentioned, this discussion will depend on the norm used to measure the error.

5. Sparse Grids: Definition and Properties

5.1. OPTIMIZATION WITH RESPECT TO THE MAXIMUM NORM

Due to (20), the L_2-norm and the L_∞-norm of $W_{\mathbf{l}}$'s contribution $u_{\mathbf{l}}$ to the hierarchical representation (18) of $u \in X_0^q(\Omega)$ are of the same order of magnitude. Therefore, we restrict ourselves to the maximum norm. According to the previous discussion, the L_∞-based local cost-benefit ratio can be written as

$$cbr_\infty(\mathbf{l}) := \frac{b_\infty(\mathbf{l})}{c(\mathbf{l})} := \frac{2^{-4 \cdot |\mathbf{l}|_1} \cdot |u|_\infty^2}{4^d \cdot 2^{|\mathbf{l} - \mathbf{1}|_1}} = \frac{1}{2^d} \cdot |u|_\infty^2 \cdot \gamma_\infty(\mathbf{l}) \qquad (28)$$

with

$$\gamma_\infty(\mathbf{l}) := 2^{-5 \cdot |\mathbf{l}|_1} . \tag{29}$$

The bigger $\gamma_\infty(\mathbf{l})$ is for some subspace $W_\mathbf{l}$, the more efficient is it to include $W_\mathbf{l}$ into our approximation space. Figure 14 shows the order of magnitude of $\log_2(\gamma_\infty(\mathbf{l}))$ in the subspace scheme from Fig. 12. Obviously, a diagonal

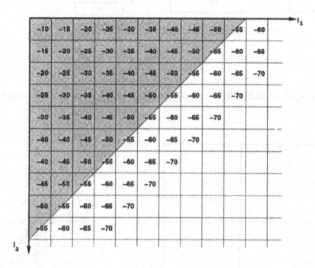

Figure 14. Order of magnitude of $\log_2(\gamma_\infty(\mathbf{l}))$ in 2 D: A smaller value means a worse cost-benefit ratio.

cut in the subspace scheme turns out to produce an optimal approximation space in our sense and results in the definition

$$V_n^{(1)} := \bigoplus_{|\mathbf{l}|_1 \leq n+d-1} W_\mathbf{l}, \tag{30}$$

which is just the standard L_∞-based sparse grid introduced in [16] and illustrated in Fig. 15.

Figure 16 gives two examples of sparse grids: a regular 2 D one and an adaptively refined 3 D one (compare with Figs. 9 and 10). Note that, although the a priori optimization process yields regular grid patterns, the hierarchical approach allows a straightforward access to adaptation, as we have already seen in Sects. 2 and 3 and as we will discuss once more when we deal with numerical examples.

Concerning the basic properties of $V_n^{(1)}$, we get the following results:

144

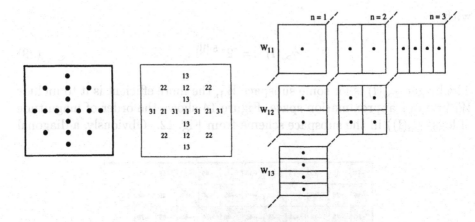

Figure 15. The sparse grid of $V_3^{(1)}$, $d = 2$, and the assignment of grid points to subspaces

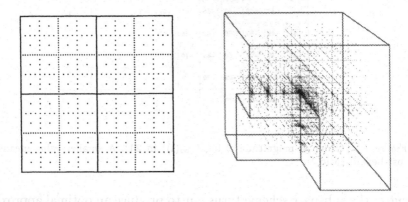

Figure 16. Sparse grids: regular example (left) and adaptive one (right)

Theorem 1 *The dimension of the space $V_n^{(1)}$, i. e. the number of degrees of freedom or inner grid points, is given by*

$$\left| V_n^{(1)} \right| = \sum_{i=0}^{n-1} 2^i \cdot \binom{d-1+i}{d-1} = O(h_n^{-1} \cdot |\log_2 h_n|^{d-1}). \quad (31)$$

For the L_∞-, the L_2-, and the energy norm, the following upper bounds for the interpolation error $u - u_n^{(1)}$ of a function $u \in X_0^q(\Omega)$ in the sparse grid space $V_n^{(1)}$ hold:

$$||u - u_n^{(1)}||_\infty \leq \frac{2 \cdot |u|_\infty}{8^d} \cdot 2^{-2n} \cdot A(d,n) = O(h_n^2 \cdot n^{d-1}),$$

$$||u - u_n^{(1)}||_2 \leq \frac{2 \cdot |u|_2}{12^d} \cdot 2^{-2n} \cdot A(d,n) = O(h_n^2 \cdot n^{d-1}), \quad (32)$$

$$||u - u_n^{(1)}||_E \leq \frac{d \cdot |u|_\infty}{2 \cdot 3^{(d-1)/2} \cdot 4^{d-1}} \cdot 2^{-n} = O(h_n).$$

The above results show the crucial improvements of the sparse grid space $V_n^{(1)}$ in comparison with $V_n^{(\infty)}$. The number of degrees of freedom is reduced significantly, whereas the accuracy is only slightly deteriorated – for the L_∞- and the L_2-norm – or stays even of the same order if the error is measured in the energy norm. This lessens the curse of dimensionality, but it does not overcome it completely.

5.2. OPTIMIZATION WITH RESPECT TO THE ENERGY NORM

Since this result is optimal with respect to both the L_∞- and the L_2-norm, a further improvement can only be expected if we turn to a setting based on another norm. Therefore, now, we study the grids that result if we optimize with respect to the energy norm. As before, we define

$$cbr_E(l) := \frac{b_E(l)}{c(l)} = \frac{3}{6^d} \cdot |u|_\infty^2 \cdot \gamma_E(l) \quad (33)$$

as the local cost-benefit ratio, where

$$\gamma_E(l) := 2^{-5 \cdot |l|_1} \cdot \sum_{j=1}^{d} 4^{l_j} \quad (34)$$

due to (20). Figure 17 shows the order of magnitude of $\log_2(\gamma_E(l))$ in our two-dimensional subspace scheme and illustrates the space $V_{30}^{(E)}$ resulting from this energy-based optimization. As we can see, we get an additional sparsening effect, and the contour lines are no longer diagonal lines. Without going into details, we summarize the main results for $V_n^{(E)}$:

Theorem 2 *The energy-based sparse grid space $V_n^{(E)}$ is a subspace of $V_n^{(1)}$, and its dimension fulfils*

$$|V_n^{(E)}| \leq 2^n \cdot \frac{d}{2} \cdot e^d = O(2^n) = O(h_n^{-1}). \quad (35)$$

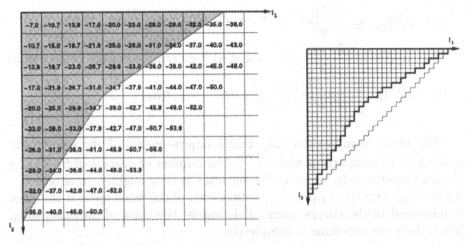

Figure 17. Order of magnitude of $\log_2(\gamma_E(1))$ in 2 D (left; a smaller value means a worse cost-benefit ratio) and resulting scheme of subspaces in $V_{30}^{(E)}$ (right)

The energy norm of the interpolation error $u - u_n^{(E)}$ of a function $u \in X_0^q(\Omega)$ in $V_n^{(E)}$ is of the order $O(h_n)$ and bounded by

$$\|u - u_n^{(E)}\|_E \leq \frac{d \cdot |u|_\infty}{3^{(d-1)/2} \cdot 4^{d-1}} \cdot \left(\frac{1}{2} + \left(\frac{5}{2}\right)^{d-1}\right) \cdot 2^{-n}. \qquad (36)$$

The crucial advantage of this energy-based approach is that the curse of dimensionality can be completely overcome. In both the bound for the number of grid points and the error bound with respect to the energy norm, the n-dependent terms are free of any d-dependencies: There is an order of $O(2^n)$ for the dimension and $O(2^{-n})$ for the interpolation error. Furthermore, it follows from Theorem 2 that the number of grid points $N(\varepsilon)$ that are necessary to compute an interpolant $u_n^{(E)} \in V_n^{(E)}$ of some $u \in X_0^q(\Omega)$ of the accuracy $\varepsilon = \varepsilon(h_n)$ is uniformly bounded in d:

$$N(\varepsilon) \leq 2012 \cdot |u|_\infty \cdot \varepsilon^{-1}. \qquad (37)$$

Though this is an asymptotic result whose practical relevance may be reduced by the influence of the factors that are constant with respect to n, it shows the potential of the sparse grid approach, especially for large d. Such an independence of d is well-known from Monte Carlo quadrature, but with a worse efficiency (i. e. an exponent of -2 in (37)).

Finally, note that, apart from the Archimedes quadrature discussed in Sect. 2, there a more historical roots of the sparse grid concept. Smolyak

[14] studied classes of quadrature formulas Q_n of resolution n of the type

$$Q_n^{(d)} f := \left(\sum_{i=0}^{n} \left(Q_i^{(1)} - Q_{i-1}^{(1)} \right) \otimes Q_{n-i}^{(d-1)} \right) f \qquad (38)$$

that are based on a tensor product $\otimes$ of lower dimensional quadrature operators and that resemble the definition of $V_n^{(1)}$, which can be written as

$$V_n^{(d,1)} := V_n^{(1)} = \sum_{|\mathbf{l}|_1 \leq n+d-1} W_\mathbf{l}^{(d)} = \sum_{l=1}^{n} W_l^{(1)} \otimes V_{n+d-1-l}^{(d-1,1)}, \qquad (39)$$

i.e. in an explicit tensor product form, too. The spaces the functions f suitable for the Smolyak approach live in are typically spaces of bounded ($\mathcal{L}^p$-integrable) mixed derivatives, too. A similar approach for the approximation of periodic multivariate functions of bounded mixed derivatives are Babenko's *hyperbolic crosses* [1]. Here, Fourier monomials or coefficients, respectively, are taken from sets of the type

$$\Gamma(n) := \left\{ \mathbf{k} \in \mathbb{Z}^d : \prod_{j=1}^{d} \max\{|k_j|, 1\} \leq n \right\}. \qquad (40)$$

For a detailed discussion of those methods, we refer to [15].

5.3. HIGHER ORDER APPROXIMATION

Up to now, we have restricted ourselves to piecewise (d-) linear hierarchical bases. However, in our tensor product setting, there is a very straightforward way to profit from higher regularity assumptions by using appropriate basis polynomials of higher order. This approach seems to be especially attractive since the effect of higher order approximation is increased further on by the reduced or even overcome curse of dimensionality.

The main idea is to define hierarchical polynomials of arbitrary degree p such that they fit into the scheme developed so far. That is, d-dimensional polynomials shall be products of d 1D ones (with degrees that may vary w.r.t. the different coordinate directions), and the character of the elements (their shape, their size, their global smoothness ($\mathcal{C}^0$), and the number of degrees of freedom living in an element) shall be preserved. Therefore, we use interpolation conditions in a grid point's ancestors, which may be located *outside* the respective element, for the construction of both the local interpolant and of the basis polynomial. The principle of this *hierarchical Lagrangian interpolation* is illustrated in Fig. 18. The resulting hierarchical interpolant is piecewise of the chosen degree p, which may vary from

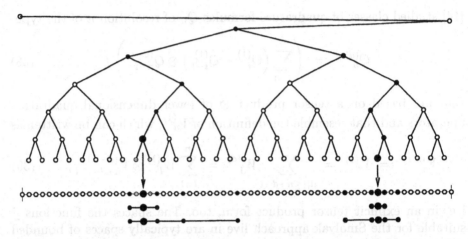

Figure 18. Hierarchical ancestors (here: boundary points of the respective basis function's support and two more; solid) used for the construction of the interpolant of degree $p = 4$ and descendants (dotted) of two grid points

element to element, but its global smoothness is still C^0, as in the piece-wise linear case. The crucial point is that the *local* interpolant on a certain element is defined with the help of hierarchical conditions, i.e. conditions that live *outside* this element.

Keeping in mind our interest in finite element methods for PDEs, we need the explicit definition of basis polynomials $\phi_{\mathbf{l},\mathbf{i}}^{(\mathbf{p})}$ of degree $\mathbf{p}$ in the d-dimensional case or, due to the tensor product approach, $\phi_{l,i}^{(p)}$ of degree p in 1 D. In contrast to the piecewise linear situation where all basis functions can be generated from the standard hat function, we now get different types of polynomials for a given degree p, depending on the position of the ancestors of the respective local grid point. As an example, all four occurring types of quartic basis polynomials are given in Fig. 19.

Figure 20 shows one of the most striking properties of our hierarchical basis polynomials: Even for higher degree p, they do not differ that much concerning their shape. In fact, an analysis of the $\phi_{l,i}^{(p)}$ reveals that this "optical" argument is true for arbitrary p. This is the main reason of the excellent numerical behaviour of the hierarchical Lagrangian polynomials which we will discuss in the PDE section.

Of course, the use of basis polynomials of degree p influences the con-tribution of a subspace W_l or, now better, $W_l^{(\mathbf{p})}$ to the subspace splitting and the corresponding interpolants, as given in (18) for the piecewise linear case. If we restart the optimization process discussed above, now for general p or $\mathbf{p}$, we essentially get the same grid patterns as in the linear case. Taking

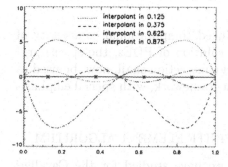 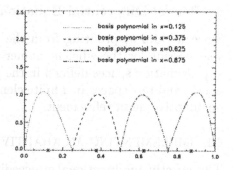

Figure 19. Hierarchical basis polynomials for $p = 4$: construction via hierarchical Lagrangian interpolation (left) and used restriction to the respective support (right)

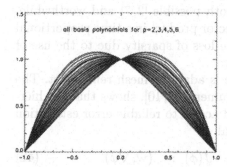 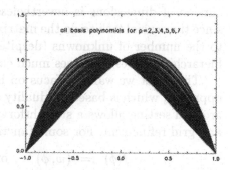

Figure 20. All hierarchical basis polynomials for $p \leq 6$ (31 different types; left) and $p \leq 7$ (63 different types; right) with respect to the support $[-1, 1]$

into account the more rigorous regularity assumptions (spaces $X_0^{q,p+1}(\Omega)$ of bounded weak derivatives up to $D^{(p+1)\cdot \mathbf{1}}$ instead of D^2), we end up with the following summary of the enhanced approximation properties of the L_∞-based sparse grid spaces $V_n^{(p,1)}$ of degree p:

Theorem 3 *For the L_∞-, the L_2-, and the energy norm, the following upper bounds for the error of the interpolant $u_n^{(p,1)} \in V_n^{(p,1)}$ of a function $u \in X_0^{q,p+1}(\Omega)$ hold:*

$$\left\| u - u_n^{(p,1)} \right\|_{L_\infty} = O(h_n^{p+1} \cdot n^{d-1}),$$

$$\left\| u - u_n^{(p,1)} \right\|_{L_2} = O(h_n^{p+1} \cdot n^{d-1}), \qquad (41)$$

$$\left\| u - u_n^{(p,1)} \right\|_{E} = O(h_n^{p}).$$

6. Partial Differential Equations

Since we aim at profiting from the reduced computational complexity of sparse grids for the efficient numerical solution of PDEs, the sparse grid approximation spaces defined in the previous section shall, now, be used as ansatz and test spaces in a finite element framework, with special emphasis on adaptive mesh refinement.

6.1. IMPLEMENTING AN ADAPTIVE FINITE ELEMENT ALGORITHM

The strictly unidirectional proceeding we have studied for the Cavalieri quadrature in Sect. 3 is much more difficult to implement for finite element methods, which is mainly due to the element-element interactions in the stiffness matrix. However, a unidirectional implementation based on a recursion of dimensions is nevertheless both possible [2, 7] and worthwhile, since the crucial task to do the matrix vector product in a time proportional to the number of unknowns (despite the loss of sparsity due to the use of hierarchical bases) becomes much easier.

The point we want to focus on here is adaptive mesh refinement. The approach, which is based on duality arguments [3, 10], shows that the hierarchical setting allows a straightforward access to reliable error estimation and grid refinement. For some functional

$$J(\phi) := (w, \phi) \quad \text{or} \quad J(\phi) := (\mathbf{w}, \nabla\phi) \qquad (42)$$

with suitable weight functions w or $\mathbf{w}$, resp., we want to estimate $J(e)$ for the error e of the original or *primal* problem's finite element solution. The choice $w := \phi/(\phi, \phi)$ yields the L_2-error, $\mathbf{w} := \nabla\phi/(\nabla\phi, \nabla\phi)$ produces the energy error, and a Dirac impulse as weight leads to the error in the respective point. Now, the *dual* problem is defined with the help of the original operator's adjoint and with J as right-hand side. For the Laplacian, we get

$$(\nabla z, \nabla\phi) = J(\phi) \qquad \forall \phi \in X_0^{q,p+1}(\Omega). \qquad (43)$$

We start from the hierarchical representation of both u, the primal problem's solution, and z, the dual problem's one,

$$u(\mathbf{x}) = \sum_{\mathbf{l}} \sum_{\mathbf{i} \in I_{\mathbf{l}}} u_{\mathbf{l},\mathbf{i}} \cdot \phi_{\mathbf{l},\mathbf{i}}^{(\mathbf{p})}(\mathbf{x}), \qquad z(\mathbf{x}) = \sum_{\mathbf{l}} \sum_{\mathbf{i} \in I_{\mathbf{l}}} z_{\mathbf{l},\mathbf{i}} \cdot \phi_{\mathbf{l},\mathbf{i}}^{(\mathbf{p})}(\mathbf{x}), \qquad (44)$$

where the hierarchical coefficients $u_{\mathbf{l},\mathbf{i}}$ and $z_{\mathbf{l},\mathbf{i}}$ decrease geometrically. Then, a short calculation provides

$$J(e) \approx \sum_{\text{n. l. o. r}} z_{\mathbf{l},\mathbf{i}} \cdot \left(\phi_{\mathbf{l},\mathbf{i}}^{(\mathbf{p})}, -\Delta e \right) =: \sum_{\text{n. l. o. r}} z_{\mathbf{l},\mathbf{i}} \cdot r_{\mathbf{l},\mathbf{i}}, \qquad (45)$$

with the summation covering the next level of refinement (n. l. o. r.), as a simple estimator for the error $J(e)$ on a certain grid (see [13] for details). Those elements with the biggest contribution to the sum (45) are subject to a refinement step in each coordinate direction.

6.2. NUMERICAL EXAMPLES

In the smooth case, numerical examples show that the finite element solution in $V_n^{(p,1)}$ is of the same quality of approximation as the interpolant is (which was to be expected for the energy norm, but is non-trivial for the maximum and L_2-norm; see [6, 7], for example). Therefore, we focus on situations with a need for adaptive refinement.

As our first example, we look at the 2 D Laplace equation with a solution that, though having a singularity only outside $\bar{\Omega}$ in $\mathbf{x} = (-0.1, 0.5)$, allows to study the impact of adaptive refinement.

Example 1 *Let* $\Delta u(\mathbf{x}) = 0$ *in* $\Omega =]0, 1[^2$ *with Dirichlet boundary conditions and the harmonic solution*

$$u(\mathbf{x}) = \frac{0.1 \cdot (x_1 + 0.1)}{(x_1 + 0.1)^2 + (x_2 - 0.5)^2}. \tag{46}$$

Figure 21 illustrates the case of the regular sparse grid spaces $V_n^{(1,1)}$ (left) and $V_n^{(2,1)}$ (right). Compared to the behaviour of the interpolation error dis-

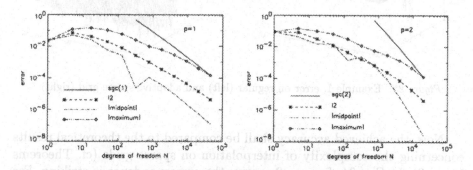

Figure 21. Example 1: convergence on $V_n^{(p,1)}$ for $p = 1$ (left) and $p = 2$ (right); the solid lines indicate the respective sparse grid convergence expected from interpolation (sgc; position of curve chosen for reasons of clarity)

cussed in the previous section, things are not too convincing for reasonable N. However, decisive progress can be made if we apply our L_2-adaptivity. Figure 22 shows a gain in accuracy with higher p that is comparable to the perfectly smooth and regular situation. In Fig. 23, we compare the error on a regular sparse grid and on an adaptively refined one. As expected, the

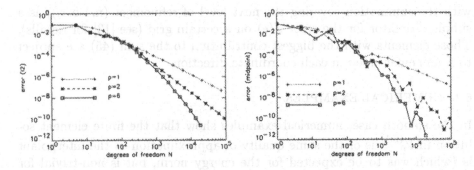

Figure 22. Example 1: L_2-error (left) and error in the midpoint $\left(\frac{1}{2}, \frac{1}{2}\right)$ for adaptive refinement, $p \in \{1, 2, 6\}$

L_2-adaptation process reduces the error equally over the whole domain. In contrast to that, regular sparse grids show large errors near the singularity.

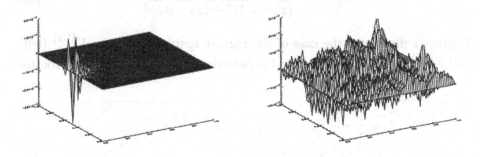

Figure 23. Example 1: error on regular (left) and adaptive sparse grid (right)

Now, the achieved accuracy shall be compared to the theoretical results concerning the complexity of interpolation on sparse grids (cf. Theorems 1 and 3). In Fig. 24, for $p = 2$, again, the correspondence is striking. For $p = 6$, it seems that the asymptotic behaviour needs bigger values of N to appear. Nevertheless, with the adaptive refinement advancing, the higher order accuracy comes to fruition.

Finally, to get an impression of the adaptation process, Fig. 25 shows two adaptively refined grids with 12 853 grid points ($p = 2$, left) and 10 965 grid points ($p = 6$, right).

As our second example, we study the three-dimensional Laplace equation with a perfectly smooth solution.

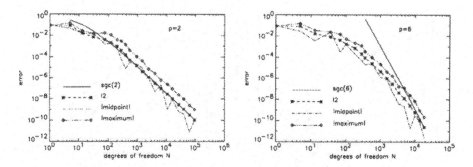

Figure 24. Example 1: convergence with adaptive mesh refinement for $p = 2$ (left) and $p = 6$ (right); the solid lines indicate the respective expected sparse grid convergence (sgc; position of curve chosen for reasons of clarity)

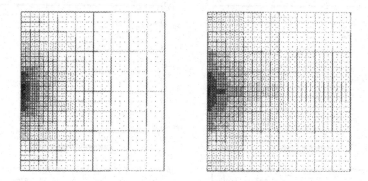

Figure 25. Example 1: adaptive grids for $p = 2$ (left), and $p = 6$ (right)

Example 2 *Let $\Delta u(\mathbf{x}) = 0$ in $\Omega =]0,1[^3$ with Dirichlet boundary conditions and the solution*

$$u(\mathbf{x}) = \frac{\sinh(\sqrt{2}\pi x_1)}{\sinh(\sqrt{2}\pi)} \cdot \sin(\pi x_2) \cdot \sin(\pi x_3). \qquad (47)$$

For the polynomial degrees $p \in \{1,\dots,4\}$, the following Fig. 26 compares the accuracy of the finite element solution with respect to the error's L_2-norm or to its absolute value in the midpoint of Ω, resp. Again, the effects of the improved approximation properties of our hierarchical polynomial bases are evident. Fig. 27 shows that we do already come quite close to the asymptotic behaviour predicted for the quality of the mere interpolation problem.

Up to now, he have studied examples on regular sparse grids and on grids that result from an refinement process driven by L_2-adaptation and

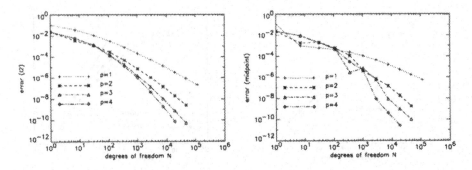

Figure 26. Example 2: L_2-error (left) and error in the midpoint $\left(\frac{1}{2}, \frac{1}{2}, \frac{1}{2}\right)$ (right) for regular sparse grid spaces $V_n^{(p,1)}$, $p \in \{1, \ldots, 4\}$

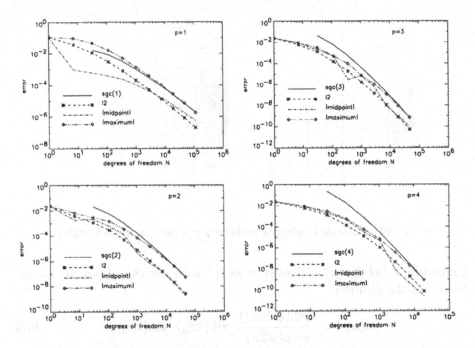

Figure 27. Example 2: convergence on $V_n^{(p,1)}$, $p \in \{1, 2\}$ (left) and $p \in \{3, 4\}$ (right); the solid lines indicate the respective expected sparse grid convergence (sgc; position of curve chosen for reasons of clarity)

aiming at a uniform reduction of the error in the whole domain. Now, as the third example to be discussed here, we consider the adaptation process with respect to the point functional aiming at an error reduction in the midpoint, especially.

Example 3 *Let* $\Delta u(\mathbf{x}) = 0$ *in* $\Omega :=]0, 1[\times]-\frac{1}{2}, \frac{1}{2}[$ *with Dirichlet boundary* *conditions and the solution*

$$u(\mathbf{x}) = Re\left((x_1 + i \cdot x_2)^{\frac{1}{2}}\right). \qquad (48)$$

We can learn from (45) that the resulting error estimator is related to the respective diagonal element of the stiffness matrix, i. e. to the *energy* of a basis function. Since we have an $O(h^p)$-behaviour of the energy norm, the energy decreases of the order $O(h^{2p})$. Consequently, a convergence behaviour of $O(h^{2p})$ for the error in the midpoint can be expected. Actually, for the quadratic case, this can bee seen from Fig. 28 for both the perfectly smooth case (the 2 D counterpart of Example 2) and the root singularity from Example 3. Note that the solid line now indicates an N^{-4}-behaviour. The results for the root singularity show that we can tackle problems with a "real" singularity, too. In Fig. 29, we present the error resulting from adap-

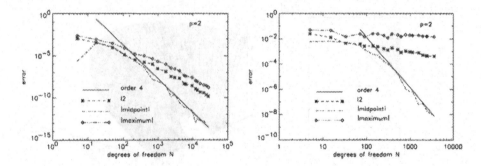

Figure 28. Reducing the error in the midpoint for $p = 2$: smooth case (left) and root singularity $Re(z^{1/2})$ (right); the solid lines indicate the expected convergence N^{-4} (position of curve chosen for reasons of clarity)

tive refinement based on the point functional for Example 1. Obviously, the error is primarily reduced near the midpoint. The right-hand side of Fig. 29 shows the underlying adaptive grid consisting of 12 767 grid points.

Finally, in Fig. 30, we want to demonstrate that our approach is not restricted to academic domains, but can be generalized with the help of mapping techniques. For details, we refer to [7, 9].

7. Concluding Remarks

In this paper, we discussed the impact of recursion and hierarchy on the design of efficient numerical algorithms. Though we presented only the main ideas with the help of simple model problems, the scope of this approach is much wider. For the case of PDEs, for example, results for more general domains or operators, resp., are very promising. More detailed studies concerning both PDEs and quadrature will be in the centre of future work.

156

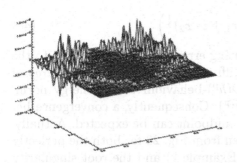
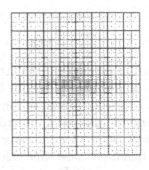

Figure 29. Reducing the error in the midpoint for $p = 2$ (smooth case): error (left) and adaptively refined grid (right)

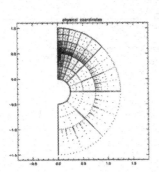
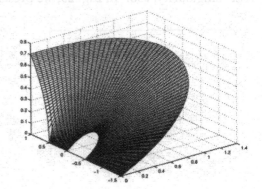

Figure 30. Laplace equation with a singularity on the boundary: adaptive sparse grid (left) and corresponding solution (right)

References

1. K. I. BABENKO, *Approximation by trigonometric polynomials in a certain class of periodic functions of several variables*, Soviet Math. Dokl., 1 (1960), pp. 672–675. Russian original in Dokl. Akad. Nauk SSSR, 132 (1960), pp. 982–985.

2. R. BALDER AND C. ZENGER, *The solution of multidimensional real Helmholtz equations on sparse grids*, SIAM J. Sci. Comp., 17 (1996), pp. 631–646.

3. R. BECKER AND R. RANNACHER, *A feed-back approach to error control in finite element methods: basic analysis and examples*, East-West J. Numer. Math., 4 (1996), pp. 237–264.

4. T. BONK, *Ein rekursiver Algorithmus zur adaptiven numerischen Quadratur mehrdimensionaler Funktionen*, Dissertation, Institut für Informatik, TU München, 1994.

5. ———, *A new algorithm for multi-dimensional adaptive numerical quadrature*, in Adaptive Methods – Algorithms, Theory, and Applications, W. Hackbusch and G. Wittum, eds., vol. 46 of NNFM, Vieweg, Braunschweig/Wiesbaden, 1994, pp. 54–68.

6. H.-J. BUNGARTZ, *Finite Elements of Higher Order on Sparse Grids*, Habilitationsschrift, Institut für Informatik, TU München, and Shaker Verlag, Aachen, 1998.

7. H.-J. BUNGARTZ AND T. DORNSEIFER, *Sparse grids: Recent developments for el-*

liptic partial differential equations. To appear in Multigrid Methods V, LNCSE, Springer, Berlin/Heidelberg, 1998. Also available as report TUM-I9702, Institut für Informatik, TU München, 1997.

8. H.-J. BUNGARTZ AND M. GRIEBEL, *A note on the complexity of solving Poisson's equation for spaces of bounded mixed derivatives.* To appear in J. Complexity, 1999.

9. T. DORNSEIFER, *Diskretisierung allgemeiner elliptischer Differentialgleichungen in krummlinigen Koordinatensystemen auf dünnen Gittern*, Dissertation, Institut für Informatik, TU München, 1997.

10. K. ERIKSSON, D. ESTEP, P. HANSBO, AND C. JOHNSON, *Adaptive Finite Elements*, Springer, Berlin/Heidelberg, 1996.

11. T. GERSTNER AND M. GRIEBEL, *Numerical integration using sparse grids*, Numerical Algorithms, 18 (1998), pp. 209–232.

12. A. R. KROMMER AND C. W. UEBERHUBER, *Numerical Integration on Advanced Computer Systems*, vol. 848 of LNCS, Springer, Berlin/Heidelberg, 1994.

13. S. SCHNEIDER AND C. ZENGER, *Multigrid methods for hierarchical adaptive finite elements.* Subm. to proc. GAMM-workshop on multigtid methods, Bonn, 1998.

14. S. A. SMOLYAK, *Quadrature and interpolation formulas for tensor products of certain classes of functions*, Soviet Math. Dokl., 4 (1963), pp. 240–243. Russian original in Dokl. Akad. Nauk SSSR, 148 (1963), pp. 1042–1045.

15. V. N. TEMLYAKOV, *Approximation of Functions with Bounded Mixed Derivative*, vol. 178 of Proc. Steklov Inst. of Math., AMS, Providence, Rode Island, 1989.

16. C. ZENGER, *Sparse grids*, in Parallel Algorithms for Partial Differential Equations, W. Hackbusch, ed., vol. 31 of NNFM, Vieweg, Braunschweig/Wiesbaden, 1991.

ORTHOGONAL MATRIX DECOMPOSITIONS IN SYSTEMS AND CONTROL

P. M. VAN DOOREN

Dept. Math. Eng., Université catholique de Louvain, Belgium

Abstract. In this paper we present several types of orthogonal matrix decompositions used in systems and control. We focus on those related to eigenvalue and singular value problems and include generalizations to several matrices.

1. Introduction

In systems and control theory, one often uses state space models to represent a dynamical system. In such models the relation between inputs $u(t) \in \mathcal{R}^m$ and outputs $y(t) \in \mathcal{R}^p$ is described via the use of a state $x(t) \in \mathcal{R}^n$ and a system of first order differential equations :

$$\begin{cases} E\dot{x}(t) &= Ax(t) + Bu(t) \\ y(t) &= Cx(t) + Du(t), \end{cases} \tag{1}$$

where E and A are real $n \times n$ matrices and B, C and D are real $n \times m$, $p \times n$ and $p \times m$ matrices, respectively. In the above model (1) the input, output and state vectors are continuous time functions. An analogous model is used for discrete time vectors functions u_k, y_k and x_k, now involving a system of first order difference equations :

$$\begin{cases} Ex_{k+1} &= Ax_k + Bu_k \\ y_k &= Cx_k + Du_k. \end{cases} \tag{2}$$

If one takes the Laplace transform of (1) and the z-transform of (2), then both can be represented by the system of algebraic equations :

$$\begin{cases} \lambda Ex(.) &= Ax(.) + Bu(.) \\ y(.) &= Cx(.) + Du(.), \end{cases} \tag{3}$$

159

H. Bulgak and C. Zenger (eds.), Error Control and Adaptivity in Scientific Computing, 159–175.
© 1999 *Kluwer Academic Publishers. Printed in the Netherlands.*

where λ stands for the differential operator and the difference operator in the two respective cases. The transfer function of this model is then obtained by eliminating the state $x(.)$:

$$T(\lambda) = C(\lambda E - A)^{-1}B + D, \qquad (4)$$

and describes directly the relation between inputs and outputs :

$$y(.) = T(\lambda)u(.).$$

Notice finally, that the transfer function (and hence the input/output behavior) is not affected by the system equivalence transformation

$$\{E, A, B, C, D\} \Longrightarrow \{\hat{E}, \hat{A}, \hat{B}, \hat{C}, \hat{D}\} \doteq \{SET, SAT, SB, CT, D\}. \qquad (5)$$

An important subclass of these models consists of the so-called standard state space models where $E = I$, in which case the state vector $x(.)$ is given explicitly by the first equation in (3). The equivalence transformation (5) now becomes a similarity transformation since $\hat{E} = ST$ must also be the identity :

$$\{A, B, C, D\} \Longrightarrow \{\hat{A}, \hat{B}, \hat{C}, \hat{D}\} \doteq \{T^{-1}AT, T^{-1}B, CT, D\}. \qquad (6)$$

Although these state space models are not the only ones used for systems and control purposes, they are the ones that have been most heavily studied as far as numerical algorithms are concerned (see e.g. [15]). We will assume for the sequel that the system under consideration is given in such a form and that the model parameters are actually known (i.e. the system has already been identified). Once the system is given, one typically has to analyze its properties (frequency response, poles/zeros, stability, robustness, ...) and then design a particular controller in order to improve some characteristics or to satisfy certain design criteria (tracking, robustness, optimality criteria, ...).

2. Condensed versus canonical forms

Many analysis and design problems are well understood these days and their theoretical solution is often described in terms of so called canonical forms, which have been defined for state space models of multivariable linear systems. These forms are typically very sparse since they are described with a minimum number of parameters. Therefore they often allow to efficiently characterize all solutions to a particular problem, which is of course very appealing. Unfortunately, it has been shown that these forms are also very sensitive to compute and amount to a coordinate transformation that can be very poorly conditioned. For most analysis and design problems encountered in linear system theory, one can as well make use of so-called *condensed forms*, which can be obtained under orthogonal transformations [25]. Such transformations have become a major tool in the development of reliable numerical linear algebra algorithms. A first reason for this is the numerical sensitivity of the problem at hand. The sensitivity (or *conditioning*) of several problems in linear algebra can be expressed in terms of norms, singular values or angles between spaces and each of these are invariant under orthogonal transformations. These transformations therefore allow to reformulate the problem in a new coordinate system which is more appropriate for solving the problem, and this without affecting its sensitivity. A second reason in the numerical stability of the algorithm used for solving the problem. Most decomposition involving orthogonal transformations can be obtained by a sequence of Givens or Householder transformations which can be performed in a numerically stable manner. The concatenation of such transformations can also be performed in a backward stable manner because numerical errors resulting from previous steps are indeed maintained in norm throughout subsequent steps since these transformations (and their inverse) have 2-norm equal to 1. Condensed forms obtained under orthogonal transformations are therefore more appropriate for solving several systems and control problems involving (generalized) state space models.

We explain this below with the special example of analyzing the poles of a single-input/single-output system given in standard state space model (i.e. $E = I$, $S = T^{-1}$ and $m = p = 1$ in the above models). For such models the poles are the eigenvalues of the matrix A and the classical form describing the fine structure of these eigenvalues is the Jordan canonical form. So we choose the similarity transformation (6) where $\hat{A}_J = T^{-1}AT$ is in Jordan canonical form. For convenience, we give the transformed system $\{\hat{A}_J, \hat{B}_J, \hat{C}_J, \hat{D}_J\}$ in the form of a compound matrix (the reason of this will become apparent later) :

$$\left[\begin{array}{c|c} \hat{B}_J & \hat{A}_J \\ \hline \hat{D}_J & \hat{C}_J \end{array}\right] \doteq \left[\begin{array}{c|cccccccc} \times & \lambda_1 & 1 & 0 & 0 & 0 & 0 & 0 \\ \times & 0 & \lambda_1 & 1 & 0 & 0 & 0 & 0 \\ \times & 0 & 0 & \lambda_1 & 0 & 0 & 0 & 0 \\ \times & 0 & 0 & 0 & \lambda_2 & 1 & 0 & 0 \\ \times & 0 & 0 & 0 & 0 & \lambda_2 & 0 & 0 \\ \times & 0 & 0 & 0 & 0 & 0 & \lambda_3 & 0 \\ \times & 0 & 0 & 0 & 0 & 0 & 0 & \lambda_4 \\ \hline \times & \times & \times & \times & \times & \times & \times & \times \end{array}\right], \qquad (7)$$

where we assume for simplicity that the eigenvalues λ_i are real. Notice that we chose our example such that there is only one Jordan block associated with each individual eigenvalue, but in general this does not have to be the case. This form not only describes the poles of the system, but contains more information, like e.g. the partial fraction description of the transfer function. But a large disadvantage of the form is that it requires a state space transformation T to put A in its Jordan form $\hat{A}_J$, and that the norms T and T^{-1} can not be bounded in general. On the other hand, when one restricts T to be orthogonal, then so is T^{-1} and both are bounded in norm. Under such transformations, one can always reduce A to triangular form, called the *Schur form*, which also has the eigenvalues on diagonal :

$$
\left[\begin{array}{c|c} \hat{B}_S & \hat{A}_S \\ \hline \hat{D}_S & \hat{C}_S \end{array} \right] \doteq \left[\begin{array}{c|ccccccc} \times & \lambda_1 & \times & \times & \times & \times & \times & \times \\ \times & 0 & \lambda_1 & \times & \times & \times & \times & \times \\ \times & 0 & 0 & \lambda_1 & \times & \times & \times & \times \\ \times & 0 & 0 & 0 & \lambda_2 & \times & \times & \times \\ \times & 0 & 0 & 0 & 0 & \lambda_2 & \times & \times \\ \times & 0 & 0 & 0 & 0 & 0 & \lambda_3 & \times \\ \times & 0 & 0 & 0 & 0 & 0 & 0 & \lambda_4 \\ \hline \times & \times & \times & \times & \times & \times & \times & \times \end{array} \right]. \tag{8}
$$

If one is only interested in computing the poles of the system, it is well known that the latter form is numerically much more reliable and actually requires less computations that the Jordan form [7].

The theorem for general $n \times n$ pencils $\lambda E - A$ is known as the generalized Schur form, and applies to regular pencils (i.e. $\det(\lambda E - A) \not\equiv 0$). Its so-called *generalized eigenvalues* are the roots of $\det(\lambda E - A) = 0$.

Theorem 1 [12]
There always exist orthogonal transformations Q and Z that transform a regular pencil $\lambda E - A$ to

$$
Q^T(\lambda E - A)Z = \lambda E_S - A_S, \tag{9}
$$

where E_S is upper triangular and A_S is block upper triangular with a 1×1 diagonal block corresponding to each real generalized eigenvalue and a 2×2 diagonal block corresponding to each pair of complex conjugate generalized eigenvalues (such matrices are called *quasi triangular*). This decomposition exists for every ordering of eigenvalues in the quasi triangular form. ∎

If $E = I$ one retrieves the standard (quasi triangular) Schur decomposition $A_S = U^T A U$ based on an orthogonal similarity transformation by taking $U = Z = Q$. Notice that if E is invertible one also has

$$
Q^T A E^{-1} Q = A_S E_S^{-1}, \quad Z^T E^{-1} A Z = E_S^{-1} A_S,
$$

which are both quasi triangular matrices. Then Q and Z of the generalized Schur form can be obtained from the standard Schur forms of AE^{-1} and

$E^{-1}A$ but this "detour" should be avoided since E may be badly conditioned in general. One of the most important uses of this form is the computation of orthogonal bases for eigenspaces. Consider any (block) triangular decomposition where we partitioned the invertible matrix X conformably :

$$X^{-1}AX = \begin{bmatrix} A_{11} & A_{12} \\ 0 & A_{22} \end{bmatrix}, \quad X = [\; X_1 \;\; X_2 \;]. \tag{10}$$

Then $AX_1 = X_1 A_{11}$ which implies that $\mathcal{X} = ImX_1$ satisfies the condition for an *invariant subspace*

$$A\mathcal{X} \subset \mathcal{X}.$$

When X is orthogonal (as in the Schur decomposition) the basis X_1 is of course orthogonal as well. The corresponding concept for the generalized eigenvalue problem $\lambda E - A$ is that of *deflating subspace* defined by the condition

$$\dim(A\mathcal{X} + E\mathcal{X}) = \dim\mathcal{X}.$$

For E invertible this is easily shown to be equivalent to $E^{-1}A\mathcal{X} \subset \mathcal{X}$ and hence each deflating subspace of $\lambda E - A$ is and invariant subspace of $E^{-1}A$. The first k columns of the right transformation Z [12] are therefore an orthogonal basis for a deflating subspace of the pencil $\lambda E - A$. We refer to [12], [7], [21] for a more rigorous discussion. The use of these eigenspaces in control shows up in the solution of several matrix equations.

We illustrate this again with a standard eigenvalue problem (i.e. $E = I$). Suppose one wants to solve the quadratic matrix equation (of dimension $q \times p$) :

$$M_{21} - X M_{11} + M_{22}X - X M_{12}X = 0 \tag{11}$$

in the $q \times p$ matrix X, then it is easily verified that this is equivalent to

$$\begin{bmatrix} I_p & 0 \\ -X & I_q \end{bmatrix} \begin{bmatrix} M_{11} & M_{12} \\ M_{21} & M_{22} \end{bmatrix} \begin{bmatrix} I_p & 0 \\ X & I_q \end{bmatrix} = \begin{bmatrix} \hat{M}_{11} & \hat{M}_{22} \\ 0 & \hat{M}_{22} \end{bmatrix}, \tag{12}$$

where $\hat{M}_{11} \doteq M_{11} + M_{12}X$, $\hat{M}_{12} = M_{12}$, $\hat{M}_{22} \doteq M_{22} - X M_{12}$, and $\hat{M}_{21} = 0$ since it is precisely equation (11). But (12) is a similarity transformation on the $(p+q) \times (p+q)$ matrix M partitioned in the 4 blocks M_{ij} $i = 1, 2$, $j = 1, 2$. The block triangular decomposition says that the eigenvalues of M are the union of those of $\hat{M}_{11}$ and of $\hat{M}_{22}$ and that the columns of $\begin{bmatrix} I_p \\ X \end{bmatrix}$ span an invariant subspace of the matrix M corresponding to the p eigenvalues of $\hat{M}_{11}$ [10]. Let us suppose for simplicity that M is simple, i.e. that it has distinct eigenvalues. Then every invariant subspace of a particular dimension p is spanned by p eigenvectors. Therefore, let $\begin{bmatrix} X_{11} \\ X_{21} \end{bmatrix}$ be a matrix whose columns are p arbitrary eigenvectors, then it is a basis for the corresponding invariant subspace. If moreover X_{11} is invertible then the columns of $\begin{bmatrix} I_p \\ X \end{bmatrix}$ with $X = X_{21}X_{11}^{-1}$ span the same subspace and hence X is a solution of the quadratic matrix equation (11). One shows that the eigenvalues corresponding to the selected eigenvectors will be the eigenvalues of $\hat{M}_{11}$ after applying the transformation (12). This approach actually yields *all solutions* X provided M is simple and the matrices X_{11} defined above are invertible. But it requires the computation of all the eigenvectors, which is obtained from a diagonalizing similarity transformation. One shows that when M has repeated eigenvalues, one should compute its Jordan canonical to find all solutions of the quadratic matrix equation (11) [15]. The disadvantage of this approach is that it involves the construction of a transformation T that may be badly conditioned.

But any invariant subspace has also an orthogonal basis, and in general these basis vectors will not be eigenvectors since eigenvectors need not be orthogonal to each other. It turns out that such a basis is exactly obtained by the Schur decomposition (8). One can always compute an orthogonal similarity transformation that quasi triangularizes the matrix M. If we then partition the triangular matrix with a $p \times p$ leading block :

$$\begin{bmatrix} U_{11} & U_{12} \\ U_{21} & U_{22} \end{bmatrix}^T \begin{bmatrix} M_{11} & M_{12} \\ M_{21} & M_{22} \end{bmatrix} \begin{bmatrix} U_{11} & U_{12} \\ U_{21} & U_{22} \end{bmatrix} = \begin{bmatrix} \tilde{M}_{11} & \tilde{M}_{22} \\ 0 & \tilde{M}_{22} \end{bmatrix} \quad (13)$$

then it follows than the columns of $\begin{bmatrix} U_{11} \\ U_{21} \end{bmatrix}$ also span an invariant subspace of M and that the columns of $\begin{bmatrix} I_p \\ X \end{bmatrix}$ with $X = U_{21}U_{11}^{-1}$ span the same subspace, provided U_{11} is invertible [10]. This approach has the advantage that it uses numerically reliable coordinate transformations in (13) but the disadvantage that only one invariant subspace is directly obtained that way. It turns out that in several applications one only needs one invariant subspace. Typical examples arise in applications involving continuous time systems :

— the algebraic Riccati equation $XBR^{-1}B^TX - XA - A^TX - Q = 0$ from optimal control. Here the relevant matrix is

$$M = \begin{bmatrix} A & -BR^{-1}B^T \\ -Q & -A^T \end{bmatrix}$$

and the matrix $\hat{M}_{11}$ must contain all eigenvalues of M in the left half plane

– the Lyapunov equation $AX + XA^T + Q = 0$ occurring in stability analysis. Here

$$M = \begin{bmatrix} -A^T & 0 \\ Q & A \end{bmatrix}$$

and $\hat{M}_{11} = -A^T$

– the Sylvester equation $AX - XB + C = 0$ where

$$M = \begin{bmatrix} B & 0 \\ C & A \end{bmatrix}$$

and $\hat{M}_{11} = B$.

In each of these cases one has a well defined spectrum in mind for the matrix $\tilde{M}_{11}$ after transformation, and so only one invariant subspace has to be computed. We point out that Lyapunov and Sylvester equations can be viewed as special linear cases of the quadratic Riccati equation, which is extensively discussed in [10] [15]. The efficient calculation of the linear equations is discussed in [1] and is based on the Schur forms of A and B. This so-called *Schur* approach was proposed in a number of papers [15] and is now the recommended basic technique to solve these problems, although improvements are still being found in this area.

The generalized eigenvalue counterpart involves deflating subspaces and arises in applications involving discrete time systems. We refer to [15] for more details. Besides the computations of eigenspaces it is also shown there that the Schur form plays an important role in simulation of dynamical systems and in the computation of system and frequency responses. In fact, any matrix function of A can be evaluated efficiently using the Schur form of A, and the same holds for matrix functions of $E^{-1}A$ using the generalized Schur form of $\lambda E - A$.

Another class of orthogonal coordinate transformations (which we do not want to elaborate on here) are the so-called Hessenberg and staircase forms. A staircase form has a typical echelon-like form for a couple of matrices as shown below for the (A, B) pair of a system with 3 inputs, 6 states and 2 outputs :

$$\left[\begin{array}{c|c} \hat{B}_c & \hat{A}_c \\ \hline \hat{D}_c & \hat{C}_c \end{array} \right] \doteq \left[\begin{array}{ccc|ccccccc} X & \times & \times & \times & \times & \times & \times & \times & \times & \times \\ 0 & X & \times & \times & \times & \times & \times & \times & \times & \times \\ 0 & 0 & X & \times & \times & \times & \times & \times & \times & \times \\ 0 & 0 & 0 & 0 & X & \times & \times & \times & \times & \times \\ 0 & 0 & 0 & 0 & 0 & X & \times & \times & \times & \times \\ 0 & 0 & 0 & 0 & 0 & 0 & 0 & X & \times & \times \\ 0 & 0 & 0 & 0 & 0 & 0 & 0 & 0 & X & \times \\ \hline \times & \times & \times & \times & \times & \times & \times & \times & \times & \times \\ \times & \times & \times & \times & \times & \times & \times & \times & \times & \times \end{array} \right]. \quad (14)$$

The X elements are nonzero and define exactly the controllability indices of the system [20]. This form is the orthogonal counterpart of the Brunovsky canonical form, which is more sparse but can only be obtained under non-orthogonal state space transformation. These have been developed in several papers and have numerous applications, including : controllability, observability, minimality of state space and generalized state space models, pole placement and robust feedback, observer design and Kalman filtering. They also exist for generalized state space systems in which case the additional matrix $\hat{E}_c$ is upper triangular. We refer to [25] [20] for more details on this form and its applications in systems and control.

A third important condensed form is the singular value decomposition of a matrix. It is given by the following theorem.

Theorem 2 [9]
Let A be a $m \times n$ real matrix of rank r. Then there exist orthogonal matrices U and V of dimension $m \times m$ and $n \times n$ respectively, such that

$$U^T A V = \Sigma = \begin{bmatrix} \Sigma_r & 0 \\ 0 & 0 \end{bmatrix} \qquad (15)$$

where $\Sigma_r = \text{diag}\{\sigma_1, \cdots, \sigma_r\}$ with $\sigma_1 \geq \cdots \geq \sigma_r > 0$. ∎

Its main uses are detecting the rank of the matrix A in a reliable fashion, finding lower rank approximations and finding orthogonal bases for the kernel and image of A. In systems and control these problems show up in identification, balancing and finding bases for various geometric concepts related to state space models. We refer to [23] [14] [13] [15] for an extensive discussion of these applications.

3. Matrix sequences

In this section we look at orthogonal decompositions of a sequence of matrices. These typically occur in the context of discrete linear time varying systems :

$$\begin{cases} E_k x_{k+1} &= A_k x_k + B_k u_k \\ y_k &= C_k x_k + D_k u_k, \end{cases} \qquad (16)$$

arising e.g. from a discretization of a continuous time system.

Let us first consider the input to state map over a period of time $[1, N]$ (notice that in the control literature one prefers to start from $k = 0$, but starting from $k = 1$ turns out to be more convenient here). If the matrices E_k are all invertible then clearly each state x_k is well defined by these equations. One is often interested in the zero input behavior (i.e. the homogeneous system) which yields an explicit expression for the final state x_{N+1} in terms of the initial state x_1 :

$$x_{N+1} = \Phi_{N,1} x_1 \qquad (17)$$

where

$$\Phi_{N,1} = E_N^{-1} A_N \cdots E_2^{-1} A_2 E_1^{-1} A_1 \qquad (18)$$

is the state transition matrix over the interval $[1, N]$. If the matrices E_k are *not* invertible, these this expression does not make sense, but still one may be able to solve the input to state map when imposing boundary conditions

$$F_N x_{N+1} - F_1 x_1 + f = 0. \qquad (19)$$

We can then rewrite these equations is the following matrix form :

$$
\begin{bmatrix}
-A_1 & E_1 & & & \\
& -A_2 & E_2 & & \\
& & \ddots & \ddots & \\
& & & -A_N & E_N \\
F_1 & & & & -F_N
\end{bmatrix}
\begin{bmatrix}
x_1 \\
x_2 \\
\vdots \\
x_N \\
x_{N+1}
\end{bmatrix}
=
\begin{bmatrix}
B_1 u_1 \\
B_2 u_2 \\
\vdots \\
B_N u_N \\
f
\end{bmatrix}, \qquad (20)
$$

which will have a solution provided the (square) matrix in the left hand side is invertible.

3.1. PERIODIC BOUNDARY VALUE PROBLEMS

A periodic system is a set of difference equations (16) where now the coefficient matrices vary periodically with time, i.e. $M_k = M_{k+K}$, $\forall k$ and for $M = E, A, B, C$ and D. The period is the smallest value of K for which this holds. It was shown in [18] that a periodic system of period K is *solvable and conditionable* (i.e. has a well defined solution for suitably chosen boundary conditions F_1, F_N) for all N, provided the pencil

$$
\lambda \mathcal{E} - \mathcal{A} \doteq
\begin{bmatrix}
-A_1 & \lambda E_1 & & \\
& \ddots & \ddots & \\
& & -A_{K-1} & \lambda E_{K-1} \\
\lambda E_K & & & -A_K
\end{bmatrix}
\qquad (21)
$$

is regular (i.e. $\det(\lambda \mathcal{E} - \mathcal{A}) \not\equiv 0$). Such periodic systems are said to be *regular*. Two point boundary value problems for regular periodic systems have thus unique solutions for *any* time interval N, provided the boundary conditions are suitably chosen.

Let us now choose a time interval equal to one period $(N = K)$ and introduce boundary conditions that allow us to define eigenvectors and eigenvalues of periodic boundary value problems. Since these concepts are typically linked to homogeneous systems, we impose :

$$u_k = 0, k = 1, \ldots, K, \quad f = 0, \tag{22}$$

and for the invariance of the boundary vectors we impose :

$$F_1 = sI_n, \ F_K = cI_n, \ \Longrightarrow \ sx_1 = cx_{K+1} \quad \text{with} \quad c^2 + s^2 = 1, \tag{23}$$

which yields :

$$\begin{bmatrix} -A_1 & E_1 & & & \\ & -A_2 & E_2 & & \\ & & \ddots & \ddots & \\ & & & -A_K & E_K \\ sI_n & & & & -cI_n \end{bmatrix} \begin{bmatrix} x_1 \\ x_2 \\ \vdots \\ x_K \\ x_{K+1} \end{bmatrix} = 0. \tag{24}$$

After some algebraic manipulations one shows that this is equivalent to :

$$(z\mathcal{E}_b - \mathcal{A}_b)\mathbf{x}(1, K) \doteq \begin{bmatrix} -A_1 & E_1 & & \\ & \ddots & \ddots & \\ & & -A_{K-1} & E_{K-1} \\ zE_K & & & -A_K \end{bmatrix} \begin{bmatrix} x_1 \\ x_2 \\ \vdots \\ x_K \end{bmatrix} = 0 \tag{25}$$

where $z = s/c$. This condition says that $\mathbf{x}(1, K)$ is an eigenvector of the pencil in the left hand side, with eigenvalue z. The pencils $z\mathcal{E}_b - \mathcal{A}_b$ and $\lambda\mathcal{E} - \mathcal{A}$ are closely related [22]. In case all matrices E_k are invertible, it follows e.g. that

$$\det(z\mathcal{E}_b - \mathcal{A}_b) = c . \det(zI_n - \Phi_{K,1}), \quad \text{and} \quad \det(\lambda\mathcal{E} - \mathcal{A}) = c . \det(\lambda^K I_n - \Phi_{K,1})$$

where $\Phi_{K,1} = E_K^{-1} A_K \cdots E_2^{-1} A_2 E_1^{-1} A_1$ is the so-called *monodromy matrix* of the periodic system. For more details on the relations between generalized eigenvectors and eigenvalues of these pencils we refer to [22]. A key decomposition for computing these generalized eigenvalues and eigenvectors is described in the next section.

3.2. PERIODIC SCHUR FORM

The role played by the generalized Schur form for time invariant systems is now replaced by a very similar orthogonal decomposition, called the *periodic Schur form*.

Theorem 3 [3]

Let the $n \times n$ matrices E_k and A_k, $k = 1, \ldots, K$ be such that the pencil $\lambda \mathcal{E} - \mathcal{A}$ is regular. Then there always exist orthogonal transformations Q_k and Z_k, $k = 1, \ldots, K$ such that

$$
\begin{bmatrix} Q_1^T & & & \\ & Q_2^T & & \\ & & \ddots & \\ & & & Q_K^T \end{bmatrix}
\begin{bmatrix} -A_1 & \lambda E_1 & & \\ & \ddots & \ddots & \\ & & -A_{K-1} & \lambda E_{K-1} \\ \lambda E_K & & & -A_K \end{bmatrix}
\begin{bmatrix} Z_1 & & & \\ & Z_2 & & \\ & & \ddots & \\ & & & Z_K \end{bmatrix}
$$

$$
= \begin{bmatrix} -\hat{A}_1 & \lambda \hat{E}_1 & & \\ & \ddots & \ddots & \\ & & -\hat{A}_{K-1} & \lambda \hat{E}_{K-1} \\ \lambda \hat{E}_K & & & -\hat{A}_K \end{bmatrix}, \tag{26}
$$

where the transformed matrices $\hat{A}_k$ and $\hat{E}_k$ are all upper triangular, except for one matrix – say, $\hat{A}_1$ – which is quasi triangular. ∎

The relation with the standard Schur form is that if the E_k matrices are invertible, then the monodromy matrix $\Phi_{K,1}$ is transformed by the orthogonal similarity Z_1 to its Schur form :

$$
\hat{\Phi}_{K,1} \doteq \hat{E}_K^{-1} \hat{A}_K \cdots \hat{E}_1^{-1} \hat{A}_1 = Z_1^T (E_K^{-1} A_K \cdots E_1^{-1} A_1) Z_1 = Z_1^T \Phi_{K,1} Z_1.
$$

Since all matrices are triangular it follows that all transformed monodromy matrices $\hat{\Phi}_{K+k-1,k}$ are quasi triangular as well, and with the same ordering of eigenvalues. From Theorem 1 it follows that the ordering of the eigenvalues can be chosen arbitrarily and hence that there exists a periodic Schur form associated with any eigenvalue ordering.

We point out here that the transformations Z_k and Q_k can also be applied directly to the system (16). Define indeed a new state $\hat{x}_k \doteq Z_k^T x_k$ and multiply the top equation of (16) by Q_k^T then we obtain an equivalent system

$$
\begin{cases} \hat{E}_k \hat{x}_{k+1} &= \hat{A}_k \hat{x}_k + \hat{B}_k u_k \\ y_k &= \hat{C}_k \hat{x}_k + D_k u_k, \end{cases} \tag{27}
$$

where $\hat{B}_k \doteq Q_k^T B_k$, $\hat{C}_k \doteq C_k Z_k$, and $\hat{E}_k$ and $\hat{A}_k$ are upper triangular, except for $\hat{A}_1$ which is quasi triangular. This is a very special coordinate system : the "bottom" equation in (27) (or the bottom 2 equations if $\hat{A}_1$ has a bottom 2×2 block) is now "decoupled" from the rest of the system. Since the ordering of the eigenvalues in the Schur form can always be chosen arbitrarily, one can choose this decoupled system to be the one with the smallest eigenvalue in absolute value and hence the "easiest" to integrate numerically [3]. Once this "bottom" component of the state has been computed, one substitutes this in the next component(s), which is then also decoupled from the rest of the system, and so on. This coordinate system is therefore very appealing for simulation purposes [11].

3.3. PERIODIC CONTROL SYSTEMS

The periodic Schur form has several other applications in control problems involving periodic discrete time systems. In optimal control of such a periodic system one considers e.g. the problem :

$$\text{Minimize } J = \sum_{k=1}^{\infty} z_k^T Q_k z_k + u_k^T R_k u_k$$
$$\text{subject to } H_k z_{k+1} = F_k z_k + G_k u_k \tag{28}$$

where the matrices Q_k, R_k, F_k, G_k, H_k are periodic with period K. In order to solve this variational problem, one needs to solve the Hamiltonian equations which is a periodic homogeneous system of difference equations (16) in the state z_k and co-state λ_k of the system [16]. The correspondences with (16) are :

$$x_k \doteq \begin{bmatrix} \lambda_k \\ z_k \end{bmatrix}, E_k \doteq \begin{bmatrix} -G_k R_k^{-1} G_k^T & H_k \\ F_k^T & 0 \end{bmatrix}, A_k \doteq \begin{bmatrix} 0 & F_k \\ H_k^T & Q_k \end{bmatrix}. \tag{29}$$

For finding the periodic solutions to the underlying periodic Riccati equation one has to find the stable invariant subspaces of the monodromy matrices $\Phi_{K+k-1,k}$ [2]. Clearly the periodic Schur form is useful here as well as the reordering of eigenvalues [3].

In pole placement of periodic systems, again the periodic Schur form and reordering is useful when one wants to extend Varga's pole placement algorithm [27] to periodic systems. Consider the system

$$E_k z_{k+1} = A_k z_k + D_k u_k$$
$$\text{with state feedback } u_k = F_k z_k + v_k \tag{30}$$

where the matrices A_k, E_k, D_k, F_k are periodic with period K. This results in the closed loop system

$$E_k z_{k+1} = (A_k + D_k F_k) z_k + D_k v_k \tag{31}$$

of which the underlying time invariant eigenvalues are those of the matrix :

$$\Phi_{K,1}^{(F)} \doteq E_K^{-1} (A_K + D_K F_K) \cdots E_2^{-1} (A_2 + D_2 F_2) E_1^{-1} (A_1 + D_1 F_1). \tag{32}$$

In the above equation it is not apparent at all how to choose the matrices F_k to assign particular eigenvalues of $\Phi_{K,1}^{(F)}$. Yet when the matrices A_k, E_k are in the triangular form (26), one can choose the F_k matrices to have only nonzero elements in the last column. This will preserve the triangular form of the matrices $A_k + D_k F_k$ and it is then trivial to choose e.g. one such column vector to assign one eigenvalue. In order to assign the other eigenvalues one needs to *reorder* the diagonal elements in the periodic Schur form and each time assign another eigenvalue with the same technique. For more details, we refer to [17].

Other applications of the periodic Schur form are the solution of periodic Lyapunov and Sylvester equations. Since these are special cases of periodic Riccati equations, they can also be solved via the periodic Schur form. These equations show up in problems of stability analysis, decoupling, balancing and so on [17], [28].

3.4. GENERALIZED QR DECOMPOSITION

The basic equation in the matrix sequences occurring in the homogeneous two point boundary value problem defined earlier is $E_k x_{k+1} = A_k x_k$. We now analyze the system of equations when both E_k and A_k are singular, and more specifically, what is needed to be able to define singular values of sequences of such equations. For this we first analyze a single equation

$$Ey = Ax. \tag{33}$$

The singular value decomposition is originally defined for a single matrix M and is closely related to its URV decomposition. If an $m \times n$ matrix M has rank $r = \min\{m, n\}$ then there exists orthogonal matrices U and V such that

$$M = U \begin{bmatrix} \tilde{M} & 0 \\ 0 & 0 \end{bmatrix} V^T$$

where $\tilde{M}$ is an $r \times r$ invertible matrix. This so-called URV decomposition can be viewed as a two sided QR decomposition and can be obtained in a finite number of operations [7]. The first r columns of U in fact span the image of M and the first r columns of V are the orthogonal complement of the kernel of M. So if in the equation

$$y = Mx$$

we constrain the vector y to $\text{Im}M$ and x to $\text{Ker}M^\perp$, then M is invertible since it essentially is represented by $\tilde{M}$. This can be applied to the implicit system (33). If A and E are invertible then

$$Ey = Ax \Leftrightarrow y = E^{-1}Ax \Leftrightarrow A^{-1}Ey = x.$$

Do these equations still make sense when the matrices are singular ? Let us consider spaces $\mathcal{X}$ and $\mathcal{Y}$ such that $\mathcal{Z} = E\mathcal{Y}$, $\mathcal{Z} = A\mathcal{X}$ and $\dim \mathcal{X} = \dim \mathcal{Y} = \dim \mathcal{Z}$. The largest such subspace Z will be $\mathcal{Z} \doteq Im A \cap Im E$. Provided we choose $x \in \mathcal{X}$ then there exists a solution $y \in \mathcal{Y}$ satisfying $Ey = Ax$, and then the mapping $y = E^{-1}Ax$ makes sense. But the solutions are not unique, unless we constrain y to be e.g. in $Ker E^{\perp}$. Conversely, provided we choose $y \in \mathcal{Y}$ then there exists a solution $x \in \mathcal{X}$ satisfying $Ey = Ax$. Again $A^{-1}Ey = x$ then make sense and x is unique e.g. if $x \in Ker A^{\perp}$.

This discussion does not change when transforming to a coordinate system

$$z_q = QEV^T y_v = QAU^T x_u$$

where U, V and Q are orthogonal. One can always find matrices Q, U, V such that in the new coordinate system this equation has the form :

$$\begin{bmatrix} z_1 \\ z_2 \\ z_3 \\ z_4 \end{bmatrix} = \begin{bmatrix} 0 & \tilde{E} & \times \\ 0 & 0 & \times \\ 0 & 0 & R_s \\ 0 & 0 & 0 \end{bmatrix} \begin{bmatrix} y_1 \\ y_2 \\ y_3 \end{bmatrix} = \begin{bmatrix} 0 & \tilde{A} & \times \\ 0 & 0 & R_t \\ 0 & 0 & 0 \\ 0 & 0 & 0 \end{bmatrix} \begin{bmatrix} x_1 \\ x_2 \\ x_3 \end{bmatrix} \quad (34)$$

where R_s, R_t are invertible matrices of dimension s and t, respectively, and $\tilde{E}$ and $\tilde{A}$ are square invertible of dimension $r = \dim Im A \cap Im E$. In the new coordinate system it is clear that we need to take $y_3 = 0$, $x_3 = 0$ in order to make the system compatible and $y_1 = 0$, $x_1 = 0$ in order to make the solution unique. Defining $\tilde{x} = x_2$ and $\tilde{y} = y_2$ we obtain the equation

$$\tilde{E}\tilde{y} = \tilde{A}\tilde{x},$$

which has a well defined solution $\tilde{y} = \tilde{E}^{-1}\tilde{A}\tilde{x}$. The singular values we are interested in are of course those of the reduced order map $\tilde{E}^{-1}\tilde{A}$. The decomposition (34) is called the generalized QR decomposition and can be extended to sequences of equations $E_k x_{k+1} = A_k x_k$. As in the single equation case, one can again extract from a possibly singular system of equations a lower dimensional one that has all matrices $\tilde{E}_k$ and $\tilde{A}_k$ nonsingular. The relevant singular values are then those of

$$\tilde{E}_K^{-1}\tilde{A}_K \cdots \tilde{E}_1^{-1}\tilde{A}_1.$$

For the details, we refer to [4]. In the next section we show how to extract from such a nonsingular sequence, the singular values by only applying orthogonal transformations to the sequence of matrices $\tilde{E}_k$, $\tilde{A}_k$.

3.5. BIDIAGONAL AND SINGULAR VALUE DECOMPOSITION

Here we consider a sequence of $n \times n$ matrices E_k, A_k which are invertible, and we want to compute the singular values of the state transition matrix over a time interval $[1, N]$.

$$\Phi_{N,1} \doteq E_N^{-1} A_N \cdots E_2^{-1} A_2 E_1^{-1} A_1.$$

It is clear that one has to perform left and right transformations on $\Phi_{N,1}$ to diagonalize $U^T \Phi_{N,1} V = \Sigma$, but these will only affect E_N and A_1. But one can insert pairs of orthogonal transformations in between the factors of this expression without altering the result. The following theorem shows how to use these degrees of freedom to find the singular values of $\Phi_{N,1}$.

Theorem 4 [8]
Let the $n \times n$ matrices E_k and A_k, $k = 1, \ldots, N$ be invertible. Then there always exist orthogonal transformations Q_k, $k = 1, \ldots, N$ and Z_k, $k = 1, \ldots, N+1$ such that

$$
\begin{bmatrix} Q_1^T & & \\ & \ddots & \\ & & Q_N^T \end{bmatrix}
\begin{bmatrix} -A_1 & E_1 & \\ & \ddots & \ddots \\ & & -A_N & E_N \end{bmatrix}
\begin{bmatrix} Z_1 & & & \\ & \ddots & & \\ & & Z_N & \\ & & & Z_{N+1} \end{bmatrix}
$$

$$
= \begin{bmatrix} -\hat{A}_1 & \hat{E}_1 & \\ & \ddots & \ddots \\ & & -\hat{A}_N & \hat{E}_N \end{bmatrix}, \tag{35}
$$

where all matrices $\hat{E}_k$ and $\hat{A}_k$ are upper triangular and the product $\hat{\Phi}_{N,1} \doteq \hat{E}_N^{-1} \hat{A}_N \cdots \hat{E}_1^{-1} \hat{A}_1$ is diagonal (alternatively, there is an algorithm of complexity $O(Nn^3)$ which *bidiagonalizes* $\hat{\Phi}_{N,1}$). ∎

The proof of this result is very simple. Choose $Z_1 = V$ and $Z_{N+1} = U$. Then alternately choose the matrices Q_k to triangularize $\hat{A}_k = Q_k^T(A_k Z_k)$ (for $k = 1, \ldots, N$) and Z_k to triangularize $\hat{E}_{k-1} = (Q_{k-1}^T E_{k-1}) Z_k$ (only for $k = 2, \ldots, N$ since Z_1 and Z_{N+1} are already fixed). All matrices but $\hat{E}_N$ in the expression for $\hat{\Phi}_{N,1}$ are now upper triangular by construction, but since all factors are invertible and the product is diagonal (or bidiagonal), $\hat{E}_N$ must be upper triangular as well. The finite algorithm for the bidiagonalization could also be derived this way, but we refer to [8] for a constructive and numerically stable algorithm.

The bidiagonalization has been shown to yield very accurate results despite the fact that the singular values of such product can become very

174

large and very small as N tends to grow [8]. The singular values of the computed bidiagonal are then computed to high relative accuracy using an appropriate technique [6]. This decomposition can e.g. be used to find the "dominant directions" of the state transition map $\Phi_{N,1}$ over a finite time interval $[1, N]$. In the case that the discrete time system (16) comes from a discretization of a nonlinear continuous time system it is known that the singular values of $\Phi_{N,1}$ are closely related to the Lyapunov exponents of the corresponding continuous time system (provided the discretization is sufficiently "fine") [5]. Singular values also show up in robustness aspects of dynamical systems [24].

4. Concluding remarks

In this paper we surveyed a number of orthogonal matrix decompositions arising in systems and control. We pointed out that they lead to numerically reliable algorithms for solving quite a large range of problems in this area. This is mainly due to the fact that orthogonal transformations are backward stable when implemented correctly, and that they do not affect the conditioning of the problem at hand. Although we did not mention all uses of orthogonal transformations in this area, we gave references for further reading on this important area of research.

Acknowledgments

This work was supported by the National Science Foundation under grant CCR-9796315. Parts of this paper presents research results of the Belgian Programme on Inter-university Poles of Attraction, initiated by the Belgian State, Prime Minister's Office for Science, Technology and Culture. The scientific responsibility rests with its authors.

References

1. R.H. Bartels and G.W. Stewart (1972), Solution of the equation $AX + XB = C$, *Comm. ACM* **15**, pp. 820–826.
2. S. Bittanti, P. Colaneri and G. de Nicolao (1988), The difference periodic Riccati equation for the periodic prediction problem, *IEEE Trans. Aut. Contr.* **33**, pp. 706–712.
3. A. Bojanczyk, G. Golub and P. Van Dooren (1992), The periodic Schur decomposition. Algorithms and applications, in *Advanced Signal Processing Algorithms, Architectures, and Implementations III*, Proceedings of SPIE, the International Society for Optical Engineering **1770**, pp. 31–42, Ed. F.T. Luk, Bellingham, Wash., USA.
4. B. De Moor and P. Van Dooren (1992), Generalizations of the singular value and QR decomposition, *SIAM Matr. Anal. & Applic.* **13**, pp. 993–1014.
5. L. Dieci, R. Russell and E. Van Vleck (1997), On the computation of Lyapunov exponents for continuous dynamical systems, *SIAM Numer. Anal.* **34**, pp. 402–423.

6. K.V. Fernando and B.N. Parlett (1994), Accurate singular values and differential qd algorithms, *Numerische Mathematik* **67**, pp. 191–229.

7. G. Golub and C. Van Loan (1989), *Matrix Computations*, 2nd edition, The Johns Hopkins University Press, Baltimore, Maryland.

8. G. Golub, K. Solna and P. Van Dooren (1995), Computing the singular values of products and quotients of matrices, in *SVD in Signal Processing III, Algorithms, Architectures and Applications*, Eds. M. Moonen, B. De Moor, Elsevier, Amsterdam, 1995, also *SIAM Matrix Anal. & Appl.*, to appear.

9. G. Golub and V. Kahan (1965), Calculating the singular values and pseudo-inverse of a matrix, *SIAM Numer. Anal.* **2**, pp. 205–224.

10. A. Laub (1990), Invariant subspace methods for the numerical solution of Riccati equations, in *The Riccati equation*, Eds. Bittanti, Laub, Willems, Springer Verlag.

11. R.M. Mattheij and S.J. Wright (1993), Parallel stabilized compactification for ODEs with parameters and multipoint conditions, *Applied Numerical Mathematics* **13**, pp. 305–333.

12. C. Moler and G. Stewart (1973), An algorithm for the generalized matrix eigenvalue problem, *SIAM J. Numer. Anal.* **10**, pp. 241–256.

13. M. Moonen, B. De Moor, L. Vandenberghe, and J. Vandewalle (1989), On- and off-line identification of linear state-space models, *Int. J. Contr.* **49**, pp. 219–232.

14. B.C. Moore (1981), Principal component analysis in linear systems: Controllability, observability, and model reduction, *IEEE Trans. Aut. Contr.* **26**, pp. 17–31.

15. R. Patel, A. Laub and P. Van Dooren (1993), *Numerical Linear Algebra Techniques for Systems and Control*, (Eds.) IEEE Press, Piscataway NJ.

16. A. Sage and C. White (1977), *Optimum Systems Control*, 2nd Ed., Prentice-Hall, New Jersey.

17. J. Sreedhar and P. Van Dooren (1994), A Schur approach for solving some periodic matrix equations, in *Systems and Networks : Mathematical Theory and Applications*, pp. 339–362, Eds. U. Helmke, R. Mennicken, J. Saurer, Mathematical Research **77**, Akademie Verlag, Berlin.

18. J. Sreedhar and P. Van Dooren (1999), Periodic descriptor systems: Solvability and conditionability, *IEEE Trans. Aut. Contr.*, to appear.

19. G.W. Stewart and J.-G. Sun (1990), *Matrix Perturbation Theory*, Academic Press, New York.

20. P. Van Dooren (1981), The generalized eigenstructure problem in linear system theory, *IEEE Trans. Aut. Contr.* **26**, pp. 111–129.

21. P. Van Dooren (1981), A generalized eigenvalue approach for solving Riccati equations, *SIAM Sci. & Stat. Comp.* **2**, pp. 121–135.

22. P. Van Dooren (1999), Two point boundary value and periodic eigenvalue problems, in *Proceedings 1999 IEEE CACSD Symposium*, Hawaii, to appear.

23. P. Van Dooren (1999), The singular value decomposition and its applications in systems and control, in *Wiley Encyclopedia of Electrical and Electronics Engineering*, Wiley and Sons, New York, to appear.

24. P. Van Dooren and V. Vermaut (1997), On stability radii of generalized eigenvalue problems, in *Proceedings European Control Conf.*, paper FR-M-H6.

25. P. Van Dooren and M. Verhaegen (1985), On the use of unitary state-space transformations, *Linear Algebra and its Role in Linear System Theory*, Ed. B.N. Datta, pp. 447–463, Providence, RI: AMS Contemporary Mathematics Series, **47**.

26. P. Van Dooren and M. Verhaegen (1988), Condensed forms for efficient time invariant Kalman filtering, *SIAM Sci. & Stat. Comp.* **9**, pp. 516–530.

27. A. Varga (1981), A Schur method for pole assignment, *IEEETrans. Aut. Contr.* **26**, pp. 517–519.

28. A. Varga (1997), Periodic Lyapunov equations : some applications and new algorithms, *Int. J. Contr.* **67**, pp. 69–87.

MODEL REDUCTION OF LARGE-SCALE SYSTEMS
RATIONAL KRYLOV VERSUS BALANCING TECHNIQUES

K. A. GALLIVAN

Comp. Sc. Dept., Florida State University, Florida, USA

E. GRIMME

Intel Corporation, Santa Clara, California, USA

AND

P. M. VAN DOOREN

Dept. Math. Eng., UniversitéCatholique de Louvain, Belgium

Abstract. In this paper, we describe some recent developments in the use of projection methods to produce a reduced-order model for a linear time-invariant dynamical system which approximates its frequency response. We give an overview of the family of Rational Krylov methods and compare them with "near-optimal" approximation methods based on balancing transformations.

1. Introduction

Physical phenomena are often modeled with linear, time-invariant (LTI) dynamical systems because of the simplicity and low complexity of the approach (both in terms of the complexity of the approximation problem and its subsequent use for simulation). Such linear models can frequently be acquired through discretizations of partial or ordinary differential equations describing the physical system. However, such physical models are becoming more complex due to either increased system size or an increased desire for detail. Large scale problems (such as the North American power grid system) and fine grid discretizations (needed in high-speed circuit designs) require models of increasing complexity. Although such models tend to accurately describe the behavior of the underlying physical system, their complexity leads to high analysis and simulation costs which are too high

177

H. Bulgak and C. Zenger (eds.), Error Control and Adaptivity in Scientific Computing, 177–190.
© 1999 *Kluwer Academic Publishers. Printed in the Netherlands.*

when one uses traditional numerical techniques. Methods which exploit the structure in the models such as sparsity have become critical. This lead to an increased interest in iterative methods for solving large sparse linear systems and/or eigenvalue problems for the simulation of such large dynamical systems.

In some cases, however, there is a need to go even further. Despite the use of efficient computational kernels, the model may still require an unacceptable amount of time to evaluate. It is then necessary to create a second model that is significantly smaller while preserving important aspects of the original system. This is the model reduction problem for linear time-invariant dynamical systems. It is assumed that the original system is described by the generalized state-space equations

$$\begin{cases} E\dot{x}(t) = Ax(t) + Bu(t) \\ y(t) = Cx(t) + Du(t). \end{cases} \tag{1}$$

The vectors $u(t) \in \mathcal{R}^m$, $y(t) \in \mathcal{R}^p$ and $x(t) \in \mathcal{R}^N$ are the vectors of input variables, output variables and state variables, respectively. For nearly all large-scale problems, it is assumed that the state transition matrices $A \in \mathcal{R}^{N \times N}$ and $E \in \mathcal{R}^{N \times N}$ are large and sparse or structured. Moreover, the input dimension m and output dimension p are assumed much smaller that the state dimension N. We point out that this system has a well-defined solution provided the pencil $(\lambda E - A)$ is *regular*, i.e. $\det(\lambda E - A) \neq 0$. A reduced-order approximation to (1) takes the corresponding form

$$\begin{cases} \hat{E}\dot{\hat{x}}(t) = \hat{A}\hat{x}(t) + \hat{B}u(t) \\ \hat{y}(t) = \hat{C}\hat{x}(t) + \hat{D}u(t). \end{cases} \tag{2}$$

The dimension n of the reduced-order is supposed to be much smaller than N. Ideally, the reduced order model would produce an output $\hat{y}(t)$ approximating well the true output $y(t)$ for all inputs $u(t)$. It is more realistic to try to match the response $y(t)$ of some "representative" input $u(t)$ and typically one chooses the response of the zero initial state impulse response $(x(0) = 0, u_j(t) = \delta(t)e_j$, where $\delta(t)$ is the Dirac impulse and e_j is the j-th column of the identity matrix). The reason for this is that the system response to an arbitrary input (with zero initial condition) can be represented as a convolution with the impulse response. Provided E is invertible, this response equals (for each input $\delta(t)e_j$) :

$$y_j(t) = \{Ce^{E^{-1}At}E^{-1}B + D\delta(t)\}e_j, \quad j = 1, \dots, m. \tag{3}$$

In the Laplace transform domain one derives the equivalent formula

$$\mathcal{L}y_j = \{C(sE - A)^{-1}B + D\}\mathcal{L}u_j, \tag{4}$$

which involves the *transfer function* $H(s) \doteq C(sE - A)^{-1}B + D$ of the system. It plays a key role in the description of the system behavior by describing the response of the system to a periodic input signal $u_j(t) \doteq e^{j\omega t} e_j$ since the corresponding output equals $y_j(t) = H(j\omega)u_j(t)$. The model reduction problem therefore reduces to an approximation of the frequency response $H(s)$ by another rational matrix of lower degree $\hat{H}(s) \doteq \hat{C}(s\hat{E} - \hat{A})^{-1}\hat{B} + \hat{D}$. Since D and $\hat{D}$ are of dimensions $p \times m$ where $p, m << N$, one typically chooses $D = \hat{D}$. Without a loss of generality, the feed-through term D of the original model can therefore be assumed to be zero since the approximation problem clearly involves only the part that depends on s and involve the large scale matrices.

Several measures of the accuracy of the reduced-order model are possible. Formally, one wants to bound the difference between the actual and low-order outputs, $y(t) - \hat{y}(t)$, given a selected input $u(t)$, and this can be characterized by a system norm. The popular H_∞ error norm measures, in the time domain, the worst ratio of output error energy to input energy. In the frequency domain, this represents the largest magnitude of the frequency-response error. A second measure of the accuracy of the approximation is to assess which properties of the original model are preserved in the reduced-order one. A common choice is modal approximation [1, 5] which preserves the dominant system's poles (eigenvalues) λ_n and corresponding residues which arise in the partial fraction expansion of the transfer function. A reduced-order model that matches (or approximates) dominant modal components of the original model is expected to approximate well its response. Unfortunately, it can be difficult to identify a priori which modes are the truly dominant modal components of the original system [23]. Alternative invariant properties that may be retained in model reduction are the Hankel singular values. Hankel singular values are related to the controllability and observability properties of a system [18]. Constructing a reduced-order model preserving the largest Hankel singular values is known as balanced truncation. A variant of this is known as optimal Hankel norm approximation [11]. These last two approximations are nearly optimal in terms of the H_∞ norm of the error and are constructed from balanced realizations. There exist sparse matrix techniques that compute these "near-optimal approximations" in an approximate fashion as well.

2. Moment matching

In this paper we focus on approximations defined from a power series expansion of the rational matrix function $H(s)$. Let σ be a point in the complex plane, then $H(s)$ has an expansion

$$H(s) = H_{-\ell}(s-\sigma)^{-\ell}+\cdots+H_{-1}(s-\sigma)^{-1}+H_0+H_1(s-\sigma)^1+H_2(s-\sigma)^2+\cdots,$$

$$(5)$$

where ℓ is the order of its pole at σ. Typically, one chooses interpolation points that are *not* a pole of $H(s)$ and then $\ell = 0$. The coefficients H_i are then easy to construct from the system model via a Neumann expansion of $H(s) = C\{(s-\sigma)E - (A-\sigma E)\}^{-1}B$ since $(A - \sigma E)$ is non-singular :

$$H_i = C(\sigma E - A)^{-1}\{E(\sigma E - A)^{-1}\}^i B = C\{(\sigma E - A)^{-1}E\}^i(\sigma E - A)^{-1}B.$$

$$(6)$$

The solution techniques proposed determine a reduced-order model that accurately matches the leading coefficients H_i – also called *moments* – arising in this power series. In general, one can produce a reduced-order model that interpolates the frequency response and its derivatives at multiple points $\{\sigma^{(1)}, \sigma^{(2)}, \ldots, \sigma^{(K)}\}$. Since

$$\left.\frac{\partial^i H(s)}{\partial s^i}\right|_{s=\sigma} \doteq i!H_i ,$$

$$(7)$$

we seek to match the moments $H_i^{(j)}$ at each interpolation point $\sigma^{(j)}, j = 1, \ldots, K$. The first $2J_1$ moments are matched at $\sigma^{(1)}$, the next $2J_2$ moments are matched at $\sigma^{(2)}$, etc., where $J_1 + J_2 + \ldots + J_K = M$. A model meeting these constraints is denoted a multipoint Padé approximation or a rational interpolant [2, 3]. By varying the location and number of interpolation points utilized with the underlying problem in mind, one can construct accurate reduced-order models in a variety of situations.

Moment matching methods are relatively old and are based on Padé approximations [6]. For the more general rational interpolation problem described here, one has to solve a system of equations involving a Loewner matrix [2]. It is important to note that the system matrices only enter the modeling problem through its moments explicitly. Unfortunately, these explicit moment-matching methods exhibit numerical instabilities, which was first pointed out in [8] and later on in [7]. The reader is referred to those papers and to [14] for a detailed discussion. The numerical difficulties come from the construction of the Hankel and Loewner matrices involved and the ill-conditioning of the associated linear systems. Both [8] and [7] point out that moment-matching via Krylov projection methods is a preferred numerical implementation.

3. Krylov projection methods

In projection methods, the M-th order reduced system is produced by applying two $N \times n$ matrices Z and V to the system matrices of the original system: $\hat{A} = Z^T AV$, $\hat{E} = Z^T EV$, $\hat{B} = Z^T B$, $\hat{C} = V^T C$. The matrices Z and V in fact define projections onto Krylov spaces

$$\mathcal{K}_j(G,g) = \text{Im}\left\{g, Gg, G^2 g, \ldots, G^{j-1} g\right\} \qquad (8)$$

for specific choices of G and g. The first connection between the Lanczos algorithm, a Krylov-based technique, and Padè approximations was given in [12]. Later work proposed related Krylov space techniques for model reduction of dynamical systems in various application areas [19, 16, 26, 24, 4, 25]. New results in the area included stability retention of the reduced order model [13] and multipoint rational Lanczos methods [10], i.e., starting from the Lanczos procedure and modifying it to produce a reduced system that matched multiple moments at multiple frequency values.

We now give a basic theorem describing the relationships between Krylov subspaces, the iterative algorithms for constructing these subspaces, and model reduction via rational interpolation. It was proven in [14] and [15] and extends [26] to multipoint approximations.

Theorem 1
If

$$\bigcup_{k=1}^{K} \mathcal{K}_{J_{b_k}}\left((\sigma^{(k)} E - A)^{-1} E, (\sigma^{(k)} E - A)^{-1} B\right) \subseteq \mathcal{V} \qquad (9)$$

and

$$\bigcup_{k=1}^{K} \mathcal{K}_{J_{c_k}}\left((\sigma^{(k)} E - A)^{-T} E^T, (\sigma^{(k)} E - A)^{-T} C^T\right) \subseteq \mathcal{Z} \qquad (10)$$

then the moments of (1) and (2) satisfy

$$H_k^{(j_k)} = C\left\{(\sigma^{(k)} E - A)^{-1} E\right\}^{j_k - 1} (\sigma^{(k)} E - A)^{-1} B = \qquad (11)$$

$$\hat{H}_k^{(j_k)} = \hat{C}\left\{(\sigma^{(k)} \hat{E} - \hat{A})^{-1} \hat{E}\right\}^{j_k - 1} (\sigma^{(k)} \hat{E} - \hat{A})^{-1} \hat{B} \qquad (12)$$

for $j_k = 1, 2, \ldots, J_{b_k} + J_{c_k}$ and $k = 1, 2, \ldots K$, provided these moments exist.
Proof: This is a trivial extension of the proof given in [14] for the case $m = p = 1$. ∎

The moments of the original system exist if one chooses interpolation points that are not poles of the system (one chooses e.g. points in the right half plane, where a stable system has no poles). But the non-singularity of the pencils $(\sigma^{(k)} \hat{E} - \hat{A})$ is not automatically guaranteed (see [14] for details on how to handle this case). Any pair of projection bases satisfying (9) and (10) achieve the desired rational interpolant. Restrictions on V or Z, such as biorthogonality or orthogonality, are implementation specific choices and lead to different variants. The Dual Rational Arnoldi method referred to later on corresponds V and Z having orthogonal columns. For algorithmic implementations and further details on the different variants we refer to [14].

4. Modal approximation

A classical method for model reduction is modal approximation. Let us assume that the system (1) has t different poles, then there exist invertible matrices Z and V transforming $\lambda E - A$ to a block diagonal form. In this new coordinate system, the model matrices $\{E, A, B, C\}$ in (1) become :

$$
\left[\begin{array}{c|c} Z^T(\lambda E - A)V & Z^T B \\ \hline CV & 0 \end{array}\right] \doteq \left[\begin{array}{ccc|c} \lambda E_1 - A_1 & & & B_1 \\ & \ddots & & \vdots \\ & & \lambda E_t - A_t & B_t \\ \hline C_1 & \cdots & C_t & 0 \end{array}\right], \quad (13)
$$

where each subpencil $\lambda E_i - A_i$ has only one eigenvalue λ_i. In general each eigenvalue λ_i can be repeated and its multiplicity k_i is the dimension of the block $\lambda E_i - A_i$. The expansion

$$
H(s) = \sum_{1}^{t} C_i(\lambda E_i - A_i)^{-1} B_i \quad (14)
$$

is essentially the partial fraction expansion of $H(s)$ since each term has only one pole (but of degree k_i which is possibly higher than 1). From a comparison of this expansion and the expansion (5) around the pole λ_i we obtain the identity

$$
C_i(\lambda E_i - A_i)^{-1} B_i = H_{-\ell}^{(i)}(\lambda - \lambda_i)^{-\ell} + \ldots + H_{-1}^{(i)}(\lambda - \lambda_i)^{-1} \quad (15)
$$

which shows indeed that (14) is the partial fraction expansion of $H(s)$.

Modal approximation now consists of keeping those terms in this expansion that correspond to "dominant" poles. The reduced order model is then obtained from keeping only the columns of Z and V corresponding to the blocks containing the "selected" dominant poles. One can describe this more formally by using the concept of deflating subspace.

Definition 1
The column spaces of the full rank matrices V_i and Z_i are called right, respectively left deflating subspaces of the regular pencil $(\lambda E - A)$ if and only if $(\lambda E - A)V_i = \hat{V}_i(\lambda E_i - A_i)$, respectively $Z_i^T(\lambda E - A) = (\lambda E_i - A_i)\hat{Z}_i^T$, and $(\lambda E_i - A_i)$ is also a regular pencil. ∎

The spectrum of a deflating subspace is that of the pencil $(\lambda E_i - A_i)$. If that spectrum is only one point (say λ_i) then $\text{Im}V_i$ and $\text{Im}Z_i$ are called deflating subspace of the eigenvalue λ_i. The largest dimension of a deflating subspace with spectrum λ_i equals the multiplicity k_i of that eigenvalue [17].

Theorem 2
Let $\text{Im}V_i$ and $\text{Im}Z_i$ be left and right invariant subspaces of the regular pencil $(\lambda E - A)$ with a given spectrum, and consider the regular reduced order system $(\lambda\hat{E} - \hat{A}) \doteq Z^T(\lambda E - A)V$. If

$$\text{Im}V_i \subseteq \text{Im}V \doteq \mathcal{V} \tag{16}$$

then $(\lambda\hat{E} - \hat{A})$ has a right invariant subspace with the same spectrum. If

$$\text{Im}Z_i \subseteq \text{Im}Z \doteq \mathcal{Z} \tag{17}$$

then $(\lambda\hat{E} - \hat{A})$ has a left invariant subspace with the same spectrum. In both cases the corresponding poles of the original system are retained in the reduced order system.

Proof: We only prove the result for the right deflating subspaces since both cases are dual. Since $\text{Im}V_i \subseteq \text{Im}V$ we have $V_i = VX_i$ for some full rank matrix X_i. Since $\text{Im}V_i$ is a deflating subspace of $(\lambda E - A)$, we have $(\lambda E - A)V_i = \hat{V}_i(\lambda E_i - A_i)$. It follows that $(\lambda\hat{E} - \hat{A})X_i = Z^T(\lambda E - A)VX_i = Z^T\hat{V}_i(\lambda E_i - A_i) = Y_i(\lambda E_i - A_i)$, whence X_i is a right deflating subspace of the reduced order pencil. ∎

Together with Theorem 1, this allows to combine moment matching with pole (or modal) matching. A simple example of this is when one has a system of differential algebraic equations (DAE's) and one wants to retain these algebraic equations in the reduced order system. For the original system this implies that E is singular with a kernel V_∞ of a particular dimension k_∞. The subscript ∞ is intentional since this kernel is in fact a deflating subspace corresponding to the eigenvalue $\lambda = \infty$. By imposing $\text{Im}V_\infty \subseteq \text{Im}V$, the reduced order system will have an $\hat{E}$ matrix with a kernel of the same dimension and hence will retain these algebraic equations. If no such condition is imposed, the reduced order system built via moment matching typically is not a DAE anymore.

We end this section by pointing out that modal matching can also be interpreted as moment matching of the transfer function $(\lambda - \lambda_i)^\ell H(s)$, which has no poles anymore at λ_i. Its first ℓ moments are indeed the matrices $H_{-\ell}^{(i)}, \ldots, H_{-1}^{(i)}$ of (15) and will be retained in the modified reduced order model $(\lambda - \lambda_i)^\ell \hat{H}(s)$, provided both V and Z contain V_i and Z_i as submatrices.

5. Near-optimal solutions

The approximation problem in the frequency domain can also be phrased in terms of the H_∞ norm of the error function $\Delta H(s) \doteq H(s) - \hat{H}(s)$. There is a well developed theory for finding nearly optimal solutions of this problem when the transfer function is given in a state-space formulation. These are based on the so-called balanced realizations of a state-space system $\{A, B, C\}$. Such realizations have Gramians that are equal and diagonal [11], and hence a diagonal product as well. We develop here the equivalent formulas for a generalized state-space $\{E, A, B, C\}$, under the assumption that E is non-singular (which can therefore be reduced to a standard state-space system).

The controllability Gramian G_c and observability Gramian G_o of a system (1) can be defined as follows :

$$G_c = \int_0^{+\infty} (e^{E^{-1}At} E^{-1} B)(e^{E^{-1}At} E^{-1} B)^T \, dt, \tag{18}$$

$$G_o = \int_0^{+\infty} (CE^{-1} e^{AE^{-1}t})^T (CE^{-1} e^{AE^{-1}t}) dt, \tag{19}$$

which by Parseval's theorem are also equal to

$$G_c = \frac{1}{2\pi} \int_0^{+\infty} (\jmath\omega E - A)^{-1} BB^T (\jmath\omega E - A)^{-*} d\omega, \tag{20}$$

$$G_o = \frac{1}{2\pi} \int_{-\infty}^{+\infty} (\jmath\omega E - A)^{-*} C^T C (\jmath\omega E - A)^{-1} d\omega. \tag{21}$$

These Gramians can be computed as the solution of the generalized state-space equations

$$AG_cE^T + EG_cA^T = -BB^T \text{ and } A^TG_oE + E^TG_oA = -C^TC. \tag{22}$$

The Gramians of the corresponding state space realization $\{E^{-1}A, E^{-1}B, C\}$ are in fact equal to G_c and E^TG_oE, respectively. It is the product $E^TG_oEG_c$ of these two matrices that one diagonalizes via a state-space similarity transformation [11]. One then truncates the smallest diagonal elements (i.e. eigenvalues of $E^TG_oEG_c$), which yields the reduced order approximation. An n-th order approximation is thus obtained from the eigenspace corresponding to the n largest eigenvalues of $E^TG_oEG_c$. An slightly better approximation to $H(s)$ is obtained from the optimal Hankel norm approximation which is also derived from the balanced realization and hence eigen-decompostion of $E^TG_oEG_c$. In practice both approximations give nearly optimal approximations in the H_∞ norm [11].

But (20,21) suggest than the rational Krylov approach tries to approximate the same objects, since the frequency response $C(\jmath\omega E - A)^{-1}B$ clearly is retrieved in the integrals describing the Gramians. Let us define the approximations to these Gramians as

$$\tilde{G}_c \doteq V\hat{G}_c V^T, \text{ and } \tilde{G}_o \doteq Z\hat{G}_o Z^T, \tag{23}$$

where $\hat{G}_c$ and $\hat{G}_o$ satisfy the projected equations

$$\hat{A}\hat{G}_c\hat{E}^T + \hat{E}\hat{G}_c\hat{A}^T = -\hat{B}\hat{B}^T \text{ and } \hat{A}^T\hat{G}_o\hat{E} + \hat{E}^T\hat{G}_o\hat{A} = -\hat{C}^T\hat{C}. \tag{24}$$

Define also the residuals

$$A\tilde{G}_c E^T + E\tilde{G}_c A^T + BB^T \doteq \Delta_c \text{ and } A^T\tilde{G}_o E + E^T\tilde{G}_o A + C^T C \doteq \Delta_o. \tag{25}$$

Then the approximations clearly satisfy the Galerkin conditions

$$Z^T\Delta_c Z = 0, \text{ and } V^T\Delta_o V = 0. \tag{26}$$

As the spaces V and Z grow, these residuals decrease and we are thus trying to improve the approximation $\tilde{G}_c \approx G_c$ and $\tilde{G}_o \approx G_o$. By choosing Z and V such that the dominant features of the transfer function $C(\jmath\omega E - A)^{-1}B$ are captured, we look for the dominant spaces of G_c and G_o, which in the balanced coordinate system are also the dominant eigenspaces of these two positive definite matrices.

6. A numerical comparison

We now compare moment matching techniques with the near-optimal approaches in terms of numerical accuracy and complexity. For the complexity, we only need to consider the most time consuming steps of each approach. We assume that the system is given in generalized state-space system. For the solution of the generalized Lyapunov equations one needs first to put the pencil $\lambda E - A$ in generalized Schur form. For a dense $N \times N$ system this requires approximately $70N^3$ flops (floating point operations). The subsequent eigendecomposition of $E^T G_o E G_c$ needed for balanced truncation takes another $30N^3$ flops, while the additional work for constructing an optimal Hankel norm approximation requires about $60N^3$ flops. Both near-optimal solutions therefore require over $100N^3$ flops. The rational Krylov approach on the other hand requires the LU factorizations of each matrix $(\sigma^{(i)}E - A)$, which is a total of $\frac{2}{3}KN^3$ flops. All other steps are less important and so this approach is of much lower complexity since the number of interpolation points K is typically small.

For sparse large-scale systems, the comparison is still in favor of rational Krylov methods. The near-optimal solutions still need the solution of the generalized Lyapunov equations but here one can also use sparse matrix

186

techniques. Efficient methods are based on the Smith iteration and multi-point accelerations of it [20] and compute a low rank approximation of the true solutions. This still has a complexity of the order of $c_1 k \alpha N$ flops where α is the average number of non-zero elements in each row of A and E, and k is the number of interpolation points used in this method. All other parts of the algorithm involve $n \times n$ matrices where n is the approximate rank of the Gramians. Notice that for this reason, the near-optimality is lost and that these approximations become much less accurate. For the rational Krylov approach one uses iterative solvers for the solution of the systems involving $(\sigma^{(i)} E - A)$, and this requires $c_2 K \alpha N$ flops, where K is the number of inter-polation points. Both approaches are therefore comparable in complexity, but the near-optimal methods tend to loose their accuracy. Moreover, the rational Krylov methods rely on independent matrix solves which can be implemented efficiently in parallel.

Now we look at the accuracy of both approaches. In order not to dis-favor the near-optimal schemes we use the full Lyapunov solvers and start from standard state-space models so that no accuracy can get lost from the inversion of E. All computations were performed on a machine with IEEE standard arithmetic and using MATLAB (which is also why all models are reasonably small). In Figures 1, 2 and 3 we compare three 15th order approximations of a single-input/single-output 120th transfer function of degree 120, used for the Compact Disc regulator [27]. The solid lines rep-resent the frequency response of the full system, the approximations are dotted for the Hankel norm approximation, dash-dotted for the balanced truncation and dashed for the rational Krylov method.

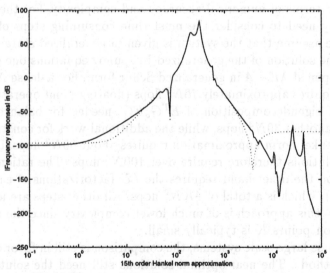

Fig. 1. Optimal Hankel norm approximation

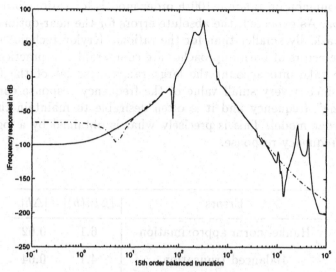

Fig. 2. Balanced truncation

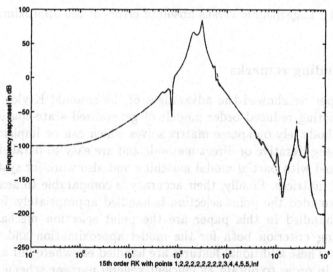

Fig. 3. Rational Krylov with 6 interpolation points

Although the near-optimal schemes should behave much better that the rational Krylov approximations, the pictures seem to indicate the opposite. This impression is due to the fact that a logarithmic scale was used. The *absolute error* of the near-optimal schemes is indeed much better that that of the Rational Krylov approach as indicated in Table 1, but for the logarithmic errors this is just the opposite. Notice that for the rational Krylov method we used 6 different interpolation points. We observed the same phenomenon on random 100th order models for various orders of approximation. As expected, the absolute errors for the near-optimal scheme are systematically smaller than for the rational Krylov technique, but the logarithmic errors of both approaches are comparable. In practice it is important to take into account the large range of scales of the frequency response, since a very small value of the frequency response corresponds to "blocking" frequency and it is often desirable to maintain this in the reduced order model. This is precisely what is obtained by a logarithmic fit of the frequency response.

Errors	$|\Delta ln(h)|$	$|\Delta h|$
Hankel norm approximation	6.1	0.02
Balanced truncation	4.1	0.04
Rational Krylov approximation	1.5	4.02

Table 1 : Logarithmic versus absolute errors of the approximations

7. Concluding remarks

In this paper we showed the advantages of the rational Krylov approach for constructing reduced order models of generalized state-space systems. These methods rely on sparse matrix solves which can be implemented efficiently using iterative or direct methods and are easy to parallelize. They can be mixed with partial modal matching and also work for systems with singular E matrices. Finally, their accuracy is comparable to near-optimal schemes provided the point selection is handled appropriately. Issues that were not handled in this paper are the point selection mechanism and the stopping criterion both for the model approximation and the iterative solves. These additional features are handled elsewhere but are equally important in order to obtain an efficient general purpose scheme.

Acknowledgments

This work was supported by a Department of Energy Fellowship, IBM Corporation, and the National Science Foundation under grant CCR-9796315. Parts of this paper presents research results of the Belgian Programme on Inter-university Poles of Attraction, initiated by the Belgian State, Prime Minister's Office for Science, Technology and Culture. The scientific responsibility rests with its authors.

References

1. L.A. Aguirre (1993), Quantitative measure of modal dominance for continuous systems, In *32nd IEEE Conf. Dec. Contr.*, San Antonio, TX.

2. B.D.O. Anderson and A.C. Antoulas (1990), Rational interpolation and state variable realizations, *Lin. Alg. & Appl.* **137/138**, pp. 479–509.

3. G.A. Baker Jr. (1975), *Essentials of Padé Approximants*, New York, Academic Press.

4. D.L. Boley (1994), *Krylov space methods on state-space control models, Circ. Syst. & Signal Process.* **13**, pp. 733–758.

5. D. Bonvin and D.A. Mellichamp (1982), A unified derivation and critical review of modal approaches to model reduction, *Int. J. Contr.* **35**, pp. 829–848.

6. C. Brezinski (1980), *Padé-Type Approximation and General Orthogonal Polynomials*, **ISNM 50**, Basel, Birkhäuser.

7. P. Feldman and R.W. Freund (1995), Efficient linear circuit analysis by Padé approximation via the Lanczos process, *IEEE Trans. Computer-Aided Design* **14**, pp. 639–649.

8. K. Gallivan, E. Grimme and P. Van Dooren (1994), Asymptotic waveform evaluation via a Lanczos method, *Appl. Math. Lett.* **7**, pp. 75–80.

9. K. Gallivan, E. Grimme and P. Van Dooren (1994), Padé Approximation of large-scale dynamic systems with Lanczos methods, *Proc. 33rd IEEE Conf. Dec. Contr.*, Lake Buena Vista, FL.

10. K. Gallivan, E. Grimme and P. Van Dooren (1995), A rational Lanczos method for model reduction, *Numer. Algorithms* **12**, pp. 33–63.

11. K. Glover (1984), All optimal Hankel norm approximations of linear time multivariable systems and their L_∞-error bounds, *Int. J. Contr.* **39**, pp. 1115–1193.

12. W.B. Gragg and A. Lindquist (1983), On the partial realization problem, *Linear Algebra & Appl.* **50**, pp. 277–319.

13. E. Grimme, D. Sorensen and P. Van Dooren (1995), Model reduction of state space systems via an implicitly restarted Lanczos method, *Numer. Algorithms* **12**, pp. 1–31.

14. E. Grimme (1997), *Krylov Projection Methods for Model Reduction*, PhD thesis, University of Illinois at Urbana-Champaign, IL.

15. E. Grimme, K. Gallivan and P. Van Dooren (1998), On Some Recent Developments in Projection-based Model Reduction, In *ENUMATH 97, 2nd European Conference on Numerical Mathematics and Advanced Applications*, H.G. Bock, F. Brezzi, R. Glowinski, G. Kanschat, Yu.A. Kuznetsov, J. Périaux, R. Rannacher (eds.), World Scientific Publishing, Singapore.

16. H.M. Kim and R.R. Craig Jr. (1998), Structural dynamics analysis using an unsymmetric block Lanczos algorithm, *Int. J. Numer. Methods in Eng.* **26**, pp. 2305–2318.

17. C.B. Moler and G.W. Stewart (1973), An algorithm for the generalized matrix eigenvalue problem, *SIAM Num. Anal.* **10**, pp. 241–256.

18. B.C. Moore (1981), Principal component analysis in linear systems : controllability, observability and model reduction, *IEEE Trans. Aut. Contr.* **26**, pp. 17–32.

19. B. Nour-Omid and R.W. Clough (1984), Dynamic analysis of structures using Lanczos coordinates, *Earthquake Eng. and Struc. Dyn.* **12**, pp. 565–577.

20. T. Penzl (1998), *A Cyclic Low Rank Smith Method for Large Sparse Lyapunov Equations with Applications in Model Reduction and Optimal Control*, T.U. Chemnitz, Dept. Mathematics, Preprint SFB393/98-6.

21. A. Ruhe (1994), Rational Krylov algorithms for nonsymmetric eigenvalue problems II: matrix pairs, *Lin. Alg. & Appl.* **197**, pp. 283–295.

22. Y. Shamash (1975), Model reduction using the Routh stability criterion and the Padé approximation technique, *Int. J. Control* **21**, pp. 475–484.

23. Y. Shamash (1981), Viability of methods for computing stable reduced-order models, *IEEE Trans. Aut. Contr.* **26**, pp.1285–1286.

24. T.-J. Su and R.R. Craig Jr. (1992), Krylov vector methods for model reduction

190

and control of flexible structures, *Advances in Control and Dynamic Systems* **54**, pp. 449–481.

25. P. Van Dooren (1992), Numerical linear algebra techniques for large scale matrix problems in systems and control, *Proc. 31st IEEE Conf. Dec. Contr.*, Tucson, AZ.

26. C.D. Villemagne and R.E. Skelton (1987), Model reduction using a projection formulation, *Int. J. Control* **46**, pp. 2141–2169.

27. P. Wortelboer (1994), *Frequency-weighted Balanced Reduction of Closed-loop Mechanical Servo-systems : Theory and Tools*, Ph.D. Thesis, T.U. Delft, The Netherlands.

ADAPTIVE SYMPLECTIC AND REVERSIBLE INTEGRATORS

B. KARASÖZEN

Department of Mathematics
Middle East Technical University
06531 Ankara-Turkey
bulent@rorqual.cc.metu.edu.tr

Abstract. The so-called structure-preserving methods which reproduce the fundamental properties like symplecticness, time reversibility, volume and energy preservation of the original model of the underlying physical problem became very important in recent years. It has been shown theoretically and experimentally, that these methods are superior to the standard integrators, especially in long term computation. In the paper the adaptivity issues are discussed for symplectic and reversible methods designed for integration of Hamiltonian systems. Molecular dynamics models and N-body problems, as Hamiltonian systems, are challenging mathematical models in many aspects; the wide range of time scales, very large number of differential equations, chaotic nature of trajectories, restriction to very small step sizes in time, etc. Recent results on variable step size integrators, multiple time stepping methods, regularization techniques with applications to classical and quantum molecular dynamics, to N- body atomic problems and planetary motion will be presented.

1. INTRODUCTION

Newton's second law of motion

$$M\frac{d^2q}{dt^2} = \nabla_q V(q) \tag{1}$$

describes the atomic motion of molecular system such as nucleic acids, proteins, biomolecules and polymers. This equation is a Hamiltonian system with the Hamiltonian $H = \frac{1}{2}p^T M^{-1} p + V(q)$, where $q \in \mathbb{R}^{3d}$ is the vector of d atomic positions, $p \in \mathbb{R}^{3d}$ is the vector of corresponding momenta, $M \in$

191

H. Bulgak and C. Zenger (eds.), Error Control and Adaptivity in Scientific Computing, 191–220.
© *1999 Kluwer Academic Publishers. Printed in the Netherlands.*

$\mathbb{R}^{3d \times 3d}$ the diagonal matrix containing atomic masses and $V(q)$ potential energy function. Equation (1) also describes the motion of planets in the solar system (N-body problems). These systems have some characteristics which make integrating them a challenging task:

- *Very large coupled system of ordinary differential equations*: the number of atoms are 10.000 - 100.000 with six equations per atom.
- *Multiple time scales*: the step size in the time variable is determined by the period associated with the high frequency modes present in all macromolecules. The most used integrator, Verlet method, sets a constraint on the maximum time step 1 fs $= 10^{-15}$ second due to the stability restrictions. In contrast to this the conformal changes (protein foldings) occur in longer time steps, 10^2 seconds. To speed up the algorithms, variable step size integration in connection with the multiple time stepping techniques must be used.
- *Long time integration*: MD simulations require efficient time stepping schemes that can faithfully approximate the dynamics of the underlying systems over many thousands of time steps.
- *Highly nonlinear and chaotic equations*: they are very sensitive to perturbations. Numerical experiments show that a tiny perturbation introduced into trajectories is doubled every pico second $= 10^{-9}$ second. After 1000 pico seconds of simulation the trajectory would be overwhelmed by the effects of roundoff error.
- *Long-time instability:* The traditional concepts of stability and accuracy for numerical integrators are of limited value for the treatment of nonlinear dynamics in long-term intervals due to the sensitive dependence on initial conditions.
- *The accurate computation of the trajectories is not desirable*: the aim of MD simulations is to obtain information like average energies, conformal distributions. For long-time integrations, accuracy with respect to the particular trajectory is neither desirable nor obtainable, only a generic or representative behavior of the systems or good sampling of the phase space is required.
- The expressions currently used for interatomic forces, the potential energy functions are only approximations that involve fitting parameters semi-empirically and the initial velocities are assigned randomly.

Standard numerical methods like Runge-Kutta and multiple time step methods will not work for MD simulations; nonstandard, unconventional numerical methods of low accuracy like the Verlet method, which is second order accurate and explicit, Lie algebraic methods are preferred. Usual multiple time stepping techniques of numerical analysis which suppress the rapidly decaying component of motion can not be applied, because of the ergodic nature of the differential equations for MD simulations; no trajectory can be smoother than its neighbors for more than a short time interval. With MD simulations and integration of Hamiltonian systems the ordinary differential equation research in numerical analyses become an interdisciplinary effort among mathematicians, physicist and chemists.

In section 2 we describe Hamiltonian systems with their properties, like symplecticness and time-reversibility. We outline the structure preserving integrals like the Verlet method and the classes of symplectic and reversible Runge-Kutta methods. Section 3 is devoted to variable step size structure preserving integrators based on a transformation of the original Hamiltonian system. In section 4 the multiple time stepping (MTS) schemes based on Verlet method is given with its different versions. Recent results of applications of MTS methods for the classical and quantum-classical MD equations are summarized in Section 5. In Section 6, application of adaptive techniques of the Section 3 and 4 to $N-$body problems are given.

2. SYMPLECTIC AND REVERSIBLE INTEGRATORS FOR HAMILTONIAN SYSTEMS

Hamiltonian systems are described by the following system of ordinary differential equations

$$\frac{dp_i}{dt} = -\frac{\partial H}{\partial q_i}(p,q), \qquad \frac{dq_i}{dt} = \frac{\partial H}{\partial p_i}(p,q), \qquad i = 1,\ldots,d \qquad (2)$$

The variables q denote generalized coordinates, p generalized momenta and d stands for the degree of freedom $q, p \in \mathbb{R}^{3d}$ and H represents the total (mechanical) energy. The Hamilton equations (2) can also be written as

$$\frac{dy}{dt} = J^{-1}\nabla H, \qquad J^{-1} = -J \qquad (3)$$

with $y = (p_1,\ldots,p_d,q_1,\ldots,q_d)$ and the skew-symmetric matrix

$$J = \begin{pmatrix} 0 & I \\ -I & 0 \end{pmatrix}$$

where ∇ is the gradient operator $\nabla = \left(\frac{\partial}{\partial p_1}, \cdots \frac{\partial}{\partial p_d}, \frac{\partial}{\partial q_1}, \cdots \frac{\partial}{\partial q_d}\right)$. Important properties of Hamiltonian systems are:

i) the Hamiltonian $H(p,q)$ is constant along the solution, i.e. the energy is conserved. Additional integrals, like momentum are also preserved.

ii) The corresponding flow is symplectic; the differential two-form $\omega = \sum_{i=1}^{d} dp_i \wedge dq_i$ is preserved, where $\wedge$ denotes the exterior product. In two dimensions symplecticness is equivalent to area preservation. In

194

higher dimensions symplecticness is still a geometric property some-
what difficult to state; a mapping $\Phi \in \mathbb{R}^{2d}$ with $\Phi_{\Delta t, H}(t)(q(t), p(t)) = (q(t+\Delta t), (p+\Delta t))$ is called symplectic if the sum of the projections of
two-dimensional oriented surfaces is preserved. A consequence of this
is volume preservation. The geometric definition of symplecticness can
be replaced by an algebraic one:

$$(\nabla\Phi)^T J (\nabla\Phi) = J,$$

where $\nabla\Phi$ denotes the Jacobian matrix of the flow Φ, which charac-
terizes the solution of the Hamiltonian system (2).

iii) the flow is time-reversible, i.e equations (2) are invariant under the
transformation $(t, p, q) \to (-t, -p, q)$. Solutions $(p(t), q(t))$ with the
initial conditions (p_0, q_0) at $t = 0$ forward in time $t \geq 0$ and solutions
$(\bar{p}(t), \bar{q}(t))$ with the initial conditions $(-p_0, q_0)$ at $t = 0$ backward in
time $t \leq 0$ satisfy the relation

$$(p(t), q(t)) = (-\bar{p}(-t), \bar{q}(-t)).$$

In other word, if one integrates forward in Δt units in time from a point
A to a point B, then replaces t by $-t$ and p by $-p$ in the differential
equations and integrates Δt units starting from the point B, one arrives
back to point A.

In the following two benchmark problems are presented, which are used
often to show the performance of the numerical methods for Hamiltonian
systems.

Kepler's problem: Kepler's problem describes the motion of a material
point in a plane that is attracted towards the origin with a force inversely
proportional to the distance squared. The motion is 2π periodic and the
distance from the center of the ellipse to the origin is called eccentricity ϵ,
$0 < \epsilon < 1$. The Hamiltonian is separable in the variables p and q and is equal
to the sum of kinetic $T(p)$ and potential $V(q)$ energies, i.e. $H = T(p) + V(q)$.

$$T(p) = \frac{1}{2}(p_1^2 + p_2^2), \quad V(q) = -\frac{1}{\sqrt{q_1^2 + q_2^2}} \tag{4}$$

Equation of motions and initial conditions are:

$$\dot{p}_1 = -\frac{q_1}{(q_1^2 + q_2^2)^{3/2}}, \quad \dot{q}_1 = p_1$$

$$\dot{p}_2 = -\frac{q_2}{(q_1^2 + q_2^2)^{3/2}}, \quad \dot{q}_2 = p_2,$$

$$q_1(0) = 1 - \epsilon, \quad q_2(0) = 0, \quad p_1(0) = 0, \quad p_2(0) = \sqrt{(1+\epsilon)/(1-\epsilon)}$$

Additional conserved quantity or first integral is momentum $I = p_1 p_2 - q_2 p_1$. Especially for high eccentricities $\epsilon \approx 1$, Kepler's equation is used as a benchmark problem for variable step size integrators.

Molecular dynamics (MD) equations: Newton's equation of motion (1)

$$\dot{q} = M^{-1}p, \qquad \dot{p} = -\nabla_q V(q) \tag{5}$$

with the kinetic energy $T(p) = \frac{1}{2}p^T M^{-1}p$ and potential energy function $V(q)$ is the most used Hamiltonian benchmark problem to evaluate the performance of different integrators. In molecular dynamics, simulation of the motion of N mutually interacting particles, require $\mathcal{O}(N^2)$ force evaluations $\nabla_q V(q)$ per time step. Therefore the number of evaluations of the vector field ∇V is taken as a cost measure of a numerical method in MD applications and N-body dynamics.

For nonseparable Hamiltonian systems, the class of Gauss-Legendre implicit Runge-Kutta (RK) methods are symplectic and reversible, including the second order implicit mid-point method

$$q_{n+1} = q_n + \Delta t M^{-1}\left(\frac{p_n + p_{n+1}}{2}\right), \; p_{n+1} = p_n - \Delta t \nabla_q V\left(\frac{q_n + q_{n+1}}{2}\right). \tag{6}$$

Explicit methods are easy to implement, they don't require solution of nonlinear equations. But they have step size restrictions due to the limited stability. Implicit methods allow the use of larger time steps because of their better stability properties. But implicit symplectic integrators like Gauss-Legendre Runge-Kutta methods with s stages increase the computational complexity enormously for large scale problem like MD equations; at each time step a system of nonlinear equations of size $2ds$ must be solved.

Another important class RK-methods for nonseparable general Hamiltonian systems are given in partitioned form:

$$\frac{dp}{dt} = f(p, q), \qquad \frac{dq}{dt} = g(p, q)$$

$$
\left|
\begin{array}{ccc}
a_{11} & \cdots & a_{1s} \\
\vdots & \ddots & \vdots \\
a_{s1} & \cdots & a_{ss} \\
\hline
b_1 & \cdots & b_s
\end{array}
\right.
\left|
\begin{array}{ccc}
\hat{a}_{11} & \cdots & \hat{a}_{1s} \\
\vdots & \ddots & \vdots \\
\hat{a}_{s1} & \cdots & \hat{a}_{ss} \\
\hline
\hat{b}_1 & \cdots & \hat{b}_s
\end{array}
\right.
\tag{7}
$$

$$P_i = p_n + \Delta t \sum_{j=1}^{s} a_{ij} f(P_j, Q_j), \qquad Q_i = q_n + \Delta t \sum_{j=1}^{s} \hat{a}_{ij} g(P_j, Q_j)$$

$$p_{n+1} = p_n + \Delta t \sum_{i=1}^{s} b_i f(P_i, Q_i), \qquad q_{n+1} = q_n + \Delta t \sum_{i=1}^{s} \hat{b}_i g(P_i, Q_i)$$

The vectors P_i, Q_i are called stage vectors and s denotes the number of stages. Symplecticness is expressed by the following relationship between the coefficients of the partitioned RK-method (7):

$$b_i \hat{a}_{ij} + \hat{b}_i a_{ji} - b_i \hat{b}_j = 0, \qquad i, j = 1, \ldots, s, \qquad \text{and} \qquad b_i = \hat{b}_i, \qquad i = 1, \ldots, s.$$

Similarly the reversibility condition is given by $\hat{b}_i = b_{s-i+1}$, $i = 1, \ldots, s$. Most of the symplectic RK-methods are reversible (symmetric). An example from this class of methods is the two stage, second order symplectic and symmetric Lobatto IIIA-B method:

$$
\begin{array}{c|cc}
0 & 0 & 0 \\
1 & 1/2 & 1/2 \\
\hline
& 1/2 & 1/2
\end{array}
\qquad
\begin{array}{c|cc}
1/2 & 1/2 & 0 \\
1/2 & 1/2 & 0 \\
\hline
& 1/2 & 1/2
\end{array}
\qquad (8)
$$

The symplectic partitioned Runge-Kutta methods for separable Hamiltonian systems $H(p, q) = T(p) + V(q)$ become explicit and are given in the following form:

$$
\begin{array}{|ccccc}
b_1 \\
b_1 & b_2 \\
b_1 & b_2 & b_3 \\
\vdots & \vdots & & \ddots \\
b_1 & b_2 & \cdots & b_{s-1} & b_s \\
\hline
b_1 & b_2 & \cdots & b_{s-1} & b_s
\end{array}
\qquad
\begin{array}{|ccccc}
0 \\
\hat{b}_1 & 0 \\
\hat{b}_1 & \hat{b}_2 & 0 \\
\vdots & \vdots & \vdots & \ddots \\
\hat{b}_1 & \hat{b}_2 & \cdots & \hat{b}_{s-1} & 0 \\
\hline
\hat{b}_1 & \hat{b}_2 & \cdots & \hat{b}_{s-1} & 0
\end{array}
\qquad (9)
$$

They have FSAL property (first same as last), which reduces the computational cost by avoiding additional force function evaluations. The simplest method of this class is the first order symplectic Euler method with $s = 1$, $b_1 = \hat{b}_1 = 1$:

$$p_{n+1} = p_n - \Delta t \nabla_q V(q_n), \qquad q_{n+1} = q_n + \Delta t M^{-1} p_{n+1} \qquad (10)$$

The Hamiltonian equations (2) can also be written in form of

$$\frac{dy}{dt} = \{y, H\} \equiv X_H(y) \qquad (11)$$

with the Poisson bracket for two real smooth functions F and G

$$\{F, G\} = \sum_{i=1}^{d} \left(\frac{\partial F}{\partial q_i} \frac{\partial G}{\partial p_i} - \frac{\partial F}{\partial p_i} \frac{\partial G}{\partial q_i} \right)$$

which is bilinear in F and G, skew-symmetric: $\{F, G\} = -\{G, F\}$, satisfies the Jacobi identity: $\{\{F, G\}, H\} + \{\{G, H\}, F\} + \{\{H, F, \}G\} = 0$, and the

Leibniz's rule: $\{F \cdot G, H\} = F \cdot \{G, H\} + G \cdot \{F, H\}$. For separable Hamiltonians $H(p, q) = V(q) + T(p)$ like Kepler's problem and MD equations, the formal solution of (11) given by

$$y(t) = \exp(t\{\cdot, H\})y(0)$$

can be approximated by a time-reversible composition of the individual flows

$$\exp\left(\frac{\Delta t}{2}\{\cdot, V\}\right) \circ \exp\left(\Delta t\{\cdot, T\}\right) \circ \exp\left(\frac{\Delta t}{2}\{\cdot, V\}\right).$$

Splitting methods are very valuable in solving Hamiltonian systems numerically. The splitted individual flows are either exactly integrable or can be approximated easily by a symplectic or time-reversible integrator. The most known algorithm for MD simulations is the Verlet/Strömer/leapfrog method, hereafter shortly Verlet method. This can be derived by the splitting of the flow corresponding to kinetic and potential energy parts of the Hamiltonian which is known as Strang splitting:

$$\Psi_{\Delta t, T+V} = \Psi_{\frac{\Delta t}{2}, V} \circ \Psi_{\Delta t, T} \circ \Psi_{\frac{\Delta t}{2}, V} + \mathcal{O}(\Delta t)^3,$$

$$\Psi_{\frac{\Delta t}{2}, V} : \quad \frac{d}{dt}q = 0, \quad \frac{d}{dt}p = -\nabla_q V(q), \quad \Psi_{\Delta t, T} : \quad \frac{d}{dt}q = M^{-1}p, \quad \frac{d}{dt}p = 0.$$

The Verlet method

$$
\begin{aligned}
p_{n+1/2} &= p_n - \frac{\Delta t}{2}\nabla_q V(q_n) & \text{half a kick} \\
q_{n+1} &= q_n + \Delta t \nabla M^{-1} p_{n+1/2} & \text{full drift} \quad (12) \\
p_{n+1} &= p_{n+1/2} - \frac{\Delta t}{2}\nabla V_q(q_{n+1}) & \text{half a kick}
\end{aligned}
$$

can also be interpreted as the consecutive application of Euler method (10) and the adjoint Euler method:

$$p_{n+1} = p_n - \Delta t \nabla_q V(q_{n+1}), \qquad q_{n+1} = q_n + \Delta t M^{-1} p_n. \quad (13)$$

For separable Hamiltonian systems it is equivalent to the second order Lobatto III A-B method (8) and belongs to the class of explicit partitioned RK-methods (9). The Verlet method can be interpreted physically as follows: two impulses of equal size are imparted to the particles just before and after each time point and between these time points the particles drift with constant velocity. It is known also as the leapfrog method, where p variables are approximated at the grid points t_n, q variables are approximated at the midpoint $\frac{t_n + t_{n+1}}{2} = t_n + \Delta t$ at staggered grids.

Higher order reversible and symplectic integrators can be constructed easily by composition methods. A well-known composition method is the Yoshida technique [54]. Suppose $\psi_{\Delta t, H}$ is a second order time reversible method (for example the implicit mid-point method or the Verlet method), then the composite method is given by the formula

$$\psi_{\alpha \Delta t} \circ \psi_{(1-2\alpha)\Delta t} \circ \psi_{\alpha \Delta t}$$

with $\alpha = \frac{1}{3}(2 + 2^{1/3} + 2^{-1/3})$ results in a fourth order time reversible method. It is possible to obtain higher order methods by this composition technique. Both ideas, splitting and composition, can be combined to integrate the Hamiltonian systems cheaply and accurately. An example for this kind of application is the integration of the solar system over a period of 10^9 years [53]. But the optimal design of splitting and composition techniques remains a challenging task.

Symplectic integrators preserve the quadratic invariants like Hamiltonians and angular momentum exactly. The backward error analysis revealed that the error of a symplectic integrator is also Hamiltonian; the energy or Hamiltonian is bounded by symplectic and reversible integrators, where for standard integrators a drift in the energy occurs. The numerical solutions satisfy the so called modified equations with a nearby Hamiltonian for symplectic integrators; they are backward stable. If a symplectic discretization is applied to approximate the Δt-flow $\Phi_{\Delta t, H}$, then the numerical map $\Psi_{\Delta t, H}$ is exactly the Δt-flow of a perturbed Hamiltonian $\Psi_{\Delta t} \equiv \Psi_{\Delta t, \widetilde{H}}$ with

$$\widetilde{H} = H + \Delta t^r H_r + \Delta t^{r+1} H_{r+1} + \cdots$$

where r denotes the order of the method and $\widetilde{H}$ is close to H, which was shown by Hairer [16] and Reich [38] for symplectic integrators. The formal series expansion of $\widetilde{H}$ doesn't need to converge in general. The errors for the trajectories are bounded for a r th order method by $\mathcal{O}(\Delta t^r)$ over long time interval $1 << T \leq \mathcal{O}(e^{c/\Delta t})$. The magnitude of energy fluctuations in actual simulation is caused by the perturbation of the Hamiltonian and it is a useful indicator of accuracy. It is impossible to have symplecticness and Hamiltonian preservation at the same time. In case of the systems with periodic solutions or integrable systems the phase error (errors in the position variables q) grows linearly for symplectic integration algorithms (SIA) and reversible integration algorithms (RIA), in contrast to quadratic error growth for standard ode integrators.

For a general introduction to Hamiltonian systems in connection with reversible and symplectic integrators see the monographs [18, Section II.16], [44], [50, Chapter 8] and for MD applications [31].

3. VARIABLE STEP SIZE SYMPLECTIC AND REVERSIBLE INTEGRATORS

Preservation of the geometric features like symplecticness and time reversibility are desirable properties of structure-preserving integrators. Symplectic and reversible integrators with conventional variable step size strategies destroy the symplecticness and time reversibility, introduce drift in the energy or Hamiltonian in long time integration. If one varies the step size in a symplectic integrator, the effect is the same as if one would solve a different Hamiltonian system over each time step. The numerical solution no longer coincides with the solutions of the corresponding nearby Hamiltonian system. Because of these, the so called reversible variable step size strategies are devised based on local error estimation of the Verlet method [27] and of symmetric Runge-Kutta methods in [7], [19], [49]. Most of the reversible step size methods are implicit; the step size becomes a function of the dynamic variables and has to be handled as a component of the solution vector.

In recent years variable step size geometric integrators are constructed by transforming the original Hamiltonian equations (11) with a time parametrization $\frac{dt}{ds} = \frac{1}{R(y)}$. The transformed equations

$$\frac{dy}{ds} = \frac{1}{R(y)} X_H(y) \tag{14}$$

are then solved with a fixed step size Δs. Time transformation leaves the orbits invariant, only the speed of the orbits is changed; in regions where the $R(y)$ and local errors is large, small step sizes Δt are required. Integration of the transformed system with the constant step size Δs corresponds to the integration of the original system with a variable step size $\Delta t = \Delta s / R$. A solution of the scaled equations (14) in the extended phase space is equivalent to that of the original system (11) up to reparametrization of time. The transformed or rescaled equations are in general not Hamiltonian, but the Poisson structure and the first integrals are preserved.

The time parametrization or monitoring function $R(q, p)$ is a smooth, positive, scalar valued function and satisfies usually the reversibility condition, i.e. $R(p, q) = R(-p, q)$. A natural choice of R is the vector field normalization $R = \|X_H(y)\|$, with $\| \cdot \|$ Euclidian 2-norm which correspond to arclength parametrization and is given for the MD equations as $R(p, q) = \sqrt{p^T M^{-1} p + V_q(q)^T M^{-1} V_q(q)}$ [2], [6], [17], [26]. Another choice is the arclength parametrization depending on the position variables q only, $R(q) = \sqrt{2(H_0 - V(q)) + V_q(q) M^{-1} V_q(q)}$ [6], [17]. In many body systems the smallest separation distance between the particles $d_{\min}$ with a positive constant α is taken as the parametrization function $R(q) = 1/d_{\min}^{2\alpha}$ [6],

[26]. There are two types of variable step size structure-preserving integrators, one is based on the reversible time transformed Hamiltonian system (*Sundman transformation*):

$$\frac{dp}{ds} = -\frac{1}{R(p,q)}\frac{\partial H}{\partial q}(p,q), \qquad \frac{dq}{ds} = \frac{1}{R(p,q)}\frac{\partial H}{\partial p}(p,q) \qquad (15)$$

and the other based on the Hamiltonian time transformed system (*Poincaré transformation*). Time transformation changes the coupling structure of the original Hamilton equations (2) and a natural generalization of the Verlet scheme (12) leads to the implicit Lobatto IIIA-B pair method (8).

We follow here the representation in [6] for the structure-preserving variable step size methods developed in the recent years. Application of the second order Lobatto IIIA-B pair (8) to the time transformed system (15) with $H(p,q) = \frac{1}{2}p^T M^{-1}p + V(q)$ results in the *implicit adaptive Verlet (IAV)* method [26]:

$$p_{n+1/2} = p_n - \frac{\Delta s}{2}\frac{1}{R(p_{n+1/2}, q_n)}\nabla V(q_n),$$

$$q_{n+1} = q_n + \frac{\Delta s}{2}\left[\frac{1}{R(p_{n+1/2}, q_n)} + \frac{1}{R(p_{n+1/2}, q_{n+1})}\right]M^{-1}p_{n+1/2},$$

$$p_{n+1} = p_{n+1/2} - \frac{\Delta s}{2}\frac{1}{R(p_{n+1/2}, q_{n+1})}\nabla V(q_{n+1}),$$

$$t_{n+1} = t_n + \frac{\Delta s}{2}\left[\frac{1}{R(p_{n+1/2}, q_n)} + \frac{1}{R(p_{n+1/2}, q_{n+1})}\right].$$

The first two equations above are implicit in the p and q variables and they have to solved by the functional iteration. If $R(q,p)$ depends on p through its norm as for archlength parametrization, then the vector iteration reduces to a scalar iteration. On the otherside if the monitoring function R depends on q only as for the minimum separation distance parametrization, the first equation is explicit, but the second is implicit in q_{n+1} and has to solved by iteration. The last equation is always explicit.

A fully explicit method was proposed in [30] by introducing a new variable ρ that represents the reciprocal of the time scaling function in a symmetric three term recurrence relation. This results in the *explicit adaptive Verlet (EAV)* method:

$$q_{n+1/2} = q_n + \frac{\Delta s}{2\rho_n}M^{-1}p_n,$$

$$p_{n+1/2} = p_n - \frac{\Delta s}{2\rho_n}\nabla V(q_{n+1/2}),$$

$$\rho_n + \rho_{n+1} = 2R(p_{n+1/2}, q_{n+1/2}),$$

$$p_{n+1} = p_{n+1/2} - \frac{\Delta s}{2\rho_{n+1}} \nabla V(q_{n+1/2}),$$

$$q_{n+1} = q_{n+1/2} + \frac{\Delta s}{2\rho_{n+1}} M^{-1} p_{n+1},$$

$$t_{n+1} = t_n + \frac{\Delta s}{2} \left[\frac{1}{\rho_n} + \frac{1}{\rho_{n+1}} \right].$$

At the first step $\rho_0 = R(p_0, q_0)$ is taken. A similar fully explicit symmetric method was formulated in [25] for general time reversible systems including the rigid body equations. If the ρ-update is replaced by

$$\rho_n + \rho_{n+1} = R(p_{n+1}, q_{n+1/2}) + R(p_n, q_{n+1/2})$$

then the method above becomes semi-explicit [26]. The superiority of the reciprocal formulation of the adaptive Verlet method was explained theoretically by examining the leading error terms in the modified equations and asymptotic expansions and it was shown there that the elimination of the dominant oscillation terms leads to an improvement of the algorithm by correcting the starting step sizes [8].

The idea of the construction of a Hamiltoninan transformed system for a symplectic variable step size interpretation was formulated independently in [17] and in [38]. The Poincaré transformed system

$$\frac{dp}{ds} = -\frac{1}{R(p,q)} H_q(p,q) - \nabla_q \left(\frac{1}{R(p,q)} \right) (H(p,q) - H_0),$$

$$\frac{dp}{ds} = \frac{1}{R(p,q)} H_p(p,q) + \nabla_p \left(\frac{1}{R(p,q)} \right) (H(p,q) - H_0) \tag{16}$$

is just a perturbation which vanishes along the solutions of the original system with initial conditions p_0, q_0, but makes it Hamiltonian. The second terms on the right hand side of (16) will vanish along a particular orbit $H = H_0$, and we are left with the Sundman transformation. However, in case of numerical integrators, there are typically some errors that are introduced in the energy and the two approaches are then quite different.

Application of the Lobatto IIIA-B pair (8) to (16) results in the *variable step symplectic method (VS)*:

$$p_{n+1/2} = p_n - \frac{\Delta s}{2} \left\{ \frac{1}{R(p_{n+1/2}, q_n)} \nabla V(q_n) , \right.$$

$$\left. + \left[H(p_{n+1/2}, q_n) - H_0 \right] \nabla_q \left(\frac{1}{R(p_{n+1/2}, q_n)} \right) \right\},$$

$$q_{n+1} = q_n + \frac{\Delta s}{2} \left\{ \frac{1}{[R(p_{n+1/2}, q_n) + R(p_{n+1/2}, q_{n+1})]} p_{n+1/2} \right.$$

$$+ \; [H(p_{n+1/2}, q_n) - H_0] \, \nabla_p \left(\frac{1}{R(p_{n+1/2}, q_n)} \right),$$

$$+ \left. [H(p_{n+1/2}, q_{n+1}) - H_0] \, \nabla_p \left(\frac{1}{R(p_{n+1/2}, q_{n+1})} \right) \right\},$$

$$p_{n+1} = p_{n+1/2} - \frac{\Delta s}{2} \left\{ \frac{1}{R(p_{n+1/2}, q_{n+1})} \nabla V(q_{n+1}) \right.$$

$$+ \left. [H(p_{n+1/2}, q_{n+1}) - H_0] \, \nabla_q \left(\frac{1}{R(p_{n+1/2}, q_{n+1})} \right) \right\},$$

$$t_{n+1} = t_n + \frac{\Delta s}{2} \left[\frac{1}{R(p_{n+1/2}, q_n)} + \frac{1}{R(p_{n+1/2}, q_{n+1})} \right].$$

If R depends on p through its norm only, again the first equation can be solved by a scalar iteration. For the minimal distance separation function depending on q, the first equation leads to a quadratic equation in $\|p_{n+1/2}^2\|$, from which $p_{n+1/2}$ can be obtained easily. The second equation is implicit in q_{n+1} and has to be solved by the functional iteration, except for the p independent monitoring functions. The methods above are all time-reversible and preserve the angular momentum. All the methods above require one force evaluation $\nabla_q V(q)$ per time step. The implicit equations in the IAV and VS algorithms have to be solved by functional iteration. The number of iterations for IAV is m and for VS is $m+1$. The adaptive Verlet methods IAV and EAV are easy to implement. The VS algorithm costs slightly more than the adaptive Verlet methods and is not applicable for all choices of R, the differentiability of $R(q,p)$ with respect to p and q variables is required. The cheapest of them all is the explicit adaptive Verlet method EAV. The variable step size structure preserving algorithms are tested for the Kepler problem (4) in [17], [26]. [38], for a three body problem [26], for rigid body equations [25] and for molecular dynamics model with the Morse potential [21]. These algorithms are tested for the Kepler problem (4) with various tolerances and eccentricities and compared with respect to accuracy and computational cost in [6].

Classical molecular dynamics equations with fast bond stretching and bond bending modes lead to Newtonian equations in form of:

$$\dot{q} = M^{-1}p, \qquad \dot{p} = -\nabla_q V(q) - G(q)^T Kg(q). \tag{17}$$

The term $G(q)^T Kg(q) = \sum_{i=1}^m G_i(q)^T k_{ii} g_i(q)$ describes the bond stretching with $g_i(q) = r - r_0$ or bond angle bending $g_i(q) = \phi - \phi_0$ [42]. The function The functions $g_i, i = 1, \dots, m$ with $m < d$ are collected in the vector

$g(q)$. The diagonal matrix $K \in \mathbb{R}^{m \times m}$ contains the force constants k_{ii}, $G_i(q) \in \mathbb{R}^{3d}$ denote the Jacobian of $g_i(q)$ and the Jacobians $\{G_i(q)\}$ are collected in the matrix $G(q) \in \mathbb{R}^{m \times 3d}$. The total energy of the system is given by the Hamiltonian:

$$H = \frac{1}{2} p^T M^{-1} p + V(q) + \frac{1}{2} g(q)^T K g(q).$$

The Verlet method (12) can be applied to the highly oscillatory system for $K >> 1$ (17) in this form, but it turns out to be inefficient due to long-range interactions. The equation (17) can be replaced for the limiting case $K \to \infty$ by the constrained system of Hamiltonian Euler-Lagrange equations:

$$\begin{aligned} \frac{dq}{dt} &= M^{-1} p, \\ \frac{dp}{dt} &= -\nabla V(q) - G(q)^T \lambda, \\ 0 &= g(q) \end{aligned} \tag{18}$$

where $g_i(q) = 0, i = 1, \ldots, m$ stand for the constraints and λ denote the Lagrange multipliers. It is assumed that the $m \times m$ matrix $G(q) M^{-1} G(q)^T$ is invertible.

The constrained system (18) was integrated using symplectic and reversible SHAKE or RATTLE ([20], Section VII.8), which correspond to a modified Verlet method [42]:

$$\begin{aligned} q_{n+1} &= q_n + \frac{\Delta t}{2} M^{-1} p_{n+1/2}, \\ p_{n+1/2} &= p_n - \frac{\Delta t}{2} \nabla V_q(q_n) - \Delta t G(q_n)^T \lambda_n, \\ p_{n+1} &= p_{n+1/2} - \frac{\Delta t}{2} \nabla V_q(q_{n+1}), \\ 0 &= g(q_{n+1}). \end{aligned}$$

Although the constrained formulation has some limitations and drawbacks [42], it can be used in MD simulation either by introducing soft constraints [39], modifying the force field [41] or correcting the forces by additional terms [37]. Similar to the unconstrained Hamiltonian systems, the constrained system (18) can be transformed using a reversible time reparametrization, leading to the time-reversible, semi-explicit (requiring only solution of algebraic equations to satisfy the constraint relations and one step force evaluation function at each time step) **VRATTLE** method [2]. The numerical experiments in [2] for the double pendulum reveal that the monitor functions based on the size of the constrained vector field give better results than those based on the size of the unconstrained vector field; peaks which occur in the energy plot for the unconstrained monitor functions are removed for monitor functions including the constrained vector field.

4. MULTIPLE TIME STEPPING METHODS

Variable step size geometric integrators transform the original system and the transformed equations are integrated with constant step sizes. Another idea opposite to variable step size integrators, is the use of different step sizes for different processes. In case of multirate methods different processes in a system of ode's are identified as subsystems. In contrast to this, in the case of the multistep methods, processes with different interaction terms are identified.

In MD applications, the forces can be decomposed as a sum of hard forces, short-range interactions; and soft forces, long-range interactions at different time scales. Therefore it is natural to reduce the overall computational work by evaluating the soft forces less often than the hard forces; this leads to the multiple time stepping (MTS) methods. This is another kind of adaptive method than the symplectic and reversible variable step methods, where different step sizes are used for different processes. In MD applications, the potential energy functions for $2, -3$ and $4-$body interactions provide a natural decomposition. The potential energy can be divided into two distinct classes:

- internal (bonded) forces: bond stretches, bond angle bends, dihedral angles with linear computational complexity,
- external (unbounded) forces: van der Waals and Coulombic forces with quadratic computational complexity.

The expense of calculating the external forces is especially severe when the step size is small. This problem has stimulated more efficient approaches that exploit the more slowly varying nature of external forces, in comparison to the internal interactions. The concept behind multiple time step methods is to increase the efficiency of a numerical integrator by decreasing the computational cost for the force calculations during the integration.

We assume that the potential energy is decomposed in two distinct classes: $V(q) = V^{hard} + V^{soft}$. A decomposition of the potential energy $V(r) = C/r$ which occur due to gravitational or electrostatic attraction is given for the Kepler problem [48] :

$$V^{soft}(r) = \begin{cases} C(2r_{cut} - r)r_{cut}^{-2}, & r \leq r_{cut}, \\ Cr^{-1}, & r \geq r_{cut}, \end{cases}$$

$$V^{hard}(r) = \begin{cases} C(r_{\text{cut}} - r)^2 r_{\text{cut}}^{-2} r^{-1}, & r \leq r_{\text{cut}}, \\ 0, & r \geq r_{\text{cut}}, \end{cases}$$

where V^{soft} is the first two terms of the Taylor series expansion of $V(r)$ for $r \leq r_{\text{cut}}$. The idea behind the MTS-SV (multiple time stepping Strömer-Verlet method) is then to sample the short-range interaction terms V^{hard} at periods of δt (micro step size) and the smooth long-range interaction V^{soft} at periods of $\Delta t = N\delta t$ (macro step size), where N is an integer.

The Verlet method can be generalized to a multiple time step method in various ways, most of them destroy symplecticness. There is a way to retain the symplecticness, the so called the Verlet-I method [3], [15], [32], [48]:

$$\tilde{p}_{iN} = p_{iN} - \frac{N\delta t}{2} V_q^{soft}(q_{iN})$$

$$\left. \begin{array}{rcl} p_{iN+k-1/2} &=& \tilde{p}_{iN+k-1/2} - \frac{\delta t}{2} V_q^{hard}(q_{iN+k-1}) \\ q_{iN+k} &=& q_{iN+k-1} + \delta t M^{-1} p_{iN+k-1/2} \\ \tilde{p}_{iN+k} &=& p_{iN+k-1/2} - \frac{\delta t}{2} V_q^{hard}(q_{iN+k}) \end{array} \right\} \quad k = 1, 2, \ldots, N$$

$$p_{iN+N} = \tilde{p}_{iN+N} - \frac{N\delta t}{2} V_q^{soft}(q_{iN+N})$$

The values (p_{iN}, q_{iN}) represent the state of the system on the macro steps $t = i\Delta t = iN\delta t$ and for $i \leq k \leq N - 1$ the computed values (p_{iN+k}, q_{iN+k}) represent the state of the system on the intermediate micro step. Further application of this procedure recursively lead to generalized multiple time stepping Verlet-Strömer (MTS-SV) methods, where the potential energy is decomposed into μ classes, which can be applied in a hierarchal form [22], [23], which will be summarized in the following. Each of the interactions between bodies is assigned to a particular class based on its strength, and the interactions in each class are evaluated at some multiple of the primary step size δt. In this way less significant interactions are computed less often in an effort to reduce the amount of work done by the numerical integrator by a factor proportional to the number of classes. The overall computational cost is reduced from $\mathcal{O}(N^2)$ to $\mathcal{O}(N^2/\mu)$ for N particles and μ classes. The potential energy function can be decomposed into μ classes: $V(q) = \sum_{i=0}^{l} V^{[i]}(q)$ where the classes are numbered by $0, 1, \ldots, l = \mu - 1$ and the force functions are computed by the formula

$$f(q) = \sum_{i=0}^{l} \omega_i(k) f^{[i]}(q), \qquad f^{[i]}(q) = -\nabla V^{[i]}(q)$$

with $\omega_i(k)$ as the scalar valued weight function whose value depend on the step number k. The weight functions $\omega_i(k)$ are defined by:

$$\omega_i(k) = \begin{cases} \pi_i & \text{if } k \equiv 0 \ (mod \ \pi_i) \\ 0 & \text{otherwise} \end{cases} \quad \text{for } i = 0, 1, \ldots, l$$

with $\pi_i = \Pi_{j=0}^{i} n_j$, for $n_0 = 1, n_j \geq 2$ be integers for $j = 0, 1, \ldots, l$. For an efficient implementation one takes usually $\pi_i = 2^i$. The weight functions are given by a divisibility condition on the step number, so that the interactions assigned to class i are evaluated every π_i steps. The stronger interactions are placed in lower classes, whereas weaker interactions are placed in higher classes. There is no correspondence between the number of classes and the number of interactions. The force component for the class i on the step number k is not evaluated if $\omega_i(k) = 0$. The MTS-SV methods with every interaction assigned to class j is equivalent to the Verlet method with step size $\pi_j \delta t$, which is known as Strömer-Verlet equivalence. If an interaction changes during the integration, it will be placed in the consecutive class, an interaction may not skip over classes. Potential energy functions for inverse square law interactions (non-bonded forces)

$$V(q) = \sum_{i<j} -\frac{c_{ij}}{r_{ij}}, \qquad r_{ij} = \|q_i - q_j\|, \qquad c_{ij} \text{ constants}$$

are decomposed into the sum of potentials which correspond to distance classes defined by a sequence of positive cutoff radii $0 < r_1 < r_2 < \cdots < r_l < \infty$ The i'th force term

$$f_i(q) = \frac{\partial V}{\partial q_i} = \sum_{j \neq i} \left(\frac{c_{ij}}{\|q_j - q_i\|^2} \right) \frac{q_j - q_i}{\|q_j - q_i\|^2}$$

is assigned to the class k if $r_{k+1} < r_{ij} < r_k$. For the gravitational forces and for the electrostatic forces the parameters become $c_{ij} = g m_i m_j$ and $c_{ij} = \frac{1}{4\epsilon_0} Q_i Q_j$ respectively. Here g, ϵ_0, m_i and Q_i denote the gravitational constant, the permitivity constant, the atomic masses, and the electrostatic charges respectively. The ij'th potential energy function of the form $V(r) = -\frac{c}{r}$ with $c = c_{ij}$, $r = r_{ij}$ can be decomposed as

$$g_k(r) = \begin{cases} S_k(r), & \text{if } 0 < r < r_k \\ V(r), & \text{if } 0 \leq r_k < r \end{cases} \quad \text{for } k = 0, 1, \ldots, l$$

where $S_k(r)$ is a Taylor series expansion for $V(r)$ around $r = r_k$ [22], [48]. The potential function is then assigned to the classes in the following way:

$$
\begin{array}{llll}
\text{class } 0: & V^{[0]} &= V(r) - g_1(r) & \text{if } \ 0 < r_{ij} < r_1, \\
\text{class } k: & V^{[k]} &= g_k(r) - g_{k+1}(r), & \text{if } \ r_k < r_{ij} < r_{k+1}, \ k = 1, \ldots, l-1, \\
\text{class } l: & V^{[l]} &= g_l(r), & \text{if } \ r_l < r_{ij}.
\end{array}
$$

The two term Taylor expansion for $V(r)$ about $r = r_k$, which ensures the C^1 continuity of the potential class functions at the joint points is

$$S_k(r) = -\frac{c}{r_k} + \frac{c}{r_k^2}(r - r_k).$$

This decomposition can be improved by taking the first two terms of the Taylor series expansion of $V(\sqrt{s})$ at $s = r_k^2$ [23], [22], [48]

$$T_k(r) = -\frac{c}{r_k} + \frac{c}{2r_k^3}(r^2 - r_k^2).$$

The forces between two particles that produce oscillatory motion are described by Hooke's law and are analogous to an idealized spring with the spring constant κ_i and equilibrium length λ_i. A decomposition of the corresponding potential energy function for a chain of n springs is given in [22]:

$$V(q) = \sum_{i=0}^{n-1} \frac{\kappa_i}{2}(q_{i+1} - q_i - \lambda_i)^2 = \sum_{j=0}^{l} V^{[j]}(q).$$

The force functions $f_i(q) = \nabla_{q_i} V(q), i = 1, \ldots, n$ with the assumption $q_i < q_{i+1}$ are assigned to classes in the following form:

$$
\begin{aligned}
f_1(q) &= \omega_{c_1}(k)\kappa_1(q_2 - q_1 - \lambda_1), \\
f_i(q) &= \omega_{c_i}(k)\kappa_i(q_{i+1} - q_i - \lambda_i), \\
&\quad - \omega_{c_{i-1}}(k)\kappa_{i-1}(q_i - q_{i-1} - \lambda_{i-1}), \quad i = 2, \ldots, n-1, \\
f_n(q) &= -\omega_{c_{n-1}}(k)\kappa_{n-1}(q_n - q_{n-1} - \lambda_{n-1}).
\end{aligned}
$$

The parameters $c_i, i = 1, \ldots, n-1$ denote the class number for the spring i and $\omega_{c_i}(k)$ the weight functions. Every spring is assigned exactly to one class, the classes 0 and l are nonempty. The stiffest spring is assigned to the lowest class . The parameters $c_i's$ are selected according to the asymptotic error considerations so that the errors introduced for the i the spring are less than for the stiffest spring, i.e. $(\pi_{c_i}\Delta t)^2\kappa_i \leq \Delta t^2\kappa_{i_0}$.

First versions of the MTS methods for MD simulations were developed in [15] and in [52], the so-called r-RESPA (reversible reference system propagator algorithm), based on operator splitting. In some cases, a clear disparity between strong and weak forces, or heavy and light particles makes the choice for slowly and rapidly varying components obvious.

There are many variants of the r-RESPA method, which result from different kinds of splittings. We will present here according to [51] two important variants of the r-RESPA splitting. For this purpose we will introduce the Liouvillian representation of the Hamiltonian system (2). Let the state $\Gamma(t)$ specifies in MD individual particles' positions and momenta

$$\Gamma(t) = (q_1(t), \ldots, q_d(t), p_1(t), \ldots, p_d(t)).$$

Then the propagation scheme or the numerical integrator for the Hamiltonian system (2) is given by $\Gamma(\Delta t) = G(\Delta t)\Gamma(0)$ where $G(t)$ represents an approximation of the classical or the true operator $e^{i\mathcal{L}t}$ with the Liouvillian for a system of d degrees of freedom with pairwise interaction terms $F_{ij} = -F_{ji}$ between the particles i and j

$$i\mathcal{L} = \{ , H\} = \sum_{i=1}^{d} \left[v_i \frac{\partial}{\partial q_i} + \frac{1}{2} \sum_{j \neq i} \frac{F_{ij}}{m_i} \frac{\partial}{\partial v_i} \right]$$

where $v_i = p_i/m_i$ denote the particle velocities. The forced based r-RESPA is the most widely used variant of the r-RESPA algorithms. If the forces act on different time scales, then the slow and fast components of the Liouvillian are:

$$i\mathcal{L}_s = \frac{1}{2} \sum_{i} \sum_{j \neq i} \frac{F_{ij}^{(s)}}{m_i} \frac{\partial}{\partial v_i} \equiv \frac{F^{(s)}}{m} \frac{\partial}{\partial v},$$

$$i\mathcal{L}_f = \sum_{i} \left[v_i \frac{\partial}{\partial q_i} + \frac{1}{2} \sum_{j \neq i} \frac{F_{ij}^{(f)}}{m_i} \frac{\partial}{\partial v_i} \right] \equiv v \frac{\partial}{\partial q} + \frac{F^{(f)}}{m} \frac{\partial}{\partial v}.$$

Application of the MTS method result in

$$v \longleftarrow v + F^{(s)} \frac{\Delta t}{2m}$$

$$\left. \begin{array}{l} v \longleftarrow v + F^{(f)} \frac{\delta t}{2m} \\ q \longleftarrow q + v\delta t \\ v \longleftarrow v + F^{(f)} \frac{\delta t}{2m} \end{array} \right\} \quad i = 1, \ldots, n$$

$$v \longleftarrow v + F^{(s)} \frac{\Delta t}{2m}$$

The inner propagator is solved by a velocity Verlet method to decouple the position and velocity operators. The velocities are updated on two different time scales, and the positions will be updated using only the smallest time step. This class of r-RESPA propagators was applied to a wide range of systems; simulation of water, organic molecules, proteins etc. (see the references in [51]).

In MD simulations some particles are significantly lighter than the others. Since the particle velocities will scale with the square root of their mass, one can use the particle based r-RESPA splitting. Heavy and light components of the Liouvillian in this case are:

$$i\mathcal{L}_h = \sum_{i \in \text{heavy}} \left[v_i \frac{\partial}{\partial q_i} + \frac{1}{2} \sum_{j \neq i} \frac{F_{ij}}{m_i} \frac{\partial}{\partial v_i} \right] \equiv v_h \frac{\partial}{\partial q_h} + \frac{F_h}{m} \frac{\partial}{\partial v_h},$$

$$i\mathcal{L}_l = \sum_{i \in \text{light}} \left[v_i \frac{\partial}{\partial q_i} + \frac{1}{2} \sum_{j \neq i} \frac{F_{ij}}{m_i} \frac{\partial}{\partial v_i} \right] \equiv v_l \frac{\partial}{\partial q_l} + \frac{F_h}{m} \frac{\partial}{\partial v_l}.$$

The MTS propagation algorithm is given then by

$$v_l \longleftarrow v_l + F_l \frac{\delta t}{4m}$$
$$q_l \longleftarrow q_l + v_l \frac{\delta t}{2}$$
$$v_l \longleftarrow v_l + F_l \frac{\delta t}{4m}$$

$$\left. \begin{array}{l} v_l \longleftarrow v_l + F_l \frac{\delta t}{2m} \\ q_l \longleftarrow q_l + v_l \delta t \\ v_l \longleftarrow v_l + F_l \frac{\delta t}{2m} \end{array} \right\} \quad i = 1, \ldots, n$$

$$v_h \longleftarrow v_h + F_h \frac{\Delta t}{4m}$$
$$q_h \longleftarrow q_h + v_h \frac{\Delta t}{2}$$
$$v_h \longleftarrow v_h + F_h \frac{\Delta t}{4m}$$

Here the positions and velocities are updated at both time scales. Propagators of this type were applied to Lennard-Jones potentials, polarizable ions etc. (see the references in [51]). The important conclusion from these different MD simulations with the r-RESPA method is that the correct choice of the splitting depends on the nature of the physics of the problem at hand. An improper choice can result in a less than optimally efficient propagator, or even one which is worse than the original single step method. Another problem of the symplectic integrators, like implicit midpoint method and Verlet methods, is the occurrence of resonance type instabilities in MD simulation for large step sizes [33], [45]. Symplectic integrators can introduce artificial coupling among the motions associated with various frequencies leading to instability. One has to distinguish between systematic errors that increase with the time steps and more erratic ones, such as caused by resonance. The resonance can be explained using backward stability arguments and looking at the perturbed, nonautonomous effective Hamiltonian obtained by the numerical integrator

$$H_{eff} = \tilde{H}(q, p; \Delta t) = H(q) + \mathcal{O}(\Delta t^2).$$

This Hamiltonian is time-periodic with the period Δt for planar systems. The trajectories generated by a symplectic algorithm correspond to some Hamiltonian that depends periodically on these steps. In such a system resonance can occur when a special relationship holds between the forcing frequency $\frac{2\pi}{\Delta t}$ and the natural frequency ω of the unforced motion. For example the resonance condition for the Morse potential is given in [33] by $n\omega = m\frac{2\pi}{\Delta t}$, where m and n are relatively prime numbers and n denotes the order of resonance. The effective frequency ω_{eff} becomes time-dependent. The exact analytical formula for ω_{eff} for the Morse potential is not known, but for relatively small energy values it is reasonable to approximate it by expressions known for the harmonic oscillator. As energy increases the resonance order increases. Resonances cause Arnold diffusion, which occurs in the form of sampling of regions for the numerical phase diagrams that the unperturbed system doesn't exhibit [33].

The highest frequencies in the highly oscillatory system (17) are due to the internal bonded interactions such as the bond stretching and bond-angle bending potentials. The step size restriction for the explicit Verlet method is $\Delta t < \epsilon$, where $\epsilon := 1/\|K\|$ stands for the period of the fastest oscillations. To overcome this step size restriction a multiple time stepping method can be applied to (17) where the fast subsystem $(M^{-1}p, -G(q)^T Kg(q))^T$ is integrated N times with the micro step size δt and the slow part $(0, -\nabla_q V(q))^T$ with the macro step size Δt [3], [15], [42], [52]. A semi-implicit method, where the highly oscillatory part is integrated by an implicit method such as implicit midpoint method or an energy conserving method was also derived in [1] to eliminate the fast oscillations. The multiple time stepping methods for the highly oscillatory system (17) become unstable at certain values of Δt due to resonance phenomena [3], [4]. The frequency of the fast bond stretching/bending modes depend on the frequencies of the slow modes. If both frequencies coincide, oscillations in the position coordinates occur, whose amplitude increase with time (resonances). A combination of the averaging and MTS, the so-called mollified impulse method, was derived in [13] to overcome the step size restrictions of the standard MTS Verlet method due to the resonance induced instabilities. In the following we will describe an extension of the MTS based on averaging the position coordinates [13]. For this purpose we consider the Newton's equation of motion (1) in the following form:

$$M\frac{d^2q}{dt^2} = F^{hard}(q) + F^{soft}(q). \tag{19}$$

The highly oscillatory part corresponding to the hard forces can be solved either exactly or approximately with very high accuracy. In the case of planetary motion with one big star and N planets, the reduced problem corresponding to the hard forces is N uncoupled Kepler problems, which can be solved exactly. In MD simulation the reduced problem can not be solved exactly, but it can be solved very accurately using a small step size δt and cheaply compared to the cost of evaluation of soft forces. The standard impulse method for the equation (19) is given by

$$\text{half a kick:} \quad \tilde{p}_n = p_n + \frac{\Delta t}{2} F^{soft}(q_n),$$

$$\text{an oscillation:} \quad \text{to advance from } (\tilde{p}_n, q_n) \rightarrow (\tilde{p}_{n+1}, q_{n+1}) \text{ integrate}$$
$$\dot{q} = M^{-1}p, \quad \dot{p} = F^{hard}(q),$$

$$\text{half a kick:} \quad p_{n+1} = \tilde{p}_n + \frac{\Delta t}{2} F^{soft}(q_{n+1}).$$

The slow part is evaluated at time averaged values of positions, and the gradient of this modified potential is used for the slow part of the force:

$$F^{slow}(\Delta t; q_n) \longrightarrow A_q(\Delta t; q_n)^T F^{slow}(\Delta t; q_n)(A(\Delta t; q_n)).$$

The averaging of F^{slow} at $A(\Delta t; q_n))$ gives more accurate description of the rapidly varying trajectories. The transpose of the Jacobian $A_q(\Delta t; q_n)^T$ has a mollifying effect; the destabilizing components of the slow forces are removed. The impulse method coincides with the Verlet method in molecular dynamics if the forces are conservative and the mollified impulse method will be symplectic. The mollified impulse method was improved in [28] by projecting the position coordinates onto the manifold in configuration space defined by the equilibrium positions for the fastest forces. and in [40],[42] by using the information about the analytical solution behavior of the fast system which is called projected MTS method. The standard MTS Verlet method for (17) becomes unstable at $\Delta t \approx 4fs$. It was shown in [28] for a box of water, that the projected and averaged MTS methods allow to increase the macro step sizes up to $\Delta t \approx 8fs$.

5. APPLICATIONS TO CLASSICAL AND QUANTUM-CLASSICAL MOLECULAR DYNAMICS

In numerical simulation of molecular dynamics one has to integrate a very large number of differential equations systems, consisting of $10^4 - 10^5$ atoms with six differential equations per atom. These systems are highly nonlinear and exhibit sensitive dependence on perturbations. The MD simulation problems involve multiple time scales; conformal changes in macromolecules occur on a time scale ranging from $10^{-12} - 10^2$ seconds [45]. Very small step size must be used for an accurate resolution of some atomic trajectories. Therefore the time step limitation is one of the major challenges of MD algorithms [9], which makes the development of efficient algorithms necessary in addition to the advances in computer hardware.

If only electrostatic forces between all pairs of atoms are considered, a direct force evaluation requires $\mathcal{O}(N^2)$ interactions, which becomes prohibitive for large N. As indicated in Section 4 the generalized MTS-Verlet algorithms reduce the computational cost to $\mathcal{O}(N^2/\mu)$ for potentials decomposed in μ-classes [22]. Multipole methods approximate the long-range forces originating from a group of point charges by truncated multipole expansion of their electrostatic potential. Multipole expansion of the potential $V(r)$ is a kind of Taylor expansion, which is accurate when r is large. The computational cost of the multipole expansions are of order $\mathcal{O}(N \log N)$ and the so-called FMA (fast multipole algorithm) reduces the computational complexity to an order of $\mathcal{O}(N)$ for large systems. The FMA is a recursive and hierarchical algorithm like the generalized MTS methods in Section 4. Multipole methods make use of regularities of Coulombic interactions in space whereas the MTS methods exploit regularities in time. A combination of both concepts, which provides additional speedups, were realized by the so called FAMUSAMM (Fast multiple-time step structure adapted multipole method) [10] and for periodic boundary conditions in [11]. The MTS methods and the fast multipole algorithms (FMA) are applied to a large-scale simulation of biomolecules, whereby the long range electrostatic interactions beyond a cutoff distance of typically $8 - 10A^0$ are neglected, which don't lead to significant drift in energy [4].

In the so-called quantum-classical Molecular Dynamics (QCMD) model, most atoms are described by classical mechanics (Newtonian equations), but an important small portion of the system by quantum mechanics (time dependent Schrödinger equations) [34], [35] [46]. The QCMD model for one quantum degree of freedom with spatial coordinate x and mass m in a bath of d classical particles with the interaction potential $V(x, q)$ is given by

$$
\begin{aligned}
\dot{\psi} &= -\frac{i}{\hbar} H(q)\psi, \\
\dot{q} &= M^{-1}p, \\
\dot{p} &= -\psi^* \nabla_q H(q)\psi - \nabla_q V_{cl}(q).
\end{aligned}
\tag{20}
$$

The purely classical potential energy function is $V_{cl}(q)$, $T = -\hbar^2 \Delta_x/2m$ denotes the kinetic energy. $H(q) = T + V(x, q)$ and ψ^* the quantum Hamiltonian operator with the complex-valued wave function $\psi \in \mathbb{C}^s$ with its' conjugate complex ψ^*. We assume that the quantum system has been truncated to a finite dimensional system by an appropriate spatial discretization. Then the total Hamiltonian

$$
\mathcal{H} = \frac{1}{2}p^T M^{-1}p + \psi^* H(q)\psi + V_{cl}(q)
$$

and the unitary propagation of the quantum part $\psi^*\psi$, are the conserved quantities. In the limiting case $\hbar \to 0$ or $M \to \infty$ the QCMD model reduces to the Born-Oppenheimer model [5], [47]; the eigenvalues of the quantum Hamiltonian $H(q)$ become adiabatic invariants. Conservation of the total energy and adiabatic invariants associated with the Born-Oppenheimer limit of the QCMD model provide a simple test for the behavior of the numerical integrator. There are different ways to consider the QCMD model as a canonical Hamiltonian system with the total Hamiltonian $\mathcal{H}$. Following [35] and [46], the complex valued wave function ψ can be decomposed into its real and imaginary parts:

$$\psi = (q_\psi + ip_\psi)/\sqrt{2}.$$

After introducing generalized positions $Q = (q_\psi^T, q^T) \in \mathbb{R}^{s+3d}$ and generalized momenta $P = (p_\psi^T, p^T) \in \mathbb{R}^{s+3d}$, the QCMD model can be written as a classical Hamiltonian system

$$\frac{dQ}{dt} = \nabla_P \mathcal{H}(Q, P), \qquad \frac{dP}{dt} = -\nabla_Q \mathcal{H}(Q, P).$$

By using the Liouville formalism $L_\mathcal{H} z_i = \{z_i, \mathcal{H}\}$, with $z = (Q_N, P_N)^T$ the formal solution $\dot{z} = L_\mathcal{H} z$ can be expressed as $z(\Delta t) = e^{\Delta t L_\mathcal{H}} z(0)$. For the separable Hamiltonian $\mathcal{H} = \mathcal{H}_1 + \mathcal{H}_2 + \cdots$ the corresponding Liouville operator can be splitted $L_\mathcal{H} = L_{\mathcal{H}_1} + L_{\mathcal{H}_2} + \cdots$ so that second order, time-reversible Strang splitting which corresponds to the Verlet method can be applied. There are two different kind of splittings. The first of them based on decomposition of the Hamiltonian in kinetic and potential energy parts $\mathcal{H} = \mathcal{H}_1 + \mathcal{H}_2$ in the following way:

$$\mathcal{H}_1 = \frac{1}{2} p^T M^{-1} p + \psi^* T \psi \quad \text{and} \quad \mathcal{H}_2 = \psi^* V(q) \psi + V_{cl}(q)$$

leads to the symplectic, time-reversible integrator PICKABACK [35], [34]:

$$q_{n+1/2} = q_n + \frac{\Delta t}{2} M^{-1} p_n,$$

$$\tilde{p}_n = p_n - \Delta t \nabla_q V_{cl}(q_{n+1/2}),$$

$$\psi_{n+1/2} = \exp(-\frac{i}{\hbar} \frac{\Delta t}{2} T) \psi_n,$$

$$p_{n+1} = \tilde{p}_n - \Delta t \psi_{n+1/2}^* \nabla_q V(q_{n+1/2}) \psi_{n+1/2},$$

$$\psi_{n+1} = \exp(-\frac{i}{\hbar} \frac{\Delta t}{2} T) \exp\left(-i \Delta t V(q_{n+1/2})\right) \psi_{n+1/2},$$

$$q_{n+1} = q_{n+1/2} + \frac{\Delta t}{2} M^{-1} p_{n+1}.$$

PICKABACK conserves the total Hamiltonian up to small fluctuations and the norm of the vector ψ exactly. The main drawback of this method is the step size restriction which is of order of the inverse of the largest eigenvalue of the quantum operator $H(q)$. If the eigenvalues of the quantum operator $H(q)$ and the gradients $\nabla V(q)$, $\nabla_q V_{cl}(q)$ are expensive due to long-range interactions, then the PICKABACK scheme becomes inefficient, i.e. the permitted step size might be much smaller than required by the classical part. The second approach is based on the splitting of classical and quantum parts of the Liouvillian $L_{\mathcal{H}} = L_{\mathcal{H}}^{cl} + L_{\mathcal{H}}^{qm}$, with

$$L_{\mathcal{H}}^{cl} = (\nabla_q \mathcal{H})^T \nabla p - (\nabla_p \mathcal{H})^T \nabla q \text{ and } L_{\mathcal{H}}^{qm} = (\nabla_{q_\psi} \mathcal{H})^T \nabla p_\psi - (\nabla_{p_\psi} \mathcal{H})^T \nabla q_\psi$$

leading to a symmetric integrator [35], [46]:

$$\psi_{n+1/2} = \exp\left(-i\frac{\Delta t}{2\hbar}H(q_n)\right)\psi_n,$$

$$q_{n+1/2} = q_n + \frac{\Delta t}{2}M^{-1}p_n,$$

$$p_{n+1} = p_n - \Delta t\psi_{n+1/2}^* \nabla_q V(q_{n+1/2})\psi_{n+1/2},$$

$$q_{n+1} = q_{n+1/2} + \frac{\Delta t}{2}M^{-1}p_{n+1},$$

$$\psi_{n+1} = \exp\left(-i\frac{\Delta t}{2\hbar}H(q_{n+1})\right)\psi_{n+1/2}.$$

Direct application of the MTS Verlet method to the QCMD equations (20) requires very small step sizes. Instead of this, a multiple time stepping scheme based on the Hamiltonian splitting for the PICKABACK algorithm by taking few macro steps with step size Δt in the slow classical part and many smaller steps with the micro step size δt in the highly oscillatory quantum subsystem [34], [42] leads to the following symplectic and time-reversible method:

$$\begin{aligned} q_{n+1/2} &= q_n + \frac{\Delta t}{2}M^{-1}p_n, \\ \tilde{p}_n &= p_n - \Delta t\nabla_q V_{cl}(q_{n+1/2}), \end{aligned}$$

$$\left.\begin{aligned} \tilde{\psi}_{n+k/j} &= \exp\left(-i\frac{\delta t}{\hbar}\frac{1}{2}T\right)\psi_{n+(k-1)/j}, \\ \tilde{p}_{n+k/j} &= \tilde{p}_{n+(k-1)/j} - \delta t\tilde{\psi}_{n+k/j}^*, \nabla V(q_{n+1/2})\tilde{\psi}_{n+k/j}, \\ \psi_{n+k/j} &= \exp\left(-i\frac{\delta t}{\hbar}\frac{1}{2}T\right)\exp\left(-\frac{i}{\hbar}\delta t V(q_{n+1/2})\right)\tilde{\psi}_{n+(k-1)/j}, \end{aligned}\right\} \quad k = 1,\ldots,j$$

$$\begin{aligned} p_{n+1} &= \tilde{p}_{n+1}, \\ q_{n+1} &= q_{n+1/2} + \frac{\Delta t}{2}M^{-1}p_{n+1}. \end{aligned}$$

A similar MTS algoritm developed in [46] which is only time-reversible. One of the main problems is to find efficient, symmetric approximations for the matrix exponentials occurring in the algorithms for QCMD models [35]. Various alternative integration schemes employing matrix functions, such as a type of mollified exponential integrators for the highly oscillatory QCMD equations were derived in [24].

The limited value of information obtained by trajectory computations in MD simulations was discussed in [9]. It was shown there using the forward and backward error analysis arguments, that symplectic methods like the Verlet method can only be used in short time trajectory computations. Instead of this it was suggested to compute time averages of physical observables or relaxation times of conformational changes using certain invariant sets of the phase space [9].

6. APPLICATIONS TO N–BODY PROBLEMS

Accurate long-term numerical integration of N–body problems play an important role in understanding the dynamics of evolution of many astronomical systems. Many problems in astrophysics have similarities with MD simulation, with respect to the large number of bodies involved and multiple time scales. The usual integration interval in astrophysics is 10^{10} years and the period of a planetary motion is typically less than one year, which corresponds to $10^{13} - 10^{14}$ numerical integration steps [12]. The longer integrations have generally revealed interesting phenomena like weak chaos in the orbits of inner and outer planets of the solar system [53]. In astronomy, for problems with larger variations in time scale, due to close encounters or high eccentricities, it is desirable to use variable time step integrators. Preserving time-symmetry plays an important role in astrophysics, because of its inherent energy conservation properties. The time-symmetry was maintained in [27] by a choice of symmetric time stepping function $\tau(t)$

$$\Delta t(t) = \frac{1}{2}(\tau(y_n) + \tau(y_{n+1}))$$

where τ depends usually on the minimum distances of particle pairs in a N-body problem. The variable step size is determined iteratively, which increases the computational cost slightly. It was shown in [27], that for N-body problems with $N = 100$, the time-symmetrized version of the Verlet algorithm with one iteration for the determination of the time-stepping function, preserves the energy, where the unsymmetrized variable step size integrators show an energy drift. The periodic orbits in Kepler's problem are also retained by symmetrized variable step size integrators.

Lee et. al. consider in [29] a time-symmetrized variable step size integrator using (18) with a multiple time scale symplectic integrator. The force function $F = \sum_{i=0}^{\infty} F_i$, $F_i = -\frac{\partial V_i}{\partial r}$, $V = -\frac{1}{r}$ is decomposed into soft and hard parts for the Kepler problem. A similar approach was also used in [36] with the time-symmetrized variable step size determination. The individual time steps for each particle are determined by a recursive algorithm, which preserves the overall reversibility of the N-body system.

Integration of an orbit with the time steps determined for another orbit seem to be somewhat artificial. This leads to the so called mixed variable step (MVS) integrators, which allows individual time steps in astrophysics applications (see [14], [29], [36], [43]). For example in the solar system the Hamiltonian can be splitted into $H = H_{Kep} + H_{int}$, where H_{Kep} describes the Keplerian motion of the Hamiltonian of bodies around the sun; H_{int} stands for interaction energy between the planets. In [43] an MVS integrator is used which switches continuously between the interplanetary motion and Kepler equations, using different time steps for each planet.

Another area of application of variable step size integrators are collision and scattering problems in atomic dynamics. To analyze the short-term behavior of collision problems in [30] the Kepler problem in polar coordinates was considered:

$$H = \frac{1}{2}p^2 - \frac{1}{q} + \frac{l^2}{2q^2}$$

where l denotes the angular momentum. Energy will fluctuate especially in the vicinity of closest approach to the fixed body and does not return to pre-collision level for the standard RK-methods, but for symplectic methods the energy returns to very near of the pre-collision level. They have always the property that the energy tends to within $\mathcal{O}(\exp(-\frac{1}{\Delta t}))$ of the pre-collision energy following a scattering event, which shows the superiority of symplectic integrators over nonsymplectic ones. But the estimate $\mathcal{O}(\exp(-\frac{1}{\Delta t}))$ is only valid for small Δt; the energy variation for large step sizes is complex and unpredictable. The Kepler problem in polar coordinates was also integrated with the implicit mid-point method using the time reparametrization function $R(p,q) = \frac{1+q^2}{q^2}$. Although the transformed equations are not Hamiltonian, the implicit mid-point method produces the same energy profile as the symplectic methods because of its time-reversible character.

Fixed step size methods lead to difficulties in neighborhoods of singularities. In such regions adaptive step size methods, based on normalization and scaling of the vector fields, must be used. This leads to a combination of symplectic or time-reversible integration with some regularization techniques for Hamiltonian systems with very steep potentials, like the Lennard-Jones potentials. Regularization techniques like Kustaanheimo-Stiefel (KS) regularization are widely used in connection with variable time-symmetrized integrators to integrate orbits having very high eccentricities [12]. In KS regularization, a Kepler orbit is transformed into a harmonic oscillator which reduces the number of integration steps significantly. The KS transformation is a coordinate and time transformation and preserves the symplectic structure. Its' effect is stabilization of the numerical solutions. The Kepler Hamiltonian in the KS-transformed variables $(Q, P) = KS(q, p)$ is

$$H = \frac{1}{8} \frac{|P|^2}{|Q|^2} - \frac{1}{|Q|^2}.$$

After a Poincaré type transformation one obtains a linear Hamiltonian system coupled with a nonlinear equation for the time transformation:

$$\frac{dQ}{ds} = \frac{1}{4} P, \qquad \frac{dP}{ds} = 2H_0 Q, \qquad \frac{dt}{ds} = |Q|^2.$$

Application of the implicit mid-point method to this system results in the so called fast Kepler solver [30]

$$Q_{n+1} = Q_n + \frac{\Delta s}{4} \frac{P_n + P_{n+1}}{2},$$

$$P_{n+1} = P_n + 2\Delta s H_0 \frac{Q_n + Q_{n+1}}{2},$$

$$\Delta t = \Delta s |Q_{n+1/2}|$$

where the time variable has to be solved by the Newton iteration. The fast Kepler solver reduces the computational cost in two-body problems by a factor of 1.5 [30]. The Hamiltonian for atomic problems

$$H(q, p) = \sum_{i=1}^{d} \frac{|p_i|^2}{2} - \sum_{i=1}^{d} \frac{k}{|q_i|} + \sum_{i=1}^{d-1} \sum_{j=i+1}^{d} \frac{1}{|q_i - q_j|}$$

can be splitted in two parts

$$H^{(i)} = \frac{|p_i|^2}{2} - \frac{k}{|q_i|}, \quad i = 1, \ldots, d, \qquad V_{e,e} = \sum_{i=1}^{d-1} \sum_{j=i+1}^{d} \frac{1}{|q_i - q_j|}$$

where the Kepler Hamiltonians $H^{(i)}$ are integrated by a fast Kepler solver. The electron-electron interaction term V_{e-e} can be integrated exactly, because it depends only on q. Due to the difficulty of Kepler problem and strong Coulombic forces for the electron-electron interaction terms, a reversible variable time step method has to be incorporated. For this purpose the perturbed Kepler problem with the Hamiltonian

$$H = \frac{1}{2} |p|^2 - \frac{1}{|q|} + \tilde{H}(q, p)$$

218

was considers in [30], where the small perturbation term for atomic systems is $\tilde{H}(q,p) = -\frac{1}{|q-q_i|}$. This system is solved by the Strang splitting of the form

$$e^{\Delta t H} = e^{\frac{\Delta t}{2} \bar{H}} \circ e^{\Delta t \tilde{H}} \circ e^{\frac{\Delta t}{2} \bar{H}}$$

where $\bar{H}$ denoting the unperturbed Kepler Hamiltonian. The first part is integrated by a fast Kepler solver incorporated with the explicit adaptive Verlet method with the time reparametrization function $R(q,p) = 1 + |q|^{-\alpha}$ with some $\alpha > 0$. This approach is called the RAR (reversible adaptive regularization) method [30]. The perturbed Hamiltonian part is solved either exactly or integrated by a symplectic integrator. Numerical experiments on several atomic systems in [30] demonstrate the performance of this approach.

References

1. U. R. Ascher and S. Reich. On some difficulties in integrating highly oscillatory Hamiltonian systems. In P. Deuflhard et al., editor, *Algorithms for Macromolecular Modelling: Challenges, Methods, Ideas*, pages 281–296. Lecture Notes in Computational Science and Engineering, Springer, 1998.
2. E. Barth, B. Leimkuhler, and S. Reich. A time-reversible variable stepsize integrator for constrained dynamics. Technical Report, SC97-53, Konrad Zuse Zentrum, Berlin, 1997. to appear in SIAM J. Sci. Comput.
3. J.J. Biesiadecki and R.D. Skeel. Dangers of multiple-time-step methods. *J. of Comp. Phys.*, 109:318–328, 1993.
4. T. C. Bishop, R. D. Skeel, and K. Schulten. Difficulties with multiple time stepping and fast multipole algorithm in molecular dynamics. *Journal of Computational Chemistry*, 18:1785–1792, 1997.
5. F. A. Bornemann and C. Schütte. On the singular limit of the Quantum-Classical molecular dynamics model. to appear in SIAM J. Appl. Math., 1998.
6. M. P. Calvo, , M. A. López-Marcos, and J. M. Sanz-Serna. Variable step implementation of geometric integrators. *Appl. Numer. Math.*, 28:1–16, 1998.
7. M.P. Calvo and J.M. Sanz–Serna. The development of variable-step symplectic integrators with application to the two-body problem. *SIAM J. Sci. Comput.*, 14:936–952, 1993.
8. S. Cirilli, E. Hairer, and B. Leimkuhler. Asymptotic error analysis of the adaptive Verlet method. *BIT*, 39:25–33,1999.
9. P. Deuflhard, M. Dellnitz, O. Junge, and C. Schütte. Computation of essential molecular dynamics by subdivision techniques. In P. Deuflhard et al., editor, *Algorithms for Macromolecular Modelling: Challenges, Methods, Ideas*, pages 98–115. Lecture Notes in Computational Science and Engineering, Springer, 1998.
10. M. Eichinger, H. Grubmüller, H. Heller, and P. Tavan. FAMUSAMM: An algorithm for rapid evaluation of electrostatic interactions in molecular dynamics. *Journal of Computational Chemistry*, 18:1729–1749, 1997.
11. F. Figueirido, R.M. Levy, Zhou, and B. J. Berne. Large scale simulation of macromolecukes in solution: Combining the periodic fast multipole method with multiple time step integrators. *J. Chem. Phys.*, 106:9835–9849, 1997.
12. Y. Funato, P. Hut, S. McMillan, and J. Makino. Time-symmetrization of Kustaanheimo-Siefel regularization. *The Astrophysics Journal*, 112:1697, 1996.
13. B. García-Archilla, J. M. Sanz-Serna, and R. D. Skeel. Long-time-steps methods for oscillatory differential equations. *SIAM J. Sci. Comput.*, 20:930–963, 1998.

14. B. Gladman, M. Duncan, and J. Candy. Symplectic integrators for long-term integration in celestial mechanics. *Celest. Mech.*, 52:221–240, 1991.
15. H. Grubmüller, H. Heller, A. Windermuth, and K. Schulten. Generalized Verlet Algorithm for Efficient Molecular Dynamics Simulations with Long-Range Interactions. *Mol. Sim.*, 6:121–142, 1991.
16. E. Hairer. Backward analysis of numerical integrators and symplectic methods. In K. Burrage, C. Baker, P. v.d. Houwen, Z. Jackiewicz, and P. Sharp, editors, *Scientific Computation and Differential Equations*, volume 1 of *Annals of Numer. Math.*, pages 107–132, Amsterdam, 1994. J.C. Baltzer. Proceedings of the SCADE'93 conference, Auckland, New-Zealand, January 1993.
17. E. Hairer. Variable time step integration with symplectic methods. *Appl. Numer. Math.*, 25:219–227, 1997.
18. E. Hairer, S.P Nørsett, and G. Wanner. *Solving Ordinary Differential Equations I, Nonstiff Problems.* Springer, 1993.
19. E. Hairer and D. Stoffer. Reversible long-term integration with variable step sizes. *SIAM J. Sci. Comput.*, 18:257–269, 1997.
20. E. Hairer and G. Wanner. *Solving Ordinary Differential Equations II, Stiff Problems and Differential-Algebraic equations.* Springer, 1996. II. Edition.
21. M. Hankel, B. Karasözen, P. Rentrop, and U. Schmitt. A Molecular Dynamics Model for Symplectic Integrators. *Mathematical Modelling of Systems*, 3(4):282–296, 1997.
22. D. J. Hardy and D. I Okunbor. Symplectic multiple time step integration. 1997.
23. D. J. Hardy, D. I. Okunbor, and R. D. Skeel. Symplectic Variable Stepsize Integration for N-Body Problems. *Appl. Numer. Math.* 29:19–30, 1999.
24. M. Hochbruck and C. Lubich. A bunch of time integrators for quantum/classical molecular dynamics. In P. Deuflhard et al., editor, *Algorithms for Macromolecular Modelling: Challenges, Methods, Ideas*, pages 421–432. Lecture Notes in Computational Science and Engineering, Springer, 1998.
25. T. Holder, B. Leimkuhler, and S. Reich. Explicit variable step-size and time-reversible integrators. Technical Report, SC98-17, Konrad Zuse Zentrum, Berlin, 1998.
26. W. Huang and B. Leimkuhler. The adaptive Verlet method. *SIAM J. Sci. Comput.*, 18(1):239, 1997.
27. P. Hut, J. Makino, and S. McMillan. Building a better leapfrog. *The Astrophysical Journal*, 443:L93–L96, 1995.
28. J.A. Izaguirre, S. Reich, and R.D. Skeel. Longer time steps for molecular dynamics. 1998.
29. M. H. Lee, M. J. Duncan, and H. F. Levinson. Variable timestep integrators for long-term orbital integrations. Technical report, Department of Physics, Queen's University, Canada, 1997. to appear in Computational Astrophysics, Proc. 12th Kingston Meeting, ed. D. A. Clarke, M. J. West.
30. B. Leimkuhler. Reversible adaptive regularization I: perturbed Kepler motion and classical atomic trajectories. to appear in Philosophical Trans. Royal Soc. A., 1997.
31. B. Leimkuhler, S. Reich, and R.D. Skeel. Integration methods for molecular dynamic. In J.P. Mesirov, K. Schulten, and D.W. Sumners, editors, *Mathematical Approaches to Biomolecular Structure and Dynamics*, pages 161–186. IMA Volumes in Mathematics and its Applications Vol. 82, Springer Verlag, 1996.
32. T. R. Littell, R. D. Skeel, and M. Zhang. Error Analysis of Symplectic Multiple Time Stepping. *SIAM J. Numer. Anal.*, 34(5), 1997.
33. M. Mandziuk and T. Schlick. Resonance in the dynamics of chemical systems simulated by the implicit midpoint scheme. *Chemical Phyics Letters*, 237:525–535, 1995.
34. P. Nettesheim and S. Reich. Symplectic multiple-time-stepping integrators for quantum-classical molecular dynamics. In P. Deuflhard et al., editor, *Algorithms for Macromolecular Modelling: Challenges, Methods, Ideas*, pages 412–420. Lecture Notes in Computational Science and Engineering, Springer, 1998.

220

35. P. Nettesheim and C. Schütte. Numerical integrators for quantum-classical molecular dynamics. In P. Deuflhard et al., editor, *Algorithms for Macromolecular Modelling: Challenges, Methods, Ideas*, pages 396–411. Lecture Notes in Computational Science and Engineering, Springer, 1998.

36. T. Quinn, N. Katz, J. Stadel, and G. Lake. Time stepping n-body simulation. Technical report, Department of Physics and Astronomy, University of Massachusetts, 1997.

37. S. Reich. Smoothed Dynamics of Highly Oscillatory Hamiltonian Systems. *Physica D*, 89:28–42, 1995.

38. S. Reich. Backward error analysis for numerical integrators. Konrad-Zuse Zentrum für Informationstechnik Berlin, 1996, to appear in SIAM J. Numer. Anal., SC 96-21

39. S. Reich. Torsions Dynamics of Molecular Dynamics. *Physical Review E*, 53:4876–4881, 1996.

40. S. Reich. Dynamical systems, numerical integration, and exponentially small estimators. Habilitationsschrift, Freie Universität Berlin, 1998.

41. S. Reich. A modified force field for constrained molecular dynamics. to appear in Numerical Algorithms, 1998.

42. S. Reich. Multiple times-scales in classical and quantum-classical molecular dynamics. to appear in J. Comput. Phys., Department of Mathematics and Statistics, University of Surrey, 1998.

43. P. Saha and Tremaine S. Long-term planetary integration with individual time steps. *The Astronomy Journal*, 108:1962–1969, 1994.

44. J.M. Sanz-Serna and M.P. Calvo. *Numerical Hamiltonian Problems*. Chapman & Hall, 1994.

45. T. Schlick, E. Barth, and M. Mandziuk. Bringing the timescale gap between simulation and experimentation. In R. M. Stroud, editor, *Annual Review of Biophyics and Biomolecular Structure*, pages 179–220, 1997.

46. U. Schmitt and J. Brickmann. Discrete time-reversible propagation scheme for mixed quantum-classical dynamics. *Chemical Physics*, 208:45–56, 1996.

47. C. Schütte and F. A. Bornemann. Approximation properies and limits of the Quantum- Classical molecular dynamics model. In P. Deuflhard et al., editor, *Algorithms for Macromolecular Modelling: Challenges, Methods, Ideas*, pages 380–395. Lecture Notes in Computational Science and Engineering, Springer, 1998.

48. R.D. Skeel and J. J. Biesiadecki. Symplectic Integration with Variable Stepsize. *Annals of Numer. Math.*, 1:191–198, 1994.

49. D. Stoffer. Variable Steps for Reversible Integration Methods. *Computing*, 55:1–22, 1995.

50. A. Stuart and A.R. Humpries. *Dynamical Systems and Numerical Analysis*. Cambridge University Press, 1996.

51. S.J. Stuart, R. Zhou, and B. J. Berne. Molecular dynamics with multiple time scales: The selection of efficient reference system propagators. *J. Chem. Phys.*, 105:1426–1436, 1996.

52. M. Tuckerman, B. J. Berne, and G. J. Martyna. Reversible Multiple Time Scale Molecular Dynamics. *J. Chem. Phys.*, 97(3):1990–2001, 1992.

53. J. Wisdom and M. Holman. Symplectic maps for the N-body problem. *Astron. J.*, 102:1528–1538, 1991.

54. H. Yoshida. Construction of higher order symplectic integrators. *Physics Letters A*, 150:262–268, 1990.

DOMAIN DECOMPOSITION METHODS FOR COMPRESSIBLE FLOWS

ALFIO QUARTERONI

Département de Mathématiques,
Ecole Polytechnique Fédérale de Lausanne,
1015 Lausanne, Switzerland and
Department of Mathematics,
Politecnico of Milano, 20133 Milano, Italy

AND

ALBERTO VALLI

Department of Mathematics,
University of Trento, 38050 Povo (Trento), Italy

1. Introduction

In these notes we consider systems of hyperbolic equations and their reformulation in the framework of multi–domain partition of the computational domain.

Our goal is at one hand to estabilish correct matching conditions at subdomain interfaces. The role of characteristics, pseudo-characteristic and Riemann invariants is analyzed in this respect. Then we devise iterative algorithms that make use of those interface conditions to split the problem into smaller subproblems that need to be faced on each subdomain. Only the mathematical principles at the differential level are addressed. These results, however, can afterwards be adapted to the numerical approximation level as well.

We start considering first order hyperbolic systems in abstract form. Then we focus on those systems that are of utmost relevance in fluid dynamics of compressible flows: the Navier-Stokes equations, the Euler equations and the full potential equation. The three types of equations can be considered in their own; however they can also play in a combined fashion in external aerodynamics. As a matter of fact, the full potential equation

H. Bulgak and C. Zenger (eds.), Error Control and Adaptivity in Scientific Computing, 221–245.
© 1999 *Kluwer Academic Publishers. Printed in the Netherlands.*

can be used in the far field where the flow is irrotational and isentropic, the Euler equations in the remaining portion of the domain where shock discontinuities may develop, except within boundary layers and wakes where the Navier-Stokes equations are the most appropriate. In particular, we discuss the coupling mechanism between the (scalar) full potential equation and the Euler system of equations. Motivations for the use of combined models in fluid dynamics are illustrated e.g. in Quarteroni and Valli (1999), Coclici (1998), Quarteroni and Stolcis (1995), Berkman and Sankart (1997), and the references therein.

2. First order hyperbolic systems

To start with, let us consider a *first-order hyperbolic system* with constant coefficients. Precisely, we consider the following one–dimensional problem

$$(2.1) \qquad \frac{\partial \mathbf{U}}{\partial t} + A \frac{\partial \mathbf{U}}{\partial x} = \mathbf{F} \qquad \text{in } Q_T = \Omega \times (0, T) \, ,$$

where Ω is an interval, $\mathbf{U} = (U_1, \cdots, U_p)$ and A is a $p \times p$ matrix with constant coefficients. This system is said to be *hyperbolic* if A is diagonalisable with real eigenvalues, so that one can consider the decomposition

$$(2.2) \qquad A = L^{-1} \Lambda L \, ,$$

where $\Lambda := \operatorname{diag}(\lambda_1, \cdots, \lambda_p)$ is the diagonal matrix of eigenvalues, and L is the matrix whose rows are given by the left eigenvectors of A, i.e.

$$\mathbf{l}^r A = \lambda_r \mathbf{l}^r \, , \quad r = 1, \ldots, p \, .$$

Introducing the *characteristic variables* $\tilde{\mathbf{U}} := L\mathbf{U}$ in (2.1) gives

$$(2.3) \qquad \frac{\partial \tilde{\mathbf{U}}}{\partial t} + \Lambda \frac{\partial \tilde{\mathbf{U}}}{\partial x} = L\mathbf{F} \, .$$

This decouples into p independent scalar advection equations

$$(2.4) \qquad \frac{\partial \tilde{U}_r}{\partial t} + \lambda_r \frac{\partial \tilde{U}_r}{\partial x} = (L\mathbf{F})_r \, , \quad r = 1, \ldots, p \, .$$

The curves $X(t) = x_0 + \lambda_r t$ of the (x, t)–plane satisfying $X'(t) = \lambda_r$ are the r–characteristics. Any characteristic variable $\tilde{U}_r$ is constant along each corresponding r–characteristic, hence the value of $\tilde{U}_r$ at any point (x, t) can be recovered from the data of the problem tracing back the characteristic line passing through (x, t).

The multi–dimensional counterpart of (2.1) reads

$$(2.5) \qquad \frac{\partial \mathbf{U}}{\partial t} + \sum_{l=1}^{d} A^{(l)} D_l \mathbf{U} = \mathbf{F} \qquad \text{in } Q_T = \Omega \times (0, T) \ .$$

Now $\Omega \subset \mathbf{R}^d$, $A^{(l)}$, $l = 1, \ldots, d$, are $p \times p$ matrices (possibly with variable coefficients) such that for each direction $\boldsymbol{\xi} \in \mathbf{R}^d$ the matrix $\sum_l \xi_l A^{(l)}$ is diagonalisable with real eigenvalues (this is surely the case if each matrix $A^{(l)}$ is symmetric).

A multi–domain version of (2.5) can be obtained upon generalising what is done for scalar advection equations (see Quarteroni and Valli (1999), Section 5.6). Let Ω be partitioned into two disjoint subdomains Ω_1 and Ω_2 whose common boundary is denoted by Γ. For any point $\mathbf{x} \in \Gamma$ define the characteristic matrix $C = C(\mathbf{n}) = \sum_l n_l A^{(l)}$, where $\mathbf{n}$ is the normal unit vector on Γ directed from Ω_1 to Ω_2. By the hyperbolicity assumption, $C(\mathbf{n})$ can be diagonalised as $\Lambda = LCL^{-1}$, where $\Lambda = \text{diag}\,(\lambda_r)$ with $\lambda_r \in \mathbf{R}$, $r = 1, \ldots, p$, and L is the matrix whose rows are the left eigenvectors of C.

The restrictions $\mathbf{U}_i$ of $\mathbf{U}$ to Ω_i, $i = 1, 2$, satisfy

$$(2.6) \qquad \frac{\partial \mathbf{U}_i}{\partial t} + \sum_{l=1}^{d} A^{(l)} D_l \mathbf{U}_i = \mathbf{F} \qquad \text{in } \Omega_i \times (0, T) \ , \quad i = 1, 2 \ ,$$

$$(2.7) \qquad C(\mathbf{n})\, \mathbf{U}_1 = C(\mathbf{n})\, \mathbf{U}_2 \qquad \text{on } \Gamma \times (0, T) \ .$$

The system (2.5) can be rewritten in the conservative form

$$\frac{\partial \mathbf{U}}{\partial t} + \sum_{l=1}^{d} D_l(A^{(l)} \mathbf{U}) = \mathbf{F} - \sum_{l=1}^{d} (D_l A^{(l)}) \mathbf{U} \qquad \text{in } Q_T \ ,$$

i.e.

$$\frac{\partial \mathbf{U}}{\partial t} + \text{div}\, \mathbf{H}(\mathbf{U}) = \mathbf{G} \ ,$$

where $\mathbf{H}(\mathbf{U})$ is the flux matrix whose l-th column is the vector $A^{(l)} \mathbf{U}$. Then the interface condition (2.7) is indeed a matching condition for the normal flux, i.e. (2.7) corresponds to $\mathbf{H}(\mathbf{U}_1) \cdot \mathbf{n} = \mathbf{H}(\mathbf{U}_2) \cdot \mathbf{n}$.

When we iterate between the subdomains, this condition ought to be split into incoming and outgoing characteristics. For this reason, for $i = 1, 2$ we introduce the variables $\mathbf{Z}_i := L\mathbf{U}_i$, that we call *pseudo-characteristics*. They coincide with the characteristic variables if C is a constant matrix. Then we distinguish among positive and negative eigenvalues. Assume for example that $\lambda_r > 0$ for $r \leq q$ and $\lambda_r < 0$ if $r > q$ for a suitable $0 \leq q \leq p$

(in particular, C is non–singular). Then (2.7) can be written equivalently as

(2.8)
$$
\begin{aligned}
Z_{1,r} &= Z_{2,r} \quad \text{for } r > q \\
Z_{2,r} &= Z_{1,r} \quad \text{for } r \leq q \,,
\end{aligned}
$$

where $Z_{i,r}$ denotes the r–th component of $\mathbf{Z}_i$, for $i = 1,2$ and $r = 1,\ldots,p$. If some eigenvalue of C is zero, (2.7) is equivalent to imposing the matching of all the components Z_r of LU for which the corresponding eigenvalue λ_r is different from zero.

If (2.6) is advanced in time from t^n to t^{n+1} by an implicit finite difference scheme (e.g. the backward Euler method), the resulting boundary value problem at the time level t^{n+1} can be solved by the following subdomain iteration method ($k \geq 0$ is the subdomain iteration counter, while we have omitted the superscript $n + 1$ which indicates the time level):

(2.9)
$$
\begin{cases}
a_0 \mathbf{U}_1^{k+1} + \displaystyle\sum_{l=1}^{d} A^{(l)} D_l \mathbf{U}_1^{k+1} = \mathbf{G}_1 & \text{in } \Omega_1 \\
Z_{1,r}^{k+1} = Z_{2,r}^{k} & \text{on } \Gamma \,, \; r > q
\end{cases}
$$

(2.10)
$$
\begin{cases}
a_0 \mathbf{U}_2^{k+1} + \displaystyle\sum_{l=1}^{d} A^{(l)} D_l \mathbf{U}_2^{k+1} = \mathbf{G}_2 & \text{in } \Omega_2 \\
Z_{2,r}^{k+1} = Z_{1,r}^{k} & \text{on } \Gamma \,, \; r \leq q \,,
\end{cases}
$$

where $a_0 := 1/\Delta t$ and $\mathbf{G}_i := \mathbf{F}_i(t^n) + a_0 \mathbf{U}_i^n$, $i = 1,2$, $\mathbf{U}_i^n$ being the approximation of $\mathbf{U}_i(t^n)$ obtained at the previous time–step, as the limit of the iteration by subdomain described above.

Note that for both problems (2.9) and (2.10), we are providing the values of the *incoming* pseudo–characteristic variables on Γ. These conditions, together with the boundary conditions prescribed on $\partial\Omega$, provide the correct number of boundary conditions to solve both subdomain problems (2.9) and (2.10).

The convergence of the sequence $\{\mathbf{U}_i^k\}$ to $\mathbf{U}_i$ as $k \to \infty$, for $i = 1,2$, has been proved in the one–dimensional case for a constant matrix A, by analysing the behaviour of the corresponding characteristic variables LU_i^k (see Quarteroni (1990); see also Bjørhus (1995)).

3. Navier–Stokes equations for compressible flows

The Navier-Stokes equations, which express the conservation of mass, momentum and energy for a compressible fluid, can be written as (see e.g.

Landau and Lifschitz (1959))

(3.1) $\qquad \dfrac{\partial \mathbf{W}}{\partial t} + \operatorname{div} \mathbf{F}(\mathbf{W}) = \operatorname{div} \mathbf{G}(\mathbf{W}) \qquad$ in $\Omega \times (0, T)$,

where $\Omega \subset \mathbf{R}^d$, $d = 2, 3$. The array $\mathbf{W}$ contains the conserved variables, $\mathbf{W} = (\rho, \rho\mathbf{u}, \rho E)$, ρ being the density, $\mathbf{u}$ the velocity vector, and E the total energy per unit mass $E = e + |\mathbf{u}|^2/2$, where e is the internal thermodynamic energy per unit mass and $|\mathbf{u}|^2/2$ is the kinetic energy per unit mass. The convective and diffusive terms $\mathbf{F}(\mathbf{W})$ and $\mathbf{G}(\mathbf{W})$ are the $(d+2) \times d$ matrices defined as

(3.2) $\qquad \mathbf{F}(\mathbf{W}) := \begin{pmatrix} \rho\mathbf{u} \\ \rho\mathbf{u} \otimes \mathbf{u} + p\mathbf{I} \\ (\rho E + p)\mathbf{u} \end{pmatrix}$, $\quad \mathbf{G}(\mathbf{W}) := \begin{pmatrix} 0 \\ \boldsymbol{\tau} \\ \boldsymbol{\tau} \cdot \mathbf{u} - \mathbf{q} \end{pmatrix}$.

Here, p is the pressure, $\mathbf{u} \otimes \mathbf{u}$ denotes the tensor whose components are $u_l u_j$, $\mathbf{I}$ is the unit tensor δ_{lj}, $\mathbf{q}$ is the heat flux, and finally $(\boldsymbol{\tau} \cdot \mathbf{u})_l := \sum_j \tau_{lj} u_j$, where $\boldsymbol{\tau}$ is the viscous stress tensor, which components are defined as

(3.3) $\qquad \tau_{lj} := \mu(D_l u_j + D_j u_l) + \left(\zeta - \dfrac{2\mu}{d}\right) \operatorname{div} \mathbf{u}\, \delta_{lj}$,

with $\mu > 0$ and $\zeta \geq 0$ being the shear and bulk viscosity coefficients, respectively. The heat flux $\mathbf{q}$ is related to the absolute temperature θ by the standard Fourier law

(3.4) $\qquad\qquad\qquad\qquad \mathbf{q} = -\kappa\nabla\theta$,

where $\kappa > 0$ is the heat conductivity coefficient.

In (3.1), the divergence of $\mathbf{F}(\mathbf{W})$ is the $(d+2)$-vector

$$\operatorname{div} \mathbf{F}(\mathbf{W}) = (\operatorname{div}(\rho\mathbf{u}), \operatorname{div}(\rho\mathbf{u} \otimes \mathbf{u} + p\mathbf{I}), \operatorname{div}[(\rho E + p)\mathbf{u}])$$

(and similarly for $\operatorname{div} \mathbf{G}(\mathbf{W})$), and the divergence of a tensor $\mathbf{T}$ is the vector with components

$$(\operatorname{div} \mathbf{T})_l := \sum_j D_j T_{lj}$$.

The entropy per unit mass s can be introduced via the second law of thermodynamics

(3.5) $\qquad\qquad\qquad\qquad \theta\, ds = de - \dfrac{p}{\rho^2} d\rho$.

For ideal polytropic gases, the pressure p and the internal energy e are related to the other thermodynamic quantities ρ and θ through the equations of state

$$p = R\rho\theta \ , \quad e = c_V \theta \ ,$$

where $R > 0$ is the difference between the specific heat at constant pressure $c_P > 0$ and the specific heat at constant volume $c_V > 0$. From these relations it follows that

$$p = (\gamma - 1)\rho e \ ,$$

with $\gamma > 1$ being the ratio of specific heats. Moreover, from (3.5) one finds

$$p = k\rho^\gamma \exp(s/c_V)$$

for a suitable constant $k > 0$.

Equations (3.1) provide an incomplete parabolic system. The hyperbolic Euler equations for inviscid and non–conductive flows can be obtained by dropping all the diffusive terms, i.e. by taking $\mu = \zeta = \kappa = 0$, or, equivalently, $\tau = 0$ and $\mathbf{q} = \mathbf{0}$ (see Section 4).

Since for compressible flows explicit methods are used as well as implicit ones, our analysis of the multi–domain formulation will be carried out without explicitly indicating the discretisation of the time derivative.

Let the domain Ω be partitioned into Ω_1 and Ω_2, as in Section 2. Denoting by $\mathbf{W}_i$ the restriction of $\mathbf{W}$ on Ω_i, $i = 1, 2$, equations (3.1) can be reformulated in the following split form

$$(3.6) \quad \begin{cases} \dfrac{\partial \mathbf{W}_i}{\partial t} + \mathrm{div}\,\mathbf{F}(\mathbf{W}_i) = \mathrm{div}\,\mathbf{G}(\mathbf{W}_i) & \text{in } \Omega_i \times (0, T) \\[2mm] \mathbf{u}_1 = \mathbf{u}_2 \ , \ E_1 = E_2 & \text{on } \Gamma \times (0, T) \\[2mm] [\mathbf{F}(\mathbf{W}_1) - \mathbf{G}(\mathbf{W}_1)] \cdot \mathbf{n} = [\mathbf{F}(\mathbf{W}_2) - \mathbf{G}(\mathbf{W}_2)] \cdot \mathbf{n} & \text{on } \Gamma \times (0, T) \ , \end{cases}$$

for $i = 1, 2$, where $\mathbf{F}(\mathbf{W}) \cdot \mathbf{n}$ denotes the $(d + 2)$–vector

$$(\rho\mathbf{u} \cdot \mathbf{n}, \rho\mathbf{u}(\mathbf{u} \cdot \mathbf{n}) + p\mathbf{n}, (\rho E + p)\mathbf{u} \cdot \mathbf{n})$$

(and similarly for $\mathbf{G}(\mathbf{W}) \cdot \mathbf{n}$).

In other words, the subdomain restrictions $\mathbf{W}_1$ and $\mathbf{W}_2$ satisfy the Navier–Stokes equations separately in Ω_1 and Ω_2, together with suitable interface conditions. Clearly, equations (3.6) inherit the same boundary and initial conditions prescribed for $\mathbf{W}$ on $\partial\Omega$ and at $t = 0$, respectively.

The interface conditions in $(3.6)_2$ are a consequence of the following fact: the unknowns $\mathbf{u}$ and θ appear in (3.1) through their second–order derivatives, hence they must be continuous across Γ. Since $E = e + |\mathbf{u}|^2/2 = c_V\theta + |\mathbf{u}|^2/2$, continuity holds also for the total energy E.

The interface condition $(3.6)_3$ states the continuity of the normal fluxes, and can be rewritten as

$$\rho_1 \mathbf{u}_1 \cdot \mathbf{n} = \rho_2 \mathbf{u}_2 \cdot \mathbf{n}$$

$$\rho_1 u_{1,l} \mathbf{u}_1 \cdot \mathbf{n} + p_1 n_l - \sum_{j=1}^{d} \tau_{1,lj} n_j$$

$$(3.7) \qquad = \rho_2 u_{2,l} \mathbf{u}_2 \cdot \mathbf{n} + p_2 n_l - \sum_{j=1}^{d} \tau_{2,lj} n_j \ , \quad l = 1,\dots,d$$

$$(\rho_1 E_1 + p_1)\mathbf{u}_1 \cdot \mathbf{n} - \sum_{j,l=1}^{d} \tau_{1,lj} u_{1,j} n_l - \kappa \nabla \theta_1 \cdot \mathbf{n}$$

$$= (\rho_2 E_2 + p_2)\mathbf{u}_2 \cdot \mathbf{n} - \sum_{j,l=1}^{d} \tau_{2,lj} u_{2,j} n_l - \kappa \nabla \theta_2 \cdot \mathbf{n} \ .$$

In particular, since $\mathbf{u}$ is continuous across Γ, from the first relation it follows that

$$(3.8) \qquad \rho_1 = \rho_2 \qquad \text{at all points on } \Gamma \times (0,T) \text{ where } \mathbf{u} \cdot \mathbf{n} \neq 0 \ .$$

Finally, the flux matching property $(3.6)_3$ is a natural consequence of the fact that the variable $\mathbf{W}$ is a distributional solution to (3.1) in Ω.

Remark. Within the frame of iterative substructuring methods, equations $(3.6)_2$ can provide the Dirichlet conditions for $\mathbf{u}$ and θ on Γ for one subdomain, while $(3.7)_2$ and $(3.7)_3$ yield the Neumann conditions on Γ for the other subdomain (see e.g. Quarteroni and Valli (1999), Section 1.3). Concerning the interface condition $(3.7)_1$ (or, equivalently, (3.8)), it must be split into a Dirichlet condition for ρ_1 on $\Gamma \cap \{\mathbf{u} \cdot \mathbf{n} < 0\}$, and a Dirichlet condition for ρ_2 on $\Gamma \cap \{\mathbf{u} \cdot \mathbf{n} > 0\}$ (see Quarteroni and Valli, Section 5.6).

4. Euler equations for compressible flows

As already noted, the Euler equations are obtained from the Navier–Stokes equations by taking the viscosity coefficients $\mu = \zeta = 0$ and the heat conductivity coefficient $\kappa = 0$, i.e. disregarding all the diffusive terms. The equations read

$$(4.1) \qquad \frac{\partial \mathbf{W}}{\partial t} + \operatorname{div} \mathbf{F}(\mathbf{W}) = 0 \qquad \text{in } \Omega \times (0,T) \ .$$

Still denoting by $\mathbf{W}_i$ the restriction of $\mathbf{W}$ to Ω_i, we have the equivalent formulation

$$(4.2) \qquad \begin{cases} \dfrac{\partial \mathbf{W}_i}{\partial t} + \operatorname{div} \mathbf{F}(\mathbf{W}_i) = 0 \quad \text{in } \Omega_i \times (0,T) \ , \quad i = 1,2 \\[2mm] \mathbf{F}(\mathbf{W}_1) \cdot \mathbf{n} = \mathbf{F}(\mathbf{W}_2) \cdot \mathbf{n} \quad \text{on } \Gamma \times (0,T) \ . \end{cases}$$

This time the only matching conditions on Γ are those prescribing the continuity of the normal inviscid flux, which can be rewritten as

$$\rho_1 \mathbf{u}_1 \cdot \mathbf{n} = \rho_2 \mathbf{u}_2 \cdot \mathbf{n}$$

(4.3)
$$\rho_1 u_{1,l} \mathbf{u}_1 \cdot \mathbf{n} + p_1 n_l = \rho_2 u_{2,l} \mathbf{u}_2 \cdot \mathbf{n} + p_2 n_l \ , \quad l = 1, \ldots, d$$

$$(\rho_1 E_1 + p_1) \mathbf{u}_1 \cdot \mathbf{n} = (\rho_2 E_2 + p_2) \mathbf{u}_2 \cdot \mathbf{n} \ .$$

As we have already noted, the first condition is equivalent to (3.7). This latter condition is in agreement with the physical properties of compressible fluid flows, which admit two types of discontinuities: shock waves or contact discontinuities. In particular, on contact discontinuities the normal velocity is zero, the pressure is continuous, but density, as well as tangential velocity and temperature, may have a non–zero jump.

If the interface Γ is not kept fixed but moves along in time, then denoting by $\sigma(t)$ its velocity at time t along the normal direction $\mathbf{n} = \mathbf{n}(t)$, the matching condition $(4.2)_2$ has to be replaced by

$$(4.4) \quad [\mathbf{W}_1(t) - \mathbf{W}_2(t)]\sigma(t) = [\mathbf{F}(\mathbf{W}_1(t)) - \mathbf{F}(\mathbf{W}_2(t))] \cdot \mathbf{n}(t) \qquad \text{on } \Gamma(t) \ .$$

If $\Gamma(t)$ intercepts (or coincides with) a shock front $\delta(t)$, then (4.4) can be easily recognised as the Rankine–Hugoniot jump condition across $\delta(t)$ (see e.g. Hirsch (1990), p. 136).

4.1. ONE-DIMENSIONAL EULER EQUATIONS

A characteristic analysis of Euler equations enlightens the role that the interface conditions could play in the framework of a substructuring iterative methods. To simplify our analysis we consider first one–dimensional flows, in which case (4.1) becomes

$$(4.5) \qquad \frac{\partial \mathbf{W}}{\partial t} + \frac{\partial \mathbf{F}(\mathbf{W})}{\partial x} = 0 \qquad \text{in } \Omega \times (0, T) \ ,$$

Ω being an interval. This time, the unknowns quantities are $\mathbf{W} = (\rho, \rho u, \rho E)$, while the inviscid flux can be written as a three–dimensional vector

$$\mathbf{F}(\mathbf{W}) = \left(\rho u, \rho u^2 + p, (\rho E + p)u \right) \ .$$

Let us recall some known facts about the characteristic form of the one–dimensional Euler equations. Using (3.5) for expressing the derivatives of e in terms of s and ρ, the quasi–linear form of (4.5) in terms of the vector of primitive variables $\mathbf{U} = (\rho, u, s)$ is given by

$$(4.6) \qquad \frac{\partial \mathbf{U}}{\partial t} + A \frac{\partial \mathbf{U}}{\partial x} = 0 \qquad \text{in } \Omega \times (0, T) \ ,$$

where

(4.7)
$$A := \begin{pmatrix} u & \rho & 0 \\ c^2/\rho & u & p_s/\rho \\ 0 & 0 & u \end{pmatrix}$$

and we have set $c := \sqrt{\frac{\partial p}{\partial \rho}}$ (the speed of sound) and $p_s := \frac{\partial p}{\partial s}$.

The eigenvalues of A are

(4.8)
$$\lambda_1 = u + c , \quad \lambda_2 = u - c , \quad \lambda_3 = u ,$$

and the matrix L of left eigenvectors is

(4.9)
$$L := \begin{pmatrix} c/\rho & 1 & p_s/(\rho c) \\ -c/\rho & 1 & -p_s/(\rho c) \\ 0 & 0 & 1 \end{pmatrix} .$$

For subsonic flows in which $0 < u < c$, the left end of the interval Ω is the upstream boundary, the two eigenvalues λ_1 and λ_3 are positive, while λ_2 is negative. If the flow is supersonic ($u > c > 0$), all the eigenvalues are positive.

In principle, system (4.6) can be diagonalised by using the eigenvectors of A, leading to a fully decoupled problem. A suitable set of variables, the *characteristic variables* $\tilde{\mathbf{U}}$, is introduced by means of the following differential form (see e.g. Hirsch (1990), p. 162)

(4.10) $d\tilde{\mathbf{U}} := L d\mathbf{U} = \left(\frac{c}{\rho} d\rho + du + \frac{p_s}{\rho c} ds, -\frac{c}{\rho} d\rho + du - \frac{p_s}{\rho c} ds, ds \right) .$

The introduction of the above relations based on variations, and not upon the quantities themselves, is motivated by the fact that the coefficients of the system, i.e. the elements of A, are not constant and depend on the solution itself.

Since $LAL^{-1} = \Lambda = \text{diag} \, (\lambda_1, \lambda_2, \lambda_3)$, owing to (4.10) equations (4.6) become

(4.11)
$$\frac{\partial \tilde{\mathbf{U}}}{\partial t} + \Lambda \frac{\partial \tilde{\mathbf{U}}}{\partial x} = 0 \qquad \text{in } \Omega \times (0, T) ,$$

which splits into three scalar independent equations

(4.12)
$$\frac{\partial \tilde{U}_r}{\partial t} + \lambda_r \frac{\partial \tilde{U}_r}{\partial x} = 0 \qquad \text{in } \Omega \times (0, T) , \quad r = 1, 2, 3 .$$

It follows that, for each $r = 1, 2, 3$, the component $\tilde{U}_r$ is constant along the *characteristic curve*

$$\{(x_r(t), t) \,|\, x_r'(t) = \lambda_r\} .$$

Notice the formal analogy between (4.12) and the characteristic system (2.4). Assuming that we know $\tilde{\mathbf{U}}$, we can then proceed as for problem (2.4), and enforce the continuity of the incoming characteristic variables on each subdomain at every iteration.

This procedure can be illustrated on the *isentropic* Euler equations, which read

$$(4.13) \qquad \frac{\partial \mathbf{V}}{\partial t} + B \frac{\partial \mathbf{V}}{\partial x} = 0 \qquad \text{in } \Omega \times (0, T) ,$$

where $\mathbf{V} = (\rho, u)$ and

$$(4.14) \qquad B := \begin{pmatrix} u & \rho \\ c^2/\rho & u \end{pmatrix} .$$

Here the entropy s is assumed to be constant, so that the pressure p depends only on the density ρ, precisely $p = K\rho^\gamma$ for a suitable constant $K > 0$.

The eigenvalues of B are

$$\lambda_1 = u + c , \quad \lambda_2 = u - c ,$$

where $c = \sqrt{K\gamma\rho^{\gamma-1}}$, and

$$L := \begin{pmatrix} c/\rho & 1 \\ -c/\rho & 1 \end{pmatrix} .$$

The characteristic variables $\tilde{\mathbf{V}}$ are defined as

$$(4.15) \qquad d\tilde{\mathbf{V}} := Ld\mathbf{V} = \left(\frac{c}{\rho}d\rho + du, -\frac{c}{\rho}d\rho + du \right) ,$$

hence by a direct integration one finds

$$\tilde{V}_1 = u + \int \frac{c}{\rho} d\rho = u + \int \sqrt{K\gamma} \rho^{\gamma/2 - 3/2} d\rho$$

$$= u + \frac{2}{\gamma - 1} c + \text{const} ,$$

and similarly for $\tilde{V}_2$. In conclusion we have

$$(4.16) \qquad \tilde{\mathbf{V}} = (R_+, R_-) , \quad R_\pm := u \pm \frac{2}{\gamma - 1} c .$$

The Riemann invariant R_+ is constant on $C_+ = \{(x(t), t) \mid x'(t) = u + c\}$, whereas R_- is constant on $C_- = \{(x(t), t) \mid x'(t) = u - c\}$.

In this case, assuming for instance that the flow is subsonic $(0 < u < c)$, the iterative method would read:

$$(4.17) \quad \begin{cases} \dfrac{\partial \mathbf{V}_1^{k+1}}{\partial t} + B\dfrac{\partial \mathbf{V}_1^{k+1}}{\partial x} = 0 & \text{in } \Omega_1 \times (0,T) \\[2mm] R_{-,1}^{k+1} = R_{-,2}^{k} & \text{at } x_\Gamma \times (0,T) \end{cases}$$

$$(4.18) \quad \begin{cases} \dfrac{\partial \mathbf{V}_2^{k+1}}{\partial t} + B\dfrac{\partial \mathbf{V}_2^{k+1}}{\partial x} = 0 & \text{in } \Omega_2 \times (0,T) \\[2mm] R_{+,2}^{k+1} = R_{+,1}^{k} & \text{at } x_\Gamma \times (0,T) \ . \end{cases}$$

Here x_Γ is the interface point separating the subintervals Ω_1 and Ω_2. For a supersonic flow, both interface conditions have to be attributed to the domain Ω_2.

Furthermore, there are the initial condition and two additional boundary conditions, one assigned at the left end of the interval Ω_1 (where the characteristic C_+ is incoming) and the other at the right end of the interval Ω_2 (where the characteristic C_- is incoming). It is easily seen that, at convergence of the iterative scheme, the continuity of R_+ and R_- at x_Γ implies that $(4.3)_1$ and $(4.3)_2$ are satisfied.

Unfortunately, for the non–isentropic system (4.6) the functions $R_\pm$ are no longer constant along the characteristic curves $C_\pm$, and in general the characteristic variables $\tilde{U}$ are not explicitly known (see e.g. Hirsch (1990), pp. 155–156), except for the entropy s, which is constant along the characteristic curve $C_0 = \{(x(t), t) \mid x'(t) = u\}$.

To circumvent the lack of knowledge of the characteristic variables, we consider the following iteration procedure. The Euler system is solved in Ω_1 and Ω_2, with given initial data at $t = 0$ and boundary data on $\partial\Omega$ (these latter conditions have to be determined by means of a suitable eigenvalue analysis, see e.g. Oliger and Sundström (1978)). Concerning the interface conditions, we can proceed as in (4.17), (4.18) and, assuming that each eigenvalue of A is different from zero, enforce the continuity of the pseudo-characteristic LU and iterate accordingly. For the Euler system, since

$$\begin{aligned} LU &= \left(u + c + \frac{p_s}{\rho c}s,\, u - c - \frac{p_s}{\rho c}s,\, s\right) \\ &=: (Z_+, Z_-, Z_0)\ , \end{aligned}$$

when all the eigenvalues are non–null we enforce the continuity of the three functions Z_+, Z_- and Z_0.

For instance, in the subsonic case $0 < u < c$, the interface conditions in the iterative process become

$$Z_{-,1}^{k+1}(x_\Gamma, t) = Z_{-,2}^{k}(x_\Gamma, t)$$

and

$$\begin{cases} Z_{+,2}^{k+1}(x_\Gamma, t) = Z_{+,1}^k(x_\Gamma, t) \\ \\ Z_{0,2}^{k+1}(x_\Gamma, t) = Z_{0,1}^k(x_\Gamma, t) \ . \end{cases}$$

For each subdomain, we are therefore prescribing a boundary condition at the interface for any variable associated with an incoming characteristic line.

Another possibility is to enforce the continuity of the Riemann 'invariants' R_+, R_- and $R_0 = s$. In both cases, it is straightforward to verify that the continuity of $LU = (Z_+, Z_-, Z_0)$ or that of (R_+, R_-, R_0) at x_Γ implies that (4.3) is satisfied.

When considering discretisation, at the interface point x_Γ one has to enforce three additional conditions (besides the other three related to Z_+, Z_- and Z_0, or R_+, R_- and R_0), in order to recover all the six interface variables. This can be accomplished through the equations

(4.19)
$$\left[\mathbf{l}^r \cdot \left(\frac{\partial \mathbf{U}_1}{\partial t} + \lambda_r \frac{\partial \mathbf{U}_1}{\partial x} \right) \right] (x_\Gamma, t) = 0 \qquad \text{for } r = 1, 3$$
$$\left[\mathbf{l}^2 \cdot \left(\frac{\partial \mathbf{U}_2}{\partial t} + \lambda_2 \frac{\partial \mathbf{U}_2}{\partial x} \right) \right] (x_\Gamma, t) = 0 \ ,$$

where

$$\mathbf{l}^1 := \left(\frac{c}{\rho}, 1, \frac{p_s}{\rho c} \right) \ , \quad \mathbf{l}^2 := \left(-\frac{c}{\rho}, 1, -\frac{p_s}{\rho c} \right) \ , \quad \mathbf{l}^3 := (0, 0, 1)$$

are the left eigenvectors of A.

In each subdomain, equations (4.19) are obtained from the Euler system (4.6) by taking the scalar product with $\mathbf{l}^r$, $r = 1, 2, 3$, at the interface point, but only for those values of r whose corresponding eigenvalue λ_r identifies a characteristic line that is outgoing at x_Γ. These equations are sometimes called the *compatibility* equations.

4.2. MULTI–DIMENSIONAL EULER EQUATIONS

Let us return now the three–dimensional Euler equations, which, with respect to the primitive variables $\mathbf{U} = (\rho, u_1, u_2, u_3, s)$, can be written as

(4.20)
$$\frac{\partial \mathbf{U}}{\partial t} + \sum_{l=1}^3 A^{(l)} D_l \mathbf{U} = 0 \qquad \text{in } \Omega \times (0, T) \ ,$$

where

$$A^{(1)} := \begin{pmatrix} u_1 & \rho & 0 & 0 & 0 \\ c^2/\rho & u_1 & 0 & 0 & p_s/\rho \\ 0 & 0 & u_1 & 0 & 0 \\ 0 & 0 & 0 & u_1 & 0 \\ 0 & 0 & 0 & 0 & u_1 \end{pmatrix}$$

$$A^{(2)} := \begin{pmatrix} u_2 & 0 & \rho & 0 & 0 \\ 0 & u_2 & 0 & 0 & 0 \\ c^2/\rho & 0 & u_2 & 0 & p_s/\rho \\ 0 & 0 & 0 & u_2 & 0 \\ 0 & 0 & 0 & 0 & u_2 \end{pmatrix}$$

$$A^{(3)} := \begin{pmatrix} u_3 & 0 & 0 & \rho & 0 \\ 0 & u_3 & 0 & 0 & 0 \\ 0 & 0 & u_3 & 0 & 0 \\ c^2/\rho & 0 & 0 & u_3 & p_s/\rho \\ 0 & 0 & 0 & 0 & u_3 \end{pmatrix}.$$

As usual, we have set $c := \sqrt{\frac{\partial p}{\partial \rho}}$ and $p_s := \frac{\partial p}{\partial s}$.

For any point $\mathbf{x} \in \Gamma$ and any time $t \in (0,T)$, denote by $C = C(\mathbf{n}) = \sum_l n_l A^{(l)}$ the characteristic matrix. The eigenvalues of C are given by

$$(4.21) \qquad \lambda_1 = \mathbf{u} \cdot \mathbf{n} + c, \quad \lambda_2 = \mathbf{u} \cdot \mathbf{n} - c, \quad \lambda_{3,4,5} = \mathbf{u} \cdot \mathbf{n}.$$

Finally, denote by L the matrix of left eigenvectors of C, which is given by

$$(4.22) \quad L := \begin{pmatrix} \dfrac{c}{\rho} & n_1 & n_2 & n_3 & \dfrac{p_s}{\rho c} \\ -\dfrac{c}{\rho} & n_1 & n_2 & n_3 & -\dfrac{p_s}{\rho c} \\ 0 & \tau_1^{(1)} & \tau_2^{(1)} & \tau_3^{(1)} & 1 \\ 0 & \tau_1^{(2)} & \tau_2^{(2)} & \tau_3^{(2)} & 1 \\ 0 & -\tau_1^{(1)} - \tau_1^{(2)} & -\tau_2^{(1)} - \tau_2^{(2)} & -\tau_3^{(1)} - \tau_3^{(2)} & 1 \end{pmatrix},$$

where $\boldsymbol{\tau}^{(1)}$ and $\boldsymbol{\tau}^{(2)}$ are two unit orthogonal vectors, spanning the plane orthogonal to $\mathbf{n}$.

Assuming that at the time t the interface Γ is not characteristic at the point $\mathbf{x}$, namely that no eigenvalue is null at $\mathbf{x}$, the matching conditions are naturally extrapolated from the one–dimensional case, and yield

$$(4.23) \qquad \sum_{q=1}^{5} L_{rq} U_{1,q} = \sum_{q=1}^{5} L_{rq} U_{2,q} \qquad \text{at } \mathbf{x} \in \Gamma, \ r = 1,\ldots,5.$$

Again, we are enforcing the continuity of the pseudo–characteristic variables LU. It is easily verified that, as a consequence of (4.23), the interface conditions (4.3) are satisfied.

The iteration–by–subdomain algorithm used for solving the multi–domain problem alternates the solution of the Euler equations (4.20) in Ω_1 and in Ω_2, with the Dirichlet boundary condition (4.23) imposed at a point $\mathbf{x}$ on Γ for all indices r corresponding to incoming characteristic lines. For instance, if we assume that at time t the interface point $\mathbf{x}$ is an outflow point for Ω_1 and that the flow is subsonic (namely, $0 < \mathbf{u} \cdot \mathbf{n} < c$), we have to impose at the $(k+1)$–th iteration

$$\sum_{q=1}^{5} L_{2q} U_{1,q}^{k+1} = \sum_{q=1}^{5} L_{2q} U_{2,q}^{k} \qquad \text{at } \mathbf{x} \in \Gamma ,$$

and

$$\sum_{q=1}^{5} L_{rq} U_{2,q}^{k+1} = \sum_{q=1}^{5} L_{rq} U_{1,q}^{k} \qquad \text{at } \mathbf{x} \in \Gamma , \ r = 1, 3, 4, 5 .$$

When a numerical discretisation is being applied, other equations, the compatibility equations, have to be imposed at $\mathbf{x}$. Proceeding as done in (4.19), in each subdomain Ω_i, $i = 1, 2$, they are obtained by taking the scalar product of (4.20) (stated for $\mathbf{U}_i^{k+1}$ instead of $\mathbf{U}$) with the r–th left eigenvector $\mathbf{l}^r$, $r = 1, \ldots, 5$, but only for those values of r for which the eigenvalue λ_r is associated with a characteristic line which is directed outward Ω_i at $\mathbf{x}$.

5. Heterogeneous models for compressible flows

As illustrated in Section 3, the Navier–Stokes equations (3.1), which express the conservation of mass, momentum and energy of a compressible fluid, provide a complete description of all flow phenomena, but need empirical laws to relate the viscous coefficients and the heat conductivity coefficient to the thermodynamic flow variables. The Euler system (4.1) shares with the Navier–Stokes system the same number of equations $d + 2$. However, the viscous stress and the heat–conduction terms are no longer considered, neither are those equations that might supplement (3.1) when the Navier–Stokes equations are Reynolds averaged through suitable turbulence models (e.g. Lesieur (1997)).

When the flow field is potential and isentropic, the Euler equations can be furtherly simplified by resorting to the so–called *full potential* equation, which is a single, scalar equation. In this case the conservation of mass reads

$$(5.1) \qquad \frac{\partial \rho}{\partial t} + \operatorname{div}(\rho \nabla \varphi) = 0 \qquad \text{in } \Omega \times (0, T) ,$$

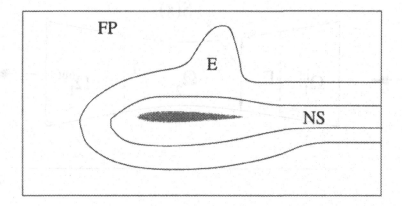

Figure 1. The three–subdomain decomposition around an airfoil.

where φ is the velocity potential, i.e. $\mathbf{u} = \nabla\varphi$, and another equation for the potential φ has to be determined by means of a suitable conservation principle (see Hirsch (1990), and Section 5.1).

Since equation (5.1) has no mechanism to generate entropy variations across discontinuities, it allows shock discontinuities that do not satisfy the Rankine–Hugoniot relations. Indeed, both mass and energy are conserved, whereas the momentum $\rho\mathbf{u}$ is not.

For the simulation of the flow field around an airfoil, a heterogeneous model based on the simultaneous use of three different sets of equations could be envisaged: the Navier–Stokes equations are used in the boundary layer and in the downstream wake, the Euler equations in the surrounding region where the shock may develop, and the full potential equation in the far field where the flow is irrotational (see Figure 1). This is a very sophisticated approach which requires the treatment of the interaction between couples of submodels such as Navier–Stokes with Euler, and Euler with full potential. For the coupling of Euler and Navier-Stokes equations we refer the reader to Quarteroni and Stolcis (1995). Here we give an account of the heterogeneous model which couples Euler equations with the full potential equation.

For ease of notation, in this section we will refer to the simple geometrical situation in which the spatial computational domain Ω is split into two disjoint subdomains Ω_1 and Ω_2, wherein the different models are used. As usual, we will denote by $\mathbf{n}$ the unit outward normal vector to Ω_1 on Γ, the latter being the interface between Ω_1 and Ω_2.

As previously pointed out, the Euler equations provide an acceptable mathematical description of compressible flows whenever the viscous stresses and the heat conducting terms can be neglected. This is the case for flows around an obstacle (e.g. an airfoil) far enough from the obstacle itself (and

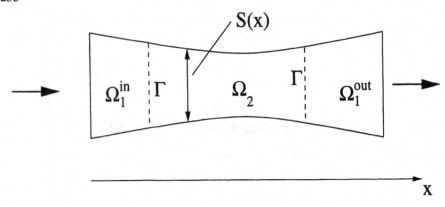

Figure 2. Converging–diverging nozzle.

from the boundary layer induced in its vicinity), or flows in ducts or pipes (e.g. the converging–diverging nozzle of Figure 2).

If there are regions of irrotational and isentropic flow, the full potential equation (5.1) can be used instead of the Euler equations, yielding two scalar equations rather than the $d + 1$ equations of the isentropic Euler system.

Before facing the issue of the heterogeneous coupling between the Euler and full potential equations, it is worthwhile to address the full potential equation alone in order to discuss the interface conditions to be used in the event of adopting a homogeneous coupling involving the full potential solely.

5.1. DOMAIN DECOMPOSITION FOR THE FULL POTENTIAL EQUATION

Assume therefore that the full potential equation is considered in the whole domain Ω; we aim at considering it separately in Ω_1 and Ω_2 and couple the corresponding solutions through suitable matching conditions on the interface Γ between Ω_1 and Ω_2.

Since the flow is potential and isentropic, the conservation of momentum now reads

$$(5.2) \qquad \frac{\partial \nabla \varphi}{\partial t} + (\mathbf{u} \cdot \nabla)\mathbf{u} + \frac{c^2}{\rho} \nabla \rho = \mathbf{0} \qquad \text{in } \Omega \times (0, T) \,,$$

where, as usual, $c := \sqrt{p'(\rho)}$ is the sound speed. On the other hand,

$$(\mathbf{u} \cdot \nabla)\mathbf{u} = \frac{1}{2} \nabla |\mathbf{u}|^2 + \text{rot } \mathbf{u} \times \mathbf{u} = \frac{1}{2} \nabla |\mathbf{u}|^2$$

and

$$\frac{c^2}{\rho}\nabla\rho = \nabla\int\frac{c^2}{\rho}d\rho .$$

Since the gas is ideal and polytropic, namely $p = R\rho\theta$ and $e = c_V\theta$, where $R = c_P - c_V$ is the difference of the specific heats, for an isentropic flow we also have $p = K\rho^\gamma$, where $K > 0$ is a suitable constant and γ is the ratio of specific heats. Hence

$$\int\frac{c^2}{\rho}d\rho = \int K\gamma\rho^{\gamma-2}d\rho = K\frac{\gamma}{\gamma-1}\rho^{\gamma-1} + \text{const}$$

$$= \frac{\gamma}{\gamma-1}\frac{p}{\rho} + \text{const} = \frac{\gamma}{\gamma-1}R\theta + \text{const}$$

$$= c_P\theta + \text{const} .$$

Introducing the enthalpy per unit mass, $h = e + \frac{p}{\rho}$, one has

$$h = c_V\theta + R\theta = c_P\theta .$$

Thus the equation of conservation of momentum finally reads

(5.3) $$\frac{\partial\varphi}{\partial t} + \frac{1}{2}|\nabla\varphi|^2 + h = H_0 \qquad \text{in } \Omega\times(0,T) ,$$

where H_0 is the (constant) stagnation total enthalpy.

In the case of isentropic flows, the second law of thermodynamics (3.5), expressed in terms of the enthalpy h, reads

$$dh - \frac{c^2}{\rho}d\rho = 0 ,$$

as the left–hand side equals θds. Hence differentiating in t equation (5.3) and using (5.1) we find

$$\frac{\partial^2\varphi}{\partial t^2} + \frac{\partial h}{\partial t} + \frac{\partial\nabla\varphi}{\partial t}\cdot\nabla\varphi$$

$$= \frac{\partial^2\varphi}{\partial t^2} + \frac{c^2}{\rho}\frac{\partial\rho}{\partial t} + \frac{\partial\nabla\varphi}{\partial t}\cdot\nabla\varphi$$

$$= \frac{\partial^2\varphi}{\partial t^2} - \frac{c^2}{\rho}\text{div}(\rho\nabla\varphi) + \frac{\partial\nabla\varphi}{\partial t}\cdot\nabla\varphi = 0 .$$

By a direct computation, differentiating equation (5.2) with respect to $\mathbf{x}$ one also has

$$-\frac{c^2}{\rho}\text{div}(\rho\nabla\varphi) = -c^2\Delta\varphi - \frac{c^2}{\rho}\nabla\varphi\cdot\nabla\rho$$

$$= -c^2\Delta\varphi - \nabla\varphi\cdot\nabla h$$

$$= -c^2\Delta\varphi + \nabla\varphi\cdot\left[\frac{\partial\nabla\varphi}{\partial t} + \frac{1}{2}\nabla|\nabla\varphi|^2\right] .$$

We have thus found the single equation for the potential φ:

(5.4) $\quad \dfrac{\partial^2 \varphi}{\partial t^2} - c^2 \Delta\varphi + \nabla\varphi \cdot [(\nabla\varphi \cdot \nabla)\nabla\varphi] + 2\nabla\varphi \cdot \dfrac{\partial \nabla\varphi}{\partial t} = 0 \qquad$ in $\Omega \times (0,T)$.

Upon introducing the vector unknown

$$\mathbf{W} := (D_t\varphi, D_1\varphi, \cdots, D_d\varphi) \ ,$$

we also obtain the following quasi–linear vector form of the full potential equation

(5.5) $\qquad \dfrac{\partial \mathbf{W}}{\partial t} + \displaystyle\sum_{l=1}^{d} A^{(l)} D_l \mathbf{W} = \mathbf{0} \qquad$ in $\Omega \times (0,T)$,

where, for $d = 3$, the Jacobian matrices $A^{(l)}$ are given by

$$A^{(1)} = \begin{pmatrix} 2u_1 & u_1^2 - c^2 & u_1 u_2 & u_1 u_3 \\ -1 & 0 & 0 & 0 \\ 0 & 0 & 0 & 0 \\ 0 & 0 & 0 & 0 \end{pmatrix}$$

$$A^{(2)} = \begin{pmatrix} 2u_2 & u_1 u_2 & u_2^2 - c^2 & u_2 u_3 \\ 0 & 0 & 0 & 0 \\ -1 & 0 & 0 & 0 \\ 0 & 0 & 0 & 0 \end{pmatrix}$$

$$A^{(3)} = \begin{pmatrix} 2u_3 & u_1 u_3 & u_2 u_3 & u_3^2 - c^2 \\ 0 & 0 & 0 & 0 \\ 0 & 0 & 0 & 0 \\ -1 & 0 & 0 & 0 \end{pmatrix} .$$

If $d = 2$, they reduce to $A^{(1)}$ and $A^{(2)}$ upon omitting the last row and column.

The hyperbolic form (5.5) suggests the coupling mechanism at the interface. As a matter of fact, following the guidelines of Section 2 we deduce that the whole vector $C(\mathbf{n})\mathbf{W}$ is continuous across the interface, where $C(\mathbf{n}) := \sum_{l=1}^{d} n_l A^{(l)}$ is the characteristic matrix.

To analyse what consequences can be drawn about the continuity of the unknown variables $\mathbf{W}$, let us investigate the spectral properties of $C(\mathbf{n})$. If $d = 3$, its eigenvalues are

$$\lambda_1 = \mathbf{u} \cdot \mathbf{n} + c \ , \quad \lambda_2 = \mathbf{u} \cdot \mathbf{n} - c \ , \quad \lambda_3 = \lambda_4 = 0 \ .$$

The first two of them coincide with those of the Euler system (see (4.21)), whereas the third and fourth are zero. The reason is that, in contrast with the Euler case, the two characteristic curves along which vorticity and entropy propagate are missing. Actually, vorticity is zero and entropy is constant for irrotational and isentropic flows.

In the one–dimensional case, the full potential equation (5.4) becomes

(5.6)
$$\frac{\partial^2 \varphi}{\partial t^2} - (c^2 - u^2)\frac{\partial^2 \varphi}{\partial x^2} + 2u\frac{\partial^2 \varphi}{\partial t \partial x} = 0 \,,$$

where $u = \frac{\partial \varphi}{\partial x}$ is the flow velocity. The eigenvalues of the characteristic matrix

$$A = \begin{pmatrix} 2u & u^2 - c^2 \\ -1 & 0 \end{pmatrix}$$

are $\lambda_1 = u + c$ and $\lambda_2 = u - c$, and the matrix of left eigenvectors is given by

$$L = \begin{pmatrix} -\dfrac{1}{c} & 1 - \dfrac{u}{c} \\ \dfrac{1}{c} & 1 + \dfrac{u}{c} \end{pmatrix} \,.$$

The characteristic variables are defined through the relations $d\tilde{\mathbf{W}} = L d\mathbf{W}$, i.e.

(5.7) $$d\tilde{W}_1 = -\frac{1}{c}dW_1 + \left(1 - \frac{u}{c}\right) dW_2 \,, \quad d\tilde{W}_2 = \frac{1}{c}dW_1 + \left(1 + \frac{u}{c}\right) dW_2 \,,$$

where $W_1 = \frac{\partial \varphi}{\partial t}$ and $W_2 = \frac{\partial \varphi}{\partial x} = u$.

For isentropic flows the sound speed can be written as follows

(5.8)
$$c^2 = p'(\rho) = \gamma\frac{p}{\rho} = \gamma R\theta$$
$$= (\gamma - 1)c_P\theta = (\gamma - 1)h = (\gamma - 1)\left(H_0 - W_1 - \frac{W_2^2}{2}\right) \,,$$

having used (5.3). Then we find

$$dW_1 = -\frac{2}{\gamma - 1}c\,dc - W_2 \,, \quad dW_2 = -\frac{2}{\gamma - 1}c\,dc - u\,du \,,$$

and therefore

$$dW_1 = \frac{2}{\gamma - 1}dc + \frac{u}{c}du + \left(1 - \frac{u}{c}\right) du = \frac{2}{\gamma - 1}dc + du$$
$$d\tilde{W}_2 = -\frac{2}{\gamma - 1}dc - \frac{u}{c}du + \left(1 + \frac{u}{c}\right) du = -\frac{2}{\gamma - 1}dc + du$$

Integrating these relations one obtains

(5.9) $$\tilde{W}_1 = u + \frac{2}{\gamma - 1}c \,, \quad \tilde{W}_2 = u - \frac{2}{\gamma - 1}c \,,$$

which coincides with the first two Riemann invariants of the one–dimensional isentropic Euler system (see (4.16)).

For simplifying the notation, let us denote for the velocity u and sound speed c the functional dependence on the potential φ as follows:

$$u = u(\varphi) := \frac{\partial \varphi}{\partial x}$$

$$c = c(\varphi) := \sqrt{(\gamma - 1)\left[H_0 - \frac{\partial \varphi}{\partial t} - \frac{1}{2}\left(\frac{\partial \varphi}{\partial x}\right)^2\right]}.$$

Repeating the analysis of Section 4.1, we obtain the following two–domain formulation for the one–dimensional full potential equation:

(5.10)
$$\begin{cases} \dfrac{\partial^2 \varphi_1}{\partial t^2} - (c(\varphi_1)^2 - u(\varphi_1)^2)\dfrac{\partial^2 \varphi_1}{\partial x^2} \\ \qquad\qquad +2u(\varphi_1)\dfrac{\partial^2 \varphi_1}{\partial t \partial x} = 0 & \text{in } \Omega_1 \times (0,T) \\ \dfrac{\partial^2 \varphi_2}{\partial t^2} - (c(\varphi_2)^2 - u(\varphi_2)^2)\dfrac{\partial^2 \varphi_2}{\partial x^2} \\ \qquad\qquad +2u(\varphi_2)\dfrac{\partial^2 \varphi_2}{\partial t \partial x} = 0 & \text{in } \Omega_2 \times (0,T) \\ u(\varphi_1) + \dfrac{2}{\gamma - 1}c(\varphi_1) = u(\varphi_2) + \dfrac{2}{\gamma - 1}c(\varphi_2) & \text{at } x_\Gamma \times (0,T) \\ u(\varphi_1) - \dfrac{2}{\gamma - 1}c(\varphi_1) = u(\varphi_2) - \dfrac{2}{\gamma - 1}c(\varphi_2) & \text{at } x_\Gamma \times (0,T) \ , \end{cases}$$

where x_Γ is the interface point separating the subintervals Ω_1 and Ω_2. As usual, this system has to be supplemented by the initial condition and the additional boundary conditions for each characteristic variable whose associated characteristic line is incoming.

An iterative procedure like (4.17), (4.18) can be set up in order to prescribing the values of the incoming Riemann invariants for the interval at hand. Assuming that u is positive at x_Γ and that the flow is subsonic, at the k–th step we prescribe

$$u(\varphi_2^{k+1}) + \frac{2}{\gamma - 1}c(\varphi_2^{k+1}) = u(\varphi_1^k) + \frac{2}{\gamma - 1}c(\varphi_1^k) \qquad \text{at } x_\Gamma \times (0,T)$$

$$u(\varphi_1^{k+1}) - \frac{2}{\gamma - 1}c(\varphi_1^{k+1}) = u(\varphi_2^k) - \frac{2}{\gamma - 1}c(\varphi_2^k) \qquad \text{at } x_\Gamma) \times (0,T) \ ,$$

whereas for supersonic flows we assign both values to the downstream domain Ω_2, i.e.

$$u(\varphi_2^{k+1}) + \frac{2}{\gamma - 1} c(\varphi_2^{k+1}) = u(\varphi_1^k) + \frac{2}{\gamma - 1} c(\varphi_1^k) \qquad \text{at } x_\Gamma \times (0, T)$$

$$u(\varphi_2^{k+1}) - \frac{2}{\gamma - 1} c(\varphi_2^{k+1}) = u(\varphi_1^k) - \frac{2}{\gamma - 1} c(\varphi_1^k) \qquad \text{at } x_\Gamma \times (0, T) \ .$$

Observe that the continuity of $\tilde{W}_1$ and $\tilde{W}_2$ at the interface ensures that of u and c, hence that of $\frac{\partial \varphi}{\partial x}$ (the potential flux) and $\frac{\partial \varphi}{\partial t}$ (through (5.8)), or else, equivalently, that of $\frac{\partial \varphi}{\partial x}$ and φ. This may encourage the use of a multi–domain formulation directly on the multi–dimensional scalar equation (5.4); in that case, the subdomain iterations should aim at enforcing the continuity of both quantities in the framework of a Dirichlet/Neumann strategy.

5.2. THE COUPLING BETWEEN THE EULER EQUATIONS AND THE FULL POTENTIAL EQUATION

The above characteristic analysis suggests how to couple the full potential equation with the Euler equations in the one–dimensional isentropic case. In fact, the characteristic lines are the same (in both cases associated with the eigenvalues $u + c$ and $u - c$), as well as the Riemann invariants $u \pm \frac{2}{\gamma - 1} c$. The heterogeneous coupling is appropriate for instance to simulating the flow field for the converging–diverging nozzle of Figure 2: in that case the full potential equation is used in the two extreme sections of the duct Ω_1^{in} and Ω_1^{out}, and the Euler equations in the inner domain Ω_2. Notice also that the Euler equations in the converging–diverging nozzle can indeed be reduced to a one–dimensional system of similar structure, by means of a suitable transformation in terms of the cross–section $S(x)$ (see e.g. Hirsch (1990), pp. 157–158).

One can proceed as follows. Let $\Omega_1 = (\alpha, x_\gamma)$ be the domain where we consider the full potential equation, and $\Omega_2 = (x_\gamma, \beta)$ the one where the Euler equations are enforced. Consider first the subsonic case $0 < u < c$. Then the iterative scheme reads (for notation see (4.13), (4.14)):

(5.11)
$$\begin{cases} \dfrac{\partial^2 \varphi_1^{k+1}}{\partial t^2} - [c(\varphi_1^{k+1})^2 - u(\varphi_1^{k+1})^2] \dfrac{\partial^2 \varphi_1^{k+1}}{\partial x^2} \\ \qquad\qquad + 2u(\varphi_1^{k+1}) \dfrac{\partial^2 \varphi_1^{k+1}}{\partial t \partial x} = 0 \qquad \text{in } \Omega_1 \times (0, T) \\ \dfrac{\partial \varphi_1^{k+1}}{\partial x} = u_2^k \qquad\qquad\qquad\qquad\qquad \text{at } x_\Gamma \times (0, T) \end{cases}$$

$$(5.12) \quad \begin{cases} \dfrac{\partial \mathbf{V}_2^{k+1}}{\partial t} + B \dfrac{\partial \mathbf{V}_2^{k+1}}{\partial x} = 0 & \text{in } \Omega_2 \times (0,T) \\[3mm] u_2^{k+1} + \dfrac{2}{\gamma - 1} c_2^{k+1} = u(\varphi_1^k) + \dfrac{2}{\gamma - 1} c(\varphi_1^k) & \text{at } x_\Gamma \times (0,T) \end{cases},$$

together with initial conditions at $t = 0$ and boundary conditions at α and β. The interface condition in Ω_1 can be substituted by

$$\frac{\partial \varphi_1^{k+1}}{\partial t} = H_0 - \frac{1}{2}(u_2^k)^2 - \frac{1}{\gamma - 1}(c_2^k)^2 \qquad \text{at } x_\Gamma \times (0,T),$$

or else by the Dirichlet condition for φ_1^{k+1} obtained by integration in time. The same iterations are used if the direction of the flow is reversed ($-c < u < 0$).

When the flow is supersonic, we have to distinguish between the inflow and the outflow case. Let us first assume that $u > c > 0$. The iterative procedure is defined as follows:

$$(5.13) \quad \begin{aligned} \frac{\partial^2 \varphi_1^{k+1}}{\partial t^2} - [c(\varphi_1^{k+1})^2 \ -u(\varphi_1^{k+1})^2] \frac{\partial^2 \varphi_1^{k+1}}{\partial x^2} \\[2mm] + 2u(\varphi_1^{k+1}) \frac{\partial^2 \varphi_1^{k+1}}{\partial t \partial x} = 0 \qquad \text{in } \Omega_1 \times (0,T) \end{aligned}$$

$$(5.14) \quad \begin{cases} \dfrac{\partial \mathbf{V}_2^{k+1}}{\partial t} + B \dfrac{\partial \mathbf{V}_2^{k+1}}{\partial x} = 0 & \text{in } \Omega_2 \times (0,T) \\[3mm] u_2^{k+1} + \dfrac{2}{\gamma - 1} c_2^{k+1} = u(\varphi_1^k) + \dfrac{2}{\gamma - 1} c(\varphi_1^k) & \text{at } x_\Gamma \times (0,T) \\[3mm] u_2^{k+1} - \dfrac{2}{\gamma - 1} c_2^{k+1} = u(\varphi_1^k) - \dfrac{2}{\gamma - 1} c(\varphi_1^k) & \text{at } x_\Gamma \times (0,T) \end{cases}.$$

Instead, when $u < -c < 0$ we consider the iterative scheme

$$(5.15) \quad \begin{cases} \dfrac{\partial^2 \varphi_1^{k+1}}{\partial t^2} - [c(\varphi_1^{k+1})^2 - u(\varphi_1^{k+1})^2] \dfrac{\partial^2 \varphi_1^{k+1}}{\partial x^2} \\[2mm] \qquad\qquad\qquad + 2u(\varphi_1^{k+1}) \dfrac{\partial^2 \varphi_1^{k+1}}{\partial t \partial x} = 0 & \text{in } \Omega_1 \times (0,T) \\[3mm] \dfrac{\partial \varphi_1^{k+1}}{\partial x} = u_2^k & \text{at } x_\Gamma \times (0,T) \\[3mm] \dfrac{\partial \varphi_1^{k+1}}{\partial t} = H_0 - \dfrac{1}{2}(u_2^k)^2 - \dfrac{1}{\gamma - 1}(c_2^k)^2 & \text{at } x_\Gamma \times (0,T) \end{cases}$$

(5.16)
$$\frac{\partial \mathbf{V}_2^{k+1}}{\partial t} + B \frac{\partial \mathbf{V}_2^{k+1}}{\partial x} = 0 \quad \text{in } \Omega_2 \times (0, T) .$$

The above procedures can also be extended to the case of a flow which is not isentropic in Ω_2. As we already noticed in Section 4.1, in this case the Riemann 'invariants' $u \pm \frac{2}{\gamma-1}c$ are no longer constant along the corresponding characteristic curves. However, we can decide to impose their continuity at the interface. Furthermore, one has to impose the continuity of the Riemann invariant s, the entropy, which is constant along the characteristic curve $C_0 = \{(x(t), t) \,|\, x'(t) = u\}$, when this curve is incoming in Ω_2.

Let us denote by $\hat{s}_1$ the constant entropy in Ω_1, and consider at first the subsonic case $0 < u < c$. The first step of the iterative scheme remains the same of the isentropic case (see (5.11)), while the second step reads (for notation see (4.6), (4.7)):

(5.17)
$$\begin{cases} \dfrac{\partial \mathbf{U}_2^{k+1}}{\partial t} + A \dfrac{\partial \mathbf{U}_2^{k+1}}{\partial x} = 0 & \text{in } \Omega_2 \times (0, T) \\[2mm] u_2^{k+1} + \dfrac{2}{\gamma-1} c_2^{k+1} = u(\varphi_1^k) + \dfrac{2}{\gamma-1} c(\varphi_1^k) & \text{at } x_\Gamma \times (0, T) \\[2mm] s_2^{k+1} = \hat{s}_1 & \text{at } x_\Gamma \times (0, T) , \end{cases}$$

together with initial conditions at $t = 0$ and boundary conditions at β.

If the direction of the flow is reversed ($-c < u < 0$), the eigenvalue u is negative and consequently the characteristic line associated with the entropy s is no longer incoming in Ω_2; the Dirichlet condition for s_2^{k+1} at the interface is therefore dropped out in (5.17). On the other hand, u is not an eigenvalue for the full potential equation, and only one interface condition for the potential φ is needed, for instance the Neumann boundary condition used in (5.11).

When the flow is supersonic and $0 < c < u$, the iteration in Ω_1 is still given by (5.13), while the step related to Ω_2 is defined as follows

(5.18)
$$\begin{cases} \dfrac{\partial \mathbf{U}_2^{k+1}}{\partial t} + A \dfrac{\partial \mathbf{U}_2^{k+1}}{\partial x} = 0 & \text{in } \Omega_2 \times (0, T) \\[2mm] u_2^{k+1} + \dfrac{2}{\gamma-1} c_2^{k+1} = u(\varphi_1^k) + \dfrac{2}{\gamma-1} c(\varphi_1^k) & \text{at } x_\Gamma \times (0, T) \\[2mm] u_2^{k+1} - \dfrac{2}{\gamma-1} c_2^{k+1} = u(\varphi_1^k) - \dfrac{2}{\gamma-1} c(\varphi_1^k) & \text{at } x_\Gamma \times (0, T) \\[2mm] s_2^{k+1} = \hat{s}_1 & \text{at } x_\Gamma \times (0, T) . \end{cases}$$

Instead, when $u < -c < 0$ no characteristic curve is incoming in Ω_2, hence all the three interface conditions in (5.18) must be dropped out. Since u is not an eigenvalue of the full potential equation, we are led to consider the iterative scheme given by (5.15) and

$$
(5.19) \qquad \frac{\partial \mathbf{U}_2^{k+1}}{\partial t} + A \frac{\partial \mathbf{U}_2^{k+1}}{\partial x} = 0 \qquad \text{in } \Omega_2 \times (0, T) .
$$

The well–posedness of all boundary value problems we have considered for the full potential equation can be analysed by the classical energy method or the Kreiss' normal mode analysis as in Oliger and Sundström (1978).

In the two– or three–dimensional case the eigenvalues of the characteristic matrix C_E of the isentropic Euler equations and those of the characteristic matrix C_{FP} of the full potential equation are different, and so are the corresponding 'characteristic' variables defined by means of the matrix of left eigenvectors of C.

Indeed, we have already seen that the eigenvalues for the Euler equations are $\mathbf{u} \cdot \mathbf{n} + c$, $\mathbf{u} \cdot \mathbf{n} - c$ and $\mathbf{u} \cdot \mathbf{n}$ (with multiplicity equal to $d - 1$), whereas for the full potential equation the characteristic matrix C_{FP} is given by (for $d = 3$)

$$
C_{FP} = \begin{pmatrix} 2\mathbf{u} \cdot \mathbf{n} & u_1 \mathbf{u} \cdot \mathbf{n} - c^2 n_1 & u_2 \mathbf{u} \cdot \mathbf{n} - c^2 n_2 & u_3 \mathbf{u} \cdot \mathbf{n} - c^2 n_3 \\ -n_1 & 0 & 0 & 0 \\ -n_2 & 0 & 0 & 0 \\ -n_3 & 0 & 0 & 0 \end{pmatrix} ,
$$

with eigenvalues $\mathbf{u} \cdot \mathbf{n} + c$, $\mathbf{u} \cdot \mathbf{n} - c$ and 0 (with multiplicity equal to $d - 1$).

Besides, a straightforward computation shows that the matrix of left eigenvectors for the Euler equations is

$$
L_E := \begin{pmatrix} c/\rho & n_1 & n_2 & n_3 \\ -c/\rho & n_1 & n_2 & n_3 \\ 0 & \tau_1^{(1)} & \tau_2^{(1)} & \tau_3^{(1)} \\ 0 & \tau_1^{(2)} & \tau_2^{(2)} & \tau_3^{(2)} \end{pmatrix} ,
$$

(see (4.22)), while the one associated with the full potential equation is given by

$$
L_{FP} := \begin{pmatrix} -1/c & -\dfrac{u_1 \mathbf{u} \cdot \mathbf{n} - c^2 n_1}{c(\mathbf{u} \cdot \mathbf{n} + c)} & -\dfrac{u_2 \mathbf{u} \cdot \mathbf{n} - c^2 n_2}{c(\mathbf{u} \cdot \mathbf{n} + c)} & -\dfrac{u_3 \mathbf{u} \cdot \mathbf{n} - c^2 n_3}{c(\mathbf{u} \cdot \mathbf{n} + c)} \\ 1/c & \dfrac{u_1 \mathbf{u} \cdot \mathbf{n} - c^2 n_1}{c(\mathbf{u} \cdot \mathbf{n} - c)} & \dfrac{u_2 \mathbf{u} \cdot \mathbf{n} - c^2 n_2}{c(\mathbf{u} \cdot \mathbf{n} - c)} & \dfrac{u_3 \mathbf{u} \cdot \mathbf{n} - c^2 n_3}{c(\mathbf{u} \cdot \mathbf{n} - c)} \\ 0 & \tau_1^{(1)} & \tau_2^{(1)} & \tau_3^{(1)} \\ 0 & \tau_1^{(2)} & \tau_2^{(2)} & \tau_3^{(2)} \end{pmatrix} ,
$$

Hence the extension of the above algorithms to these multi–dimensional cases is not straightforward and needs some further investigation.

Acknowledgements

This work has been supported by the Swiss National Science Foundation, Project N. 21-54139.98, by MURST 40% and by the C.N.R. Special Project "Mathematical Methods in Fluid Dynamics and Molecular Dynamics.

References

1. Berkman, M.E. and Sankart, L.N. (1997). Navier-Stokes/full potential/free-wake method for rotor flows. *Journal of Aircraft*, **34**, pp. 635–640.
2. Bjørhus, M. (1995). *On domain decomposition, subdomain iteration, and waveform relaxation*. Dr.Ing. Thesis, Department of Mathematical Sciences, University of Trondheim.
3. Coclici, C.A. (1998). *Domain decomposition methods and far–field boundary conditions for two–dimensional compressible flows around airfoils*. Ph.D. Thesis, University of Stuttgart.
4. Hirsch, C. (1990). *Numerical computation of internal and external flows. Volume 2: Computational methods for inviscid and viscous flows*. John Wiley & Sons, Chicester.
5. Landau, L.D. and Lifshitz, E.M. (1959). *Fluid mechanics*. Pergamon Press, Oxford. Russian 2nd edn: Gosudarstv. Izdat. Tehn.-Teor. Lit., Moscow, 1953.
6. Lesieur, M. (1997). *Turbulence in fluids* (3rd edn). Kluwer, Dordrecht.
7. Oliger, J. and Sundström, A. (1978). Theoretical and practical aspects of some initial boundary value problems in fluid dynamics. *SIAM J. Appl. Math.*, **35**, pp. 419–446.
8. Quarteroni, A. (1990). Domain decomposition methods for systems of conservation laws: spectral collocation approximations. *SIAM J. Sci. Stat. Comput.*, **11**, pp. 1029–1052.
9. Quarteroni, A. and Stolcis, L. (1995). Heterogeneous domain decomposition for compressible flows. In *Proceedings of the ICFD conference on numerical methods for fluid dynamics*, M. Baines and W.K. Morton eds., Oxford University Press, Oxford, pp. 113–128.
10. Quarteroni, A. and Valli, A. (1999). *Domain decomposition methods for partial differential equations*. Oxford University Press, Oxford (in press).

ERROR CONTROL IN FINITE ELEMENT COMPUTATIONS
An introduction to error estimation and mesh-size adaptation

R. RANNACHER
Institut für Angewandte Mathematik
Universität Heidelberg
INF 293, D-69120 Heidelberg, Germany
e-mail: rannacher@iwr.uni-heidelberg.de
URL: http://gaia.iwr.uni-heidelberg.de

Summary. We present a general paradigm for a posteriori error control and adaptive mesh design in finite element Galerkin methods. The conventional strategy for controlling the error in finite element methods is based on a posteriori estimates for the error in the global energy or L^2-norm involving local residuals of the computed solution. Such estimates contain constants describing the local approximation properties of the finite element spaces and the stability properties of a linearized dual problem. The mesh refinement then aims at the equilibration of the local error indicators. However, meshes generated via controlling the error in a global norm may not be appropriate for local error quantities like point values or line integrals and in case of strongly varying coefficients. This deficiency may be overcome by introducing certain weight-factors in the a posteriori error estimates which depend on the dual solution and contain information about the relevant error propagation. This way, optimally economical meshes may be generated for various kinds of error measures. This is systematically developed first for a simple model case and then illustrated by results for more complex problems in fluid mechanics, elasto-plasticity and radiative transfer.

Recommended literature and references: The basics on the mathematical theory of finite element methods used in this paper can be found in the books of Johnson [14] and Brenner, Scott [10]. An introduction into the general concept of residual-based error control for finite element methods has been given in the survey article by Eriksson, Estep, Hansbo, Johnson [11], and with some modifications in the papers by Becker, Rannacher [6], [7]. Surveys of the traditional approach to a posteriori error estimation are given by Verfürth [23] and Ainsworth, Oden [1]. The material of this paper has mainly been collected from the papers of Becker, Rannacher [6], [7], [8], Becker [4], Führer, Kanschat [12], Kanschat [17], Rannacher, Suttmeier [19], [20], Becker, Braack, Rannacher [5], and Rannacher [18]. For results of computations for special applications, we refer to the PhD theses of Becker [3] (incompressible flows), Kanschat [16] (radiative transfer problems), Suttmeier [22] (elasto-plasticity problems), and Braack [9], Waguet [24] (flows with chemical reactions).

H. Bulgak and C. Zenger (eds.), Error Control and Adaptivity in Scientific Computing, 247–278.

1 Introduction

In this chapter, first, we introduce a general concept for a posteriori error estimation and mesh-size selection in finite element Galerkin methods, and then describe this approach for a simple model situation. The goal is to develop techniques for reliable estimation of the discretization error for quantities of physical interest and based on this criteria for mesh adaptation. The use of Galerkin discretization provides the appropriate framework for a mathematically rigorous a posteriori error analysis.

1.1 Discretization error

The total discretization error in a mesh cell T splits into two components, the locally produced error (truncation error) and the transported error (pollution error)

$$E_T^{tot} = E_T^{loc} + E_T^{trans}. \tag{1.1}$$

The effect of the cell residual ϱ_T on the local error $E_{T'}$, at another cell T', is essentially governed by the Green's function of the continuous problem. This is the general philosophy underlying our approach to error control.

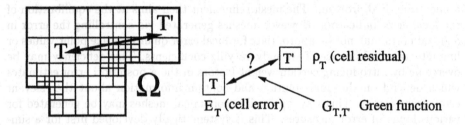

Figure 1: Scheme of error propagation

This rises the following questions:

1) How can we detect and use the interplay of the various error propagation effects for the design of economical meshes in solving coupled problems: Given N cells, what is the "best" distribution of grid cells?

2) How can we achieve a posteriori accuracy control for quantities of physical interest (e.g., drag/lift in flows around bodies, mean values of species concentrations in chemically reacting flows, point values of stresses in loaded bodies, etc.)?

We know from a priori error analysis that in finite element approximations the error propagation (information transport) is governed by different mechanisms related to the physical features of the problem. An effective error estimation has to take these properties into account.

- *Diffusion:* The term $-\varepsilon \Delta u$ causes slow isotropic decay, but global error pollution may occur from local irregularities.

- *Transport:* The term $\beta \cdot \nabla u$ leads to "no" decay in β-direction, but exponential error decay occurs in $\beta^{\perp}$-direction.

- *Reaction:* The term $\partial_t u + \alpha u$ leads to isotropic exponential decay, but "stiff" behavior may occur in some components.

1.2 Principles of error estimation

I) A priori error analysis: The classical a priori error estimation aims at estimating the error to be expected in a computation to be done. These bounds are expressed in terms of powers of a mesh size h and involve constants which depend on the (unknown) exact solution. In this way, only asymptotic (as $h \to 0$) information about the error behavior is provided, but no quantitatively useful error bound. In particular, no criterion for local mesh adaptation is obtained.

II) A posteriori error analysis: The local a posteriori error analysis generates error estimates in the course of the computation. Accordingly, these bounds are in terms of computable local residuals of the approximate solution and do not require information about the exact solution. However, a posteriori error analysis usually does not provide a priori information about the convergence as $h \to 0$ of the discretization.

We illustrate the basic principles underlying error estimation by considering the perturbation of linear algebraic systems. Let $A, \tilde{A} \in \mathbb{R}^{n \times n}$, $b, \tilde{b} \in \mathbb{R}^n$ be given and solve $Ax = b$, $\tilde{A}\tilde{x} = \tilde{b}$ (perturbed problem). For estimating the error $e := x - \tilde{x}$, there are several approaches. The a priori method uses the "truncation error" $\tau := \tilde{A}x - \tilde{b} = \tilde{A}(x - \tilde{x})$,

$$e = \tilde{A}^{-1}\tau \quad \Rightarrow \quad \|e\| \leq \tilde{c}_s \|\tau\|, \tag{1.2}$$

with the "discrete" stability constant $\tilde{c}_s := \|\tilde{A}^{-1}\|$. The a posteriori method uses the "residual" (or "defect") $\varrho := b - A\tilde{x} = A(x - \tilde{x})$,

$$e = A^{-1}\varrho \quad \Rightarrow \quad \|e\| \leq c_s \|\varrho\|, \tag{1.3}$$

with the "continuous" stability constant $c_s := \|A^{-1}\|$. Alternatively, we may use the solution z of the "dual problem" $A^* z = \|e\|^{-1} e$, to obtain

$$\|e\| = (e, A^* z) = (b - A\tilde{x}, z) = (\varrho, z) \leq \|\varrho\| \|z\| \leq c_s^* \|\varrho\|, \tag{1.4}$$

with the "dual" stability constant $c_s^* := \|A^{*-1}\|$. Of course, this approach does not yield a new result in estimating the error in the l_2-norm. But it also gives us the possibility to bound other error quantities, e.g., single error components e_i. Below, we will use this duality technique for generating a posteriori error estimates in finite element Galerkin methods for differential equations.

An analogous argument can also be applied in the case of nonlinear equations. Let $F, \tilde{F} : \mathbb{R}^n \to \mathbb{R}^n$ be (differentiable) vector functions and solve $F(x) = 0$ and $\tilde{F}(\tilde{x}) = 0$ *(perturbed problem)*. Then, the residual $\varrho := -F(\tilde{x})$ satisfies

$$\varrho = F(x) - F(\tilde{x}) = \left(\int_0^1 F'(\tilde{x} + se) \, ds \right) e =: L(x, \tilde{x})e, \qquad (1.5)$$

with the Jacobian F'. The term in parentheses defines a linear operator $L(x, \tilde{x}) : \mathbb{R}^n \to \mathbb{R}^n$ which depends on the (unknown) solution x. It follows that $\|e\| \leq c_s \|\varrho\|$, with the (nonlinear) stability constant $c_s := \|L(x, \tilde{x})^{-1}\|$.

1.3 A model problem

For illustration, we consider the model problem

$$-\Delta u = f \quad \text{in } \Omega, \quad u = 0 \text{ on } \partial\Omega. \qquad (1.6)$$

The variational formulation of this problem uses the function space $V := H_0^1(\Omega)$ and the L^2 product $(\cdot, \cdot)$,

$$u \in V : \qquad (\nabla u, \nabla \varphi) = (f, \varphi) \quad \forall \varphi \in V. \qquad (1.7)$$

The finite element approximation uses subspaces $V_h := \{v \in V : v_{|T} \in Q_1(T), \, T \in \mathbb{T}_h\}$ defined on decompositions $\mathbb{T}_h$ of Ω into quadrilaterals T ("cells") of width $h_T := diam(T)$. Furthermore, we write $h := \max_{T \in \mathbb{T}_h} h_T$ for the *global* mesh width. The discrete problem reads

$$u_h \in V_h : \qquad (\nabla u_h, \nabla \varphi_h) = (f, \varphi_h) \quad \forall \varphi_h \in V_h. \qquad (1.8)$$

The essential feature of this approximation scheme is the "Galerkin orthogonality" of the error $e := u - u_h$,

$$(\nabla e, \nabla \varphi_h) = 0, \quad \varphi_h \in V_h. \qquad (1.9)$$

a) A priori error analysis: We begin with a brief discussion of the a priori error analysis for the scheme (1.8). With the nodal interpolant $I_h u \in V_h$, there holds

$$\|\nabla(u - I_h u)\|_T \leq c_i h_T \|\nabla^2 u\|. \qquad (1.10)$$

This yields an a priori error estimate in the "energy norm",

$$\|\nabla e\| = \inf_{\varphi_h \in V_h} \|\nabla(u - \varphi_h)\| \leq c_i h \|\nabla^2 u\|. \qquad (1.11)$$

Employing a duality argument ("Aubin-Nitsche trick"),

$$-\Delta z = \|e\|^{-1} e \quad \text{in } \Omega, \quad z = 0 \text{ on } \partial\Omega, \qquad (1.12)$$

we obtain the improved L^2-error estimate

$$\|e\| = (e, -\Delta z) = (\nabla e, \nabla z) = (\nabla e, \nabla(z - I_h z)) \leq c_i c_s h \|\nabla e\|, \qquad (1.13)$$

where the "stability constant" c_s is defined by the a priori bound $\|\nabla^2 z\| \leq c_s$.

b) A posteriori error analysis: Next, we seek to derive a posteriori error estimates. Let $J(\cdot)$ be an arbitrary "error functional" defined on V, and $z \in V$ the solution of the corresponding dual problem

$$(\nabla \varphi, \nabla z) = J(\varphi) \quad \forall \varphi \in V. \qquad (1.14)$$

Taking $\varphi = e$ in (1.14) and using the Galerkin orthogonality, cell-wise integration by parts results in the error representation

$$
\begin{aligned}
J(e) &= (\nabla e, \nabla z) = (\nabla e, \nabla(z - I_h z)) \\
&= \sum_{T \in T_h} \left\{ (-\Delta u + \Delta u_h, z - I_h z)_T - (\partial_n u_h, z - I_h z)_{\partial T} \right\} \qquad (1.15) \\
&= \sum_{T \in T_h} \left\{ (f + \Delta u_h, z - I_h z)_T - \tfrac{1}{2}([\partial_n u_h], z - I_h z)_{\partial T} \right\},
\end{aligned}
$$

where $[\partial_n u_h]$ is the jump of $\partial_n u_h$ across the inter-element boundaries. This gives us the a posteriori error estimate

$$|J(e)| \leq \eta(u_h) := \sum_{T \in T_h} \alpha_T \varrho_T(u_h) \omega_T(z), \qquad (1.16)$$

with the cell parameters $\alpha_T = c_i h_T^4$, the cell residuals

$$\varrho_T(u_h) := h_T^{-1} \|f + \Delta u_h\|_T + \tfrac{1}{2} h_T^{-3/2} \|[\partial_n u_h]\|_{\partial T},$$

and the weights

$$\omega_T(z) := \max \left\{ h_T^{-3} \|z - I_h z\|_T, h_T^{-5/2} \|z - I_h z\|_{\partial T} \right\}.$$

The interpretation of this relation is that the weights $\omega_T(z)$ describe the dependence of $J(e)$ on variations of the cell residuals $\varrho_T(u_h)$,

$$\frac{\partial J(e)}{\partial \varrho_T} \approx \alpha_T \omega_T(z) \approx h_T^{-2} \max_T |z - I_h z| \approx \max_T |\nabla^2 z|.$$

We remark that in a finite difference discretization of the model problem (1.6) the corresponding "influence factors" behave like $\omega_T(z) \approx h_T^{-2} \max_T |z|$.

In practice the weights $\omega_T(z)$ have to be determined computationally. Let $z_h \in V_h$ be the finite element approximation of z,

$$(\nabla \varphi_h, \nabla z_h) = J(\varphi_h) \quad \forall \varphi_h \in V_h. \qquad (1.17)$$

We can estimate

$$\omega_T(z) \leq c_i h_T^{-1} \|\nabla^2 z\|_T \approx c_i \max_T |\nabla_h^2 z_h|, \qquad (1.18)$$

where $\nabla_h^2 z_h$ is a suitable difference quotient approximating $\nabla^2 z$. The interpolation constant is usually in the range $c_i \approx 0.1 - 1$ and can be determined by calibration. Alternatively, we may construct from $z_h \in V_h$ a patchwise biquadratic interpolation $I_h^{(2)} z_h$ and replace $z - I_h z$ in the weight $\omega_T(z)$ by $I_h^{(2)} z_h - z_h$. This gives an approximation to $\omega_T(z)$ which is free of any interpolation constant. The quality of these approximations for the model problem will be discussed below.

By the same type of argument, we can also derive the traditional global error estimates in the energy and the L^2-norm.

i) Energy-norm error bound: Using the functional $J(\varphi) := \|\nabla e\|^{-1}(\nabla e, \nabla \varphi)$ in the dual problem, we obtain the estimate

$$\|\nabla e\| \leq \sum_{T \in T_h} \alpha_T \, \varrho_T(u_h) \, \omega_T(z) \leq c_i \sum_{T \in T_h} h_T^2 \, \varrho_T(u_h) \, \|\nabla z\|_{\tilde{T}}, \qquad (1.19)$$

where $\tilde{T}$ is the union of all cells neighboring T. In view of the a priori bound $\|\nabla z\| \leq c_s = 1$, this implies the a posteriori error estimate

$$\|\nabla e\| \leq \eta_E := c_i c_s \Big(\sum_{T \in T_h} h_T^4 \, \varrho_T(u_h)^2 \Big)^{1/2}. \qquad (1.20)$$

ii) L^2-norm error bounds: Using the functional $J(\varphi) := \|e\|^{-1}(e, \varphi)$ in the dual problem, we obtain the estimate

$$\|e\| \leq \sum_{T \in T_h} \alpha_T \, \varrho_T(u_h) \, \omega_T(z) \leq c_i \sum_{T \in T_h} h_T^3 \, \varrho_T(u_h) \, \|\nabla^2 z\|_T. \qquad (1.21)$$

In view of the a priori bound $\|\nabla^2 z\| \leq c_s$ ($c_s = 1$, if Ω is convex), this implies the a posteriori error bound

$$\|e\| \leq \eta_{L^2} := c_i c_s \Big(\sum_{T \in T_h} h_T^6 \, \varrho_T(u_h)^2 \Big)^{1/2}. \qquad (1.22)$$

2 Model case analysis

In the following, we want to investigate the mechanism underlying a posteriori error estimation and mesh adaptation as introduced above in some more detail.

2.1 Strategies for adaptive mesh design

We use the notation introduced above: u is the solution of the variational problem posed on a d-dimensional domain Ω, u_h is its finite element approximation of order $m = 2$ (piecewise P_1- or Q_1-shape functions), and $\varrho_T(u_h)$ is the corresponding residual on mesh cell T. Further, $e = u - u_h$ is the discretization error and $J(\cdot)$ a linear error functional for measuring e. We suppose that there holds an a posteriori error estimate of the form

$$|J(e)| \leq \eta := \sum_{T \in \mathbb{T}_h} \alpha_T \, \varrho_T(u_h) \, \omega_T(z), \qquad (2.1)$$

with the local mesh parameters and weights $\alpha_T := h_T^{d+2}$, $\omega_T(z)$, and the local error indicators $\eta_T := \alpha_T \, \varrho_T(u_h) \, \omega_T(z)$. The mesh design strategies are oriented towards a prescribed tolerance TOL for the error quantity $J(e)$ and the number of mesh cells N which measures the complexity of the computational model. Usually the admissible complexity is constrained by some maximum value $N_{\max}$.

There are various strategies for organizing a mesh adaptation process on the basis of the a posteriori error estimate (2.1).

i) Error balancing strategy: Cycle through the mesh and equilibrate the local error indicators,

$$\eta_T \approx \frac{TOL}{N} \quad \Rightarrow \quad \eta \approx TOL.$$

This process requires iteration with respect to the number of mesh cells N.

ii) Fixed fraction strategy: Order cells according to the size of η_T and refine a certain percentage (say 30%) of cells with largest η_T and coarsen those cells with smallest η_T. By this strategy, we may achieve a prescribed rate of increase of N (or keep it constant as desirable in nonstationary computations).

iii) Optimized mesh strategy: Use the representation

$$\eta := \sum_{T \in \mathbb{T}_h} \alpha_T \, \varrho_T(u_h) \, \omega_T(z) \approx \int_\Omega h(x)^2 A(x) \, dx$$

for generating a formula of an optimal mesh-size distribution $h_{\mathrm{opt}}(x)$.

We want to discuss the strategy for deriving an optimal mesh-size distribution in more detail. As a side-product, we will also obtain the justification of the error equilibration strategy. Let $N_{\max}$ and TOL be prescribed. We assume that for $TOL \to 0$, the cell residuals and the weights approach certain limits,

$$\varrho_T(u_h) = h_T^{-d/2}\|f + \Delta u_h\|_T + \tfrac{1}{2}h_T^{-(d-1)/2}\|h_T^{-1}[\partial_n u_h]\|_{\partial T} \;\to\; |D^2 u(x_T)|,$$

$$\omega_T(z) = \max\left\{ h_T^{-2-d/2}\|z - I_h z\|_T,\, h_T^{-3/2-d/2}\|z - I_h z\|_{\partial T} \right\} \;\to\; |D^2 z(x_T)|.$$

These properties can be proven on uniformly refined meshes by exploiting super-convergence effects, but still need theoretical justification on locally refined meshes as constructed by the strategies described above. This suggests the relations

$$\eta \approx \int_\Omega h(x)^2 A(x)\,dx, \qquad N = \sum_{T \in \mathcal{T}_h} h_T^d h_T^{-d} \approx \int_\Omega h(x)^{-d}\,dx. \tag{2.2}$$

Consider the mesh optimization problem $\eta \to min!$, $N \leq N_{\max}$. Applying the usual Lagrange approach yields

$$\frac{d}{dt}\left[\int_\Omega (h + t\varphi)^2 A\,dx + (\lambda + t\mu)\Big((h + t\mu)^{-d}\,dx - N_{\max}\Big)\right]_{t=0} = 0,$$

implying

$$2h(x)A(x) - d\lambda h(x)^{-d-1} = 0, \qquad \int_\Omega h(x)^{-d}\,dx - N_{\max} = 0.$$

Consequently,

$$h(x) = \left(\frac{2}{\lambda d}A(x)\right)^{-1/(2+d)} \quad \Rightarrow \quad \eta \approx h^{2+d}A = \frac{2}{\lambda d},$$

and

$$\left(\frac{2}{d\lambda}\right)^{d/(2+d)} \int_\Omega A(x)^{d/(2+d)}\,dx = N_{\max}, \quad W := \int_\Omega A(x)^{d/(2+d)}\,dx.$$

From this, we infer a formula for the "optimal" mesh-size distribution,

$$\lambda = \frac{2}{d}\left(\frac{W}{N_{\max}}\right)^{(2+d)/d} \quad \Rightarrow \quad h_{\text{opt}}(x) \equiv \left(\frac{W}{N_{\max}}\right)^{1/d} A(x)^{-1/(2+d)}. \tag{2.3}$$

In an analogous way, we can also treat the adjoint optimization problem $N \to min!$, $\eta \leq TOL$. We note that for "regular" functionals $J(\cdot)$ the quantity W is bounded, e.g.,

$$J(e) = \partial_i e(0) \quad \Rightarrow \quad A(x) \approx |x|^{-d-1} \quad \Rightarrow \quad W = \int_\Omega |x|^{d/(d+2)-d}\,dx < \infty.$$

Hence, the optimization approach is valid. However, the evaluation of *hyper-singular* error functionals (e.g., higher derivatives) may require regularization.

2.2 Computational tests

For our computational test, we consider again the model problem

$$-\Delta u = f \quad \text{in } \Omega, \quad u = 0 \text{ on } \partial\Omega, \tag{2.4}$$

defined on the rectangular domain $\Omega = (-1, 1) \times (-1, 3)$ with slit at $(0, 0)$. In the presence of a reentrant corner, here a slit, with angle $\omega = 2\pi$, the solution involves a "corner singularity", i.e., it can be written in the form $u = \psi r^{1/2} + \tilde{u}$, with r being the distance to the corner point and $\tilde{u} \in H^2(\Omega)$. We want to illustrate how the singularity introduced by the weights interacts with the pollution effect caused by the slit singularity. Let the point value $J(u) = u(P)$ to be computed at $P = (0.75, 2.25)$. In this case the dual solution z behaves like the Green's function,

$$|\nabla^2 z(x)| \approx d(x)^{-2} + r(x)^{-3/2}.$$

This implies that

$$|e(P)| \approx c_i \sum_{T \in T_h} h_T^4 \varrho_T(u_h) \left\{ d_T^{-2} + r_T^{-3/2} \right\}, \tag{2.5}$$

and, consequently, $N_{\text{opt}} \approx TOL^{-1}$, as suggested by a priori analysis.

Next, we evaluate derivative values, $J(u) = \partial_1 u(P)$. In this case the dual solution z behaves like the derivative Green's function

$$|\nabla^2 z(x)| \approx d(x)^{-3} + r(x)^{-3/2}.$$

This implies that

$$|\partial_1 e(P)| \approx c_i \sum_{T \in T_h} h_T^4 \varrho_T(u_h) \left\{ d_T^{-3} + r_T^{-3/2} \right\}. \tag{2.6}$$

Equilibrating the local error indicators yields

$$\eta_T \approx \frac{h_T^4}{d_T^3} \approx \frac{TOL}{N} \quad \Rightarrow \quad h_T^2 \approx d_T^{3/2} \left(\frac{TOL}{N} \right)^{1/2},$$

and, consequently,

$$N = \sum_{T \in T_h} h_T^2 h_T^{-2} = \left(\frac{N}{TOL} \right)^{1/2} \sum_{T \in T_h} h_T^2 d_T^{-3/2} \approx \left(\frac{N}{TOL} \right)^{1/2}.$$

This implies that again $N_{\text{opt}} \approx TOL^{-1}$. We note that in this case, the dual solution does not exist in the sense of $H_0^1(\Omega)$, such that for practical use, we have to regularize the functional $J(u) = \partial_1 u(P)$. Notice that the energy estimator leads to a mesh efficiency like $J(e) \sim N^{-1/2}$. This predicted asymptotic behavior is well confirmed by the results shown in Figure 2. Second derivatives $J(u) = \partial_1^2 u(P)$ can also be calculated on $N_{\text{opt}} \approx TOL^{-1}$ mesh cells to accuracy TOL. For more details, we refer to [7]. In Figure 3, we show optimized meshes generated by the weighted error estimator compared to the standard energy-error estimator. We see that the former one concentrates mesh cells at the evaluation point but also

256

at the slit in order to suppress the pollution effect of the corner singularity, while the energy-norm estimator induces a significantly stronger refinement at the slit.

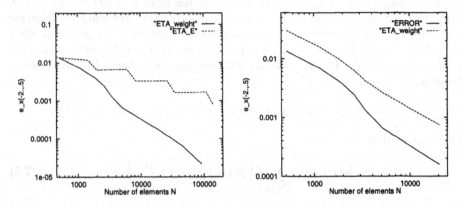

Figure 2: Comparison of efficiency between η_E and η_{weight} on the slit domain

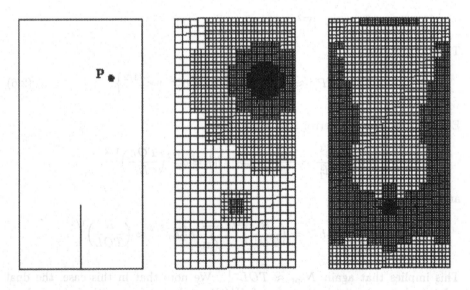

Figure 3: Refined meshes with about $5,000$ elements for computing $\partial_1 u(P)$ using the weighted error estimator η_{weight} (middle) and the energy error estimator η_E (right)

At the end of this introductory discussion, we pose an exercise which illustrates the essential points of our approach to a posteriori error estimation. For the model problem (2.4), we consider the functional

$$J(u) := \int_{\partial\Omega} \partial_n u \, ds \quad \left(= \int_{\Omega} f \, dx \right).$$

What is an optimal mesh-size distribution for computing $J(u)$? The answer is based on the observation that the corresponding dual problem

$$(\nabla\varphi, \nabla z)\text{"} = \text{"} \int_{\partial\Omega} \partial_n \varphi \, ds \quad \forall\varphi \in V \cup W^{2,1}(\Omega),$$

has a measure solution of the form $z\text{"} \equiv \text{"}1$ in Ω, $z = 0$ on $\partial\Omega$. In order to avoid the use of measures, we may use regularization. Setting $\varepsilon = TOL$ and $S_\varepsilon := \{x \in \Omega, dist\{x, \partial\Omega\} < \varepsilon\}$, we have

$$J_\varepsilon(\varphi) = \frac{1}{\varepsilon} \int_{S_\varepsilon} \partial_r \varphi \, dx \quad \rightarrow \quad \int_{\partial\Omega} \partial_r \varphi \, ds = J(\varphi) \quad (\varepsilon \to 0).$$

The corresponding regularized dual solution is

$$z_\varepsilon = 1 \quad \text{in } \Omega \setminus S_\varepsilon, \quad z_\varepsilon(x) = \frac{1}{\varepsilon}(1 - |x|) \quad \text{on } S_\varepsilon.$$

This implies that

$$J_\varepsilon(e) = \sum_{T \in T_h} h_T^{-1} \varrho_T(u_h) \|\nabla^2 z_\varepsilon\|_T \approx \sum_{T \subset S_\varepsilon} \cdots,$$

i.e., there is no contribution to the error from the interior of Ω. Hence, the optimal strategy is to refine the elements adjacent to the boundary and to leave the others unchanged, assuming that the force term f is integrated exactly.

2.3 Theoretical and experimental backup

In the following, we discuss the practical evaluation of the a posteriori error bounds developed above in some more detail.

i) The best error bound can be obtained by direct evaluation of the identity

$$J(e) = \eta(u_h) := \sum_{T \in T_h} \left\{ (f + \Delta u_h, z - I_h z)_T - \frac{1}{2}([\partial_n u_h], z - I_h z)_{\partial T} \right\},$$

without separating its components by use of the triangle inequality. We introduce the "effectivity index" $I_{\text{eff}} := |J(e)/\eta(u_h)|$. In [7] three possibilities for evaluating $\eta(u_h)$ have been compared:

a) Approximating $z \approx z_h^{(2)}$ by its *biquadratic Ritz projection* actually yields the asymptotically optimal behavior $\lim_{TOL \to 0} I_{\text{eff}} = 1$.

b) Approximating $z \approx I_h^{(2)} z_h$ by patchwise (four cells) *biquadratic interpolation* of the bilinear Ritz projection z_h yields $\lim_{TOL \to 0} I_{\text{eff}} < 1$ ($\approx 0.5 - 0.9$).

c) Approximating $z - I_h z \approx c_{i,T} h_T^2 \nabla_h^2 z_h$ by appropriate second-order finite difference quotients yields $\lim_{TOL \to 0} I_{\text{eff}} < 1$ ($\approx 0.5 - 0.9$).

The two "cheap" procedures (b) and (c) have an acceptable asymptotic accuracy and are used in most of the computations described below. Notice that procedure (b) does not require to specify any interpolation constants c_i.

ii) For deriving mesh refinement criteria, we may use the a posteriori error estimate

$$|J(e)| \le c_i \sum_{T \in T_h} h_T^{2+d} \varrho_T(u_h) \max_T |\nabla_h^2 z_h|,$$

derived by approximating $h_T^{-1} \|\nabla^2 z\|_T \approx \max_T |\nabla_h^2 z_h|$, with the finite element approximation $z_h \in V_h$. One may try to further improve the quality of the error estimate by solving local (patchwise) defect equations, either Dirichlet problems (à la Babuška, Miller) or Neumann problems (à la Bank, Weiser), for details we refer to [2]. References for these approaches are Verfürth [23] and Ainsworth, Oden [1]. Comparison with simpler mesh adaptation techniques, e.g., refinement criteria based on difference quotients of the computed solution, local gradient recovery techniques "ZZ-technique" (à la Zienkiewicz, Zhu [25]), or other local "ad hoc" criteria will be given in some of the applications below.

3   Extensions

So far, we have considered error estimation only for a simple model problem. In the following section, we will discuss various extensions of this approach to other types of partial differential problems. Finally, we will prepare for its application to general nonlinear situations.

3.1   More general elliptic problems

We consider a diffusion problem with variable coefficients,

$$-\nabla \cdot (a(x)\nabla u) = f \quad \text{in } \Omega, \quad u = 0 \quad \text{on } \partial\Omega.$$

The related dual problem for estimating the L^2 error is

$$-\nabla \cdot (a(x)\nabla z) = \|e\|^{-1} e \quad \text{in } \Omega, \quad z = 0 \quad \text{on } \partial\Omega.$$

Recalling a posteriori error estimates (1.20) and (1.22), we obtain:

i) Standard L^2-error estimate:

$$\|e\| \le c_i c_s \left(\sum_{T \in T_h} h_T^6 \varrho_T(u_h)^2 \right)^{1/2}, \quad c_s \approx \|\nabla_h^2 z_h\|, \tag{3.1}$$

with the cell residuals $\varrho_T(u_h) := h_T^{-1}\|f + \nabla \cdot (a\nabla u_h)\|_T + \frac{1}{2}h_T^{-3/2}\|[n \cdot a\nabla u_h]\|_{\partial T}.$

ii) Weighted L^2-error estimate:

$$\|e\| \leq \sum_{T\in\mathbb{T}_h} h_T^3 \, \varrho_T(u_h)\, \omega_T(z), \quad \omega_T(z) \approx c_i\|\nabla_h^2 z_h\|_T, \qquad (3.2)$$

with the same cell residuals as above.

The quality of these two a posteriori error estimates is compared for a test case defined on the square $\Omega = (-1,1) \times (-1,1)$ with the coefficient function $a(x) = 0.1 + e^{3(x_1+x_2)}$. A reference solution is generated by a computation on a very fine mesh. Figure 4 shows the results for computations using the standard L^2-error estimate ($\tilde{\eta}_2(u_h)$) and the weighted L^2-error estimate ($\tilde{\eta}_{weight}(u_h)$), with approximate evaluation of $J(\cdot)$ by taking $\tilde{e} := I_h^{(2)}u_h - u_h$. The weights $\tilde{\omega}_K$ are computed by taking second difference quotients of the discrete dual solution $z_h \in V_h$, the interpolation constant being chosen as $c_i = 0.2$.

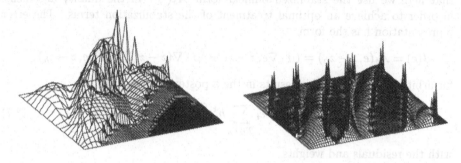

Figure 4: Errors on meshes with $N \sim 10000$ obtained by the L^2-error estimator $\tilde{\eta}_2(u_h)$ (left, scaled by 1:30) and the weighted error estimator $\tilde{\eta}_{weight}^{approx}(u_h)$ (right, scaled by 1:10)

3.2 Stationary transport problems

As a simple model, we consider the scalar transport problem

$$\beta \cdot \nabla u = f \quad \text{in } \Omega, \quad u = g \text{ on } \Gamma_+, \qquad (3.3)$$

where $\Gamma_+ = \{x \in \partial\Omega, \, n \cdot \beta \leq 0\}$ is the "inflow boundary". Accordingly, $\Gamma_- = \partial\Omega \setminus \Gamma_+$ is the "outflow boundary". This problem is discretized using the finite element Galerkin method with least-squares or streamline diffusion stabilization (called SDFEM). On regular quadrilateral meshes $\mathbb{T}_h$ (one irregular "hanging" node is allowed per edge), we define again subspaces $V_h = \{v \in H^1(\Omega), \, v_{|T} \in \tilde{Q}_1(T), \, T \in \mathbb{T}_h\}$, where $\tilde{Q}_1$ means the space of "isoparametric" bilinear functions on cell T. The discrete solution $u_h \in V_h$ is defined by

$$(\beta \cdot \nabla u_h, \Phi) - (\beta_n u_h, \varphi)_+ = (f, \Phi) - (\beta_n g, \varphi)_+ \quad \forall \varphi \in V_h, \qquad (3.4)$$

where $\Phi := \varphi + \delta\beta \cdot \nabla\varphi$, and the stabilization parameter is determined locally by $\delta_T = \kappa h_T$. In the formulation (3.4) the inflow boundary condition is imposed in the weak sense. This facilitates the use of a duality argument in generating a posteriori error estimates. The right-hand and left-hand side of (3.4) define a bilinear form $A_\delta(\cdot, \cdot)$ and a linear form $F_\delta(\cdot)$, respectively. Using this notation, (3.4) may be written as

$$A_\delta(u_h, \varphi) = F_\delta(\varphi) \quad \forall \varphi \in V_h. \tag{3.5}$$

Let $J(\cdot)$ be a given functional defined on V with respect to which the error $e = u - u_h$ is to be controlled. Following our general approach, we consider the corresponding dual problem

$$A_\delta(\varphi, z) = J(\varphi) \quad \forall \varphi \in V, \tag{3.6}$$

which is a transport problem with transport in the negative β-direction. We note that here we use the stabilized bilinear form $A_\delta(\cdot, \cdot)$ in the duality argument, in order to achieve an optimal treatment of the stabilization terms. The error representation has the form

$$J(e) = A_\delta(e, z - z_h) = (\beta \cdot \nabla e, z - z_h + \delta\beta \cdot \nabla(z - z_h)) - (\beta_n e, z - z_h)_+,$$

for arbitrary $z_h \in V_h$. This results in the a posteriori error estimate

$$|J(e)| \leq \eta(u_h) := c_i \sum_{T \in \mathcal{T}_h} h_T^4 \left\{ \varrho_T^{(1)} \omega_T^{(1)} + \varrho_T^{(2)} \omega_T^{(2)} \right\}, \tag{3.7}$$

with the residuals and weights

$$\varrho_T^{(1)} = h_T^{-1} \| f - \beta \cdot \nabla u_h \|_T, \quad \omega_T^{(1)} = h_T^{-3} \{ \| z - z_h \|_T + \delta_T \| \beta \cdot \nabla(z - z_h) \|_T \},$$

$$\varrho_T^{(2)} = h_T^{-3/2} \| \beta_n(u_h - g) \|_{\partial T \cap \Gamma_+}, \quad \omega_T^{(2)} = h_T^{-5/2} \| z - z_h \|_{\partial T \cap \Gamma_+}.$$

This a posteriori error bound explicitly contains the mesh size h_T and the stabilization parameter δ_T as well. This gives us the possibility to simultaneously adapt both parameters, which may be particularly advantageous in capturing sharp layers in the solution.

We want to illustrate the performance of the error estimator (3.7) by two thought experiments. Let $f = 0$. First, we take the functional

$$J(e) := (1, \beta_n e)_-.$$

The corresponding dual solution is $z \equiv 1$, so that $J(e) = 0$. This reflects the global conservation property of the SDFEM. Next, we set

$$J(e) := (1, e) + (1, \delta\beta_n e)_-.$$

The corresponding dual problem reads

$$(-\beta \cdot \nabla z, \varphi - \delta\beta \cdot \nabla\varphi) + (\beta_n z, \varphi)_- = (1, \varphi) + (\delta 1, \beta_n \varphi)_-.$$

Assuming that $\delta \equiv const.$, this dual problem has the same solution as

$$-\beta \cdot \nabla z = 1 \quad \text{in } \Omega, \quad z = \delta \text{ on } \Gamma_-.$$

Consequently, z is linear almost everywhere, i.e., the weights in the a posteriori bound (3.7) are non-zero only along the characteristic line $\{x \in \Omega, x_1 = x_2\}$. Therefore, the mesh refinement will be restricted to this critical region although the cell residuals $\varrho_T(u_h)$ may be non-zero everywhere.

3.3 Parabolic problems

We consider the parabolic problem (diffusion or heat-transfer problem)

$$u_t - a\Delta u = 0 \quad \text{in } \Omega \times I, \quad u_{|t=0} = u^0 \text{ in } \Omega, \quad u_{|\partial\Omega} = 0 \text{ on } I, \tag{3.8}$$

where $\Omega \subset \mathbb{R}^2$ is a bounded domain and $I := [0, T]$. This problem is discretized by a Galerkin method in space-time using standard (continuous) bilinear finite elements in space and discontinuous shape functions of degree $r \geq 0$, in time (so-called "dG(r) method"). We split the time interval $[0, T]$ into subintervals I_n according to

$$0 = t_0 < \dots < t_n < \dots < t_N = T, \quad k_n := t_n - t_{n-1}, \quad I_n = (t_{n-1}, t_n].$$

At each time level t_n, we define a regular finite element mesh $\mathbb{T}_h^n$, where the local mesh width is again $h_T = diam(T)$, $T \in \mathbb{T}_h^n$. Extending the spatial mesh to the corresponding space-time slab $\Omega \times I_n$, we obtain a global space-time mesh consisting of prisms $Q_T^n := T \times I_n$. On this mesh, we define the global finite element space

$$V_h^k := \{v \in H_0^1(\Omega) \times L^2([0, T]), \ v_{|Q_T^n} \in \tilde{Q}_1(T) \times P_r(I_n), \ T \in \mathbb{T}_h^n, \ n = 1, \dots, N\}.$$

For functions from this space (and their continuous analogous) we use the notation

$$U_n^+ := \lim_{t \to t_n+0} U(t), \quad U_n^- := \lim_{t \to t_n-0} U(t), \quad [U]_n := U_n^+ - U_n^-.$$

The Galerkin discretization of problem (3.8) is based on a variational formulation which allows the use of piecewise discontinuous functions in time. Then, the dG(r) method determines approximations $U \in V_h^k$ by requiring

$$\sum_{n=1}^{N} \int_{I_n} \left\{ (U_t, V) + (a\nabla U, \nabla V) \right\} dt + (U_0^+, V_0^+) + \sum_{n=2}^{N} ([U]_{n-1}, V_{n-1}^+) \tag{3.9}$$

$$= (u_0, V_0^+) + \int_0^T (f, V) \, dt \quad \forall V \in V_h^k.$$

Since the test functions $V \in V_h^k$ may be discontinuous at times t_n, this global system decouples and can be written in form of a time-stepping scheme,

$$\int_{I_n} \left\{ (U_t, V) + (a\nabla U, \nabla V) \right\} dt + ([U]_{n-1}, V_{n-1}^+) = \int_{I_n} (f, V) \, dt \quad \forall V \in V_h^n,$$

for $n = 1, ..., N$. In the following, we consider only the lowest-order case $r = 0$ (dG(0) method, corresponding to the backward Euler scheme). In this case, we have the final-time a priori error estimate

$$\|e_N\| \leq c \max_{1 \leq n \leq N} \left\{ h^2 \|\nabla^2 u\|_{\Omega \times I_n} + k_n \|u_t\|_{\Omega \times I_n} \right\}.$$

In the a posteriori error analysis, we concentrate on the control of the spatial L^2 error at the end time T. We use a duality argument in space-time,

$$z_t - a\Delta z = 0 \quad \text{in } \Omega \times I, \tag{3.10}$$

$$z_{|t=0} = \|e_N\|^{-1} e_N \text{ in } \Omega, \quad z_{|\partial\Omega} = 0 \text{ on } I,$$

to obtain the error representation

$$\|e_N\| = \sum_{n=1}^{N} \sum_{T \in T_h^n} \left\{ (f + a\Delta U - U_t, z - Z)_{T \times I_n} - \tfrac{1}{2}(a[\partial_n U], z - Z)_{\partial T \times I_n} \right.$$

$$\left. - ([U]_{n-1}, (z - Z)_{n-1}^+)_T \right\}. \tag{3.11}$$

From this, we infer the following a posteriori error estimate for the $dG(0)$ method,

$$\|e_N\| \leq c_i \sum_{n=1}^{N} \sum_{T \in T_h^n} \sum_{j=1}^{3} \left\{ \varrho_{T,k}^{n,j}(U)\, \omega_{T,k}^{n,j}(z) + \varrho_{T,h}^{n,j}(U)\, \omega_{T,h}^{n,j}(z) \right\}, \tag{3.12}$$

with cell residuals and weights defined by ($R(U) := f + \Delta U - \partial_t U$)

$$\varrho_{T,k}^{n,1}(U) := \|R(U)\|_{T \times I_n}, \quad \omega_{T,k}^{n,j}(z) := k_n \|\partial_t z\|_{T \times I_n},$$

$$\varrho_{T,h}^{n,1}(U) := \|R(U)\|_{T \times I_n}, \quad \omega_{T,h}^{n,j}(z) := h_T^2 \|\nabla^2 z\|_{T \times I_n},$$

$$\varrho_{T,k}^{n,2}(U) := 0,$$

$$\varrho_{T,h}^{n,2}(U) := \tfrac{1}{2} h_T^{-1/2} \|[\partial_n U]\|_{\partial T \times I_n}, \quad \omega_{T,h}^{n,2}(z) := h_T^2 \|\nabla^2 z\|_{\tilde{T} \times I_n},$$

$$\varrho_{T,k}^{n,3}(U) := k_n^{-1/2} \|[U]_{n-1}\|_T, \quad \omega_{T,k}^{n,3}(z) := k_n \|\partial_t z\|_{T \times I_n},$$

$$\varrho_{T,h}^{n,3}(U) := k_n^{-1/2} \|[U]_{n-1}\|_T, \quad \omega_{T,h}^{n,3}(z) := h_T^2 \|\nabla^2 z_{n-1}\|_{\tilde{T} \times I_n}.$$

The performance of this error estimator is illustrated by a simple test where the (known) exact solution represents a smooth rotating bulb on the unit square (for details see [13]). For a sequence of time levels, we compare the meshes obtained by controlling, first, the spatial L^2 error at the final time, corresponding to an initial condition in the dual problem, $z_{|t=T} = \|e_N\|^{-1} e_N$, and, second, the global space-time L^2 error, corresponding to a right-hand side in the dual problem, $\partial_t z - a\Delta z = \|e\|_{\Omega \times I}^{-1} e$. Figure 5 shows the development of the mesh refinement for the end-time error in contrast to that for the global error. We see clear differences which are explained by the different behavior of the dual solutions related to the two error measures considered.

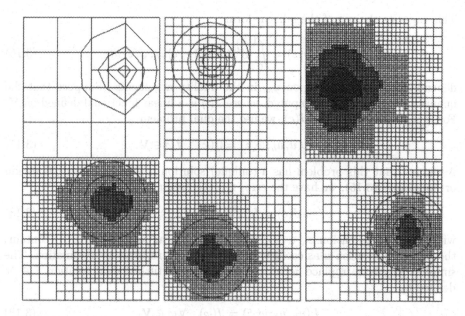

Figure 5: Sequence of refined meshes for controlling the end-time error $\|e_N\|_\Omega$ (upper row), and the global error $\|e\|_{\Omega \times I}$ (lower row)

3.4 A general paradigm for a posteriori error estimation

The approach to residual-based error estimation described above can be extended to general nonlinear problems. We outline the general concept in an abstract setting following the paradigm introduced by Eriksson, Estep, Hansbo, Johnson [11]. Let $\mathbf{V}$ be a Hilbert space with product $(\cdot, \cdot)$ and corresponding norm $\|\cdot\|$, and $A(\cdot; \cdot)$ a semi-linear form continuously defined on $\mathbf{V} \times \mathbf{V}$. We seek a solution to the abstract variational problem

$$u \in \mathbf{V}: \quad A(u; \varphi) = 0 \quad \forall \varphi \in \mathbf{V}. \tag{3.13}$$

This problem is approximated by a Galerkin method using a sequence of finite dimensional subspaces $\mathbf{V}_h \subset \mathbf{V}$ parameterized by a discretization parameter h. The discrete problems read

$$u_h \in \mathbf{V}_h: \quad A(u_h; \varphi) = 0 \quad \forall \varphi \in \mathbf{V}_h. \tag{3.14}$$

With the tangent form $A'(\cdot; \cdot, \cdot)$, we have the following orthogonality relation for the error $e = u - u_h$:

$$\int_0^1 A'(tu + (1-t)u_h; e, \varphi_h)\, dt = A(u; \varphi) - A(u_h; \varphi) = 0, \quad \varphi \in \mathbf{V}_h. \tag{3.15}$$

This suggests the use of the bilinear form

$$L(u, u_h; \varphi, z) = \int_0^1 A'(tu + (1-t)u_h; \varphi, z) \, dt, \qquad (3.16)$$

depending on u as well as on u_h, in the duality arguments. Suppose that the quantity $J(u)$ has to be computed, where $J(\cdot)$ is a linear functional defined on $\mathbf{V}$. For representing the error $J(e)$, we use the dual problem

$$L(u, u_h; \varphi, z) = J(\varphi) \quad \forall \varphi \in \mathbf{V}. \qquad (3.17)$$

Assuming that this problem has a (unique) solution $z \in \mathbf{V}$, and using Galerkin orthogonality (3.15), we have the error representation

$$J(e) = L(u, u_h; e, z - z_h), \qquad (3.18)$$

with any approximation $z_h \in \mathbf{V}_h$. Since the bilinear form $L(u, u_h; \cdot, \cdot)$ contains the unknown solution u as coefficient, its evaluation requires approximation. The simplest way is to replace u by u_h yielding a perturbed dual solution $\tilde{z} \in \mathbf{V}$ defined by

$$L(u_h, u_h; \varphi, \tilde{z}) = J(\varphi) \quad \forall \varphi \in \mathbf{V}. \qquad (3.19)$$

Controlling the effect of this perturbation on the accuracy of the resulting error estimator may be a delicate task and depends strongly on the particular problem considered. Our experience with the stationary Navier-Stokes equations indicates that this problem seems to be less critical if the continuous solution is stable. The crucial problem is the numerical computation of the perturbed dual solution $\tilde{z}$ by solving a discretized dual problem

$$L(u_h, u_h; \varphi, \tilde{z}_h) = J(\varphi) \quad \forall \varphi \in \mathbf{V}_h. \qquad (3.20)$$

This results in a practically useful error estimator $J(e) \approx \tilde{\eta}(u_h)$, in which the difference $z - z_h$ is replaced by some approximation, e.g., $(z - z_h)_{|T} \approx c_i h_T^2 \nabla_h^2 z_h$, obtained by using local interpolation estimates.

3.5 A nested solution approach

For solving the nonlinear problem (3.13) by a finite element Galerkin method (3.14), we employ the following iterative scheme. Starting from a coarse initial mesh $\mathbb{T}_0$, a hierarchy of refined meshes $\mathbb{T}_i$, $i \geq 1$, and corresponding finite element spaces $\mathbf{V}_i$ is generated by a nested solution process.

(0) Initialization $i = 0$: Start on coarse mesh $\mathbb{T}_0$ with $U_0^{(0)} = U_{-1} \in \mathbf{V}_0$.

(1) Defect correction iteration: For $i \geq 1$, start with $U_i^{(0)} = U_{i-1} \in \mathbf{V}_i$.

(2) Iteration step: Evaluate the defect

$$(d_i^{(j)}, \varphi) := f(\varphi) - A(U_i^{(j)}; \varphi), \quad \varphi \in \mathbf{V}_i,$$

and solve the correction equation

$$\tilde{A}'(U_i^{(j)}; V_i^{(j)}, \varphi) = (d_i^{(j)}, \varphi) \quad \forall \varphi \in \mathbf{V}_i,$$

by Krylov-space or multigrid iterations using the hierarchy of precedingly constructed meshes $\{\mathbb{T}_i, ..., \mathbb{T}_0\}$. Update $U_i^{(j+1)} = U_i^{(j)} + V_i^{(j)}$, set $j = j + 1$ and go back to (2). This process is repeated until a limit $U_i \in \mathbf{V}_i$, is reached with some required accuracy.

(3) Error estimation: Solve the (linearized) discrete dual problem

$$Z_i \in \mathbf{V}_i : \quad A'(U_i; \varphi, Z_i) = J(\varphi) \quad \forall \varphi \in \mathbf{V}_i$$

and evaluate the a posteriori error estimate

$$|J(e_i)| \approx \tilde{\eta}(U_i).$$

If $\tilde{\eta}(U_i) \leq TOL$ or $N_i \geq N_{max}$, then stop. Otherwise,cell-wise mesh adaptation yields the new mesh $\mathbb{T}_{i+1}$. Then, set $i = i + 1$ and go back to (1).

This kind of nested solution process is employed in all applications presented in the next section.

4 Applications

In this section, we present several applications of the method described above for error control and mesh adaptation. We have chosen four prototypical examples with different characteristic features. Example 1 is concerned with a problem from fluid mechanics. We compute drag and lift coefficients of a blunt body in an incompressible fluid modeled by the Navier-Stokes equations. Example 2 extends this example to weakly compressible flows using the low-Mach approximation and includes chemical reactions. Example 3 deals with a strongly nonlinear application in structural mechanics, elasto-plastic deformations described by Hencky's law. Finally, in Example 4, we consider a nonstandard problem of mixed differential-integral type arising in radiative transfer problems in astrophysics.

In all these applications the solution process is organized according to the general strategy described in the preceding chapter. Its characteristics are:

- Galerkin approximation with (conforming) bilinear finite elements for all physical quantities.

- Galerkin least-squares stabilization of velocity-pressure coupling and transport terms.

- Linearization by an outer fixed-point defect correction iteration, if necessary combined with pseudo-time stepping for generating starting values.

266

- Solution of the linear sub-problems by Krylov-space methods with optimized multigrid preconditioning (using block Gauss-Seidel or ILU-smoothing).

- Error control and mesh adaptation for the full (generally nonlinear) problem by the fixed fraction strategy.

4.1 Viscous incompressible flow

The results in this section are taken from Becker [3], [4], and Becker, Rannacher [6]. We consider a viscous incompressible Newtonian fluid modeled by the classical (stationary) Navier-Stokes equations

$$-\nu \Delta v + v \cdot \nabla v + \nabla p = f, \quad \nabla \cdot v = 0 \quad \text{in } \Omega, \qquad (4.1)$$

on a bounded domain $\Omega \subset \mathbb{R}^2$. Here, vector functions are also denoted by normal type, and no distinction is made in the notation of the corresponding products and norms. The unknowns are the velocity field $v = (v_1, v_2)$ and the pressure p, while ν is the normalized viscosity (density $\varrho = 1$), and f a prescribed body force. At the boundary $\partial\Omega$, the usual no-slip condition is posed along rigid parts together with suitable inflow and free-stream outflow conditions,

$$v|_{\Gamma_{\text{rigid}}} = 0, \quad v|_{\Gamma_{\text{in}}} = v_{\text{in}}, \quad (\nu \partial_n v - pn)|_{\Gamma_{\text{out}}} = 0. \qquad (4.2)$$

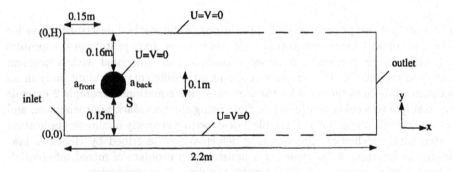

Figure 6: Geometry of the benchmark problem "Flow around a Cylinder" in 2D

As an example, we consider the flow around the cross section of a cylinder in a channel shown in Figure 6. This is part of a set of benchmark problems discussed in [21]. Quantities of physical interest are, for example,

$$\text{pressure drop:} \qquad J_{\Delta p}(v, p) = p(a_{\text{front}}) - p(a_{\text{back}}),$$

$$\text{drag/lift coefficient:} \qquad J_{\text{drag/lift}}(v, p) = \frac{2}{\bar{U}^2 D} \int_S n \cdot \sigma(v, p) e_{1/2} \, ds,$$

where S is the surface of the cylinder, D its diameter, $\bar{U}$ the reference velocity, and $\sigma(v,p) = \frac{1}{2}\nu(\nabla v + \nabla v^T) + pI$ the stress force acting on S. In our example, the Reynolds number is $Re = \bar{U}^2 D/\nu = 20$, such that the flow is stationary. Below, we present some detailed results for the pressure drop calculation. Corresponding results for drag and lift coefficients are similar and can be found in [4].

For discretizing this problem, we use a finite element method based on the quadrilateral Q_1/Q_1-Stokes element with globally continuous (isoparametric) bi-linear shape functions for both unknowns, velocity and pressure. The trial spaces for the velocity are denoted by V_h and those for the pressure by L_h. In the following, we use the compact notation $u := \{v, p\}$ for the continuous and $u_h := \{v_h, p_h\} \in V_h \times L_h$ for the discrete solution couple. Accordingly, the Navier-Stokes system can be written in the vector form

$$Lu := \{-\nu\Delta v + v \cdot \nabla v + \nabla p, \nabla \cdot v\} = \{f, 0\} =: F.$$

Further, for tuples $u = \{v, p\}$, $\varphi = \{\psi, \chi\}$, we define the semi-linear form

$$A(u; \varphi) := \nu(\nabla v, \nabla \psi) + (u \cdot \nabla u, \psi) - (p, \nabla \cdot \psi) + (\nabla \cdot v, \chi),$$

and for vectors v, w the weighted bilinear form

$$(v, w)_\delta := \sum_{T \in \mathbf{T}_h} \delta_T (v, w)_T.$$

Then, with a suitable finite element approximation $v_{h,\text{in}} \approx v_{\text{in}}$, the discrete problem seeks to determine $v_h \in V_h + v_{h,\text{in}}$, $p_h \in L_h$, such that

$$A(u_h; \varphi_h) + (Lu_h, S\varphi_h)_\delta = (F, \varphi_h) + (F, S\varphi_h)_\delta \quad \forall \varphi_h \in V_h \times L_h,$$

where

$$S\varphi_h := \{\nu\Delta\psi + v \cdot \nabla\psi + \nabla\chi, 0\}, \quad \delta_T = ch_T \min\left\{\frac{h_T}{\nu}, \frac{1}{\min_T |v_h|}\right\}.$$

This formulation simultaneously contains the least-squares terms for achieving velocity-pressure as well as transport stabilization. The resulting nonlinear problem is solved by the nested multilevel techniques described above. For details, we refer to [3].

For this discretization the global energy-norm *a posteriori* error estimate reads (neglecting boundary terms along the cylinder contour)

$$\|\nabla e_v\| + \|e_p\| \leq c_i c_s \Big(\sum_{T \in \mathbf{T}_h} \Big\{ (h_T^2 + \delta_T)\|R(u_h)\|_T^2 + \|\operatorname{div} v_h\|_K^2 \qquad (4.3)$$
$$+ \nu h_T \|[\partial_n v_h]\|_{\partial T}^2 \Big\} \Big)^{1/2},$$

with the residual $R(u_h) = f + \nu\Delta v_h - v_h \cdot \nabla v_h - \nabla p_h$.

Corresponding *weighted a posteriori* error estimates can be obtained following the general line of argument described in the preceding sections. In computing the pressure drop, the (approximate) dual problem seeks a couple $z := \{w, q\} \in V \times L^2$ satisfying

$$A'(u_h; \varphi, z) + (L'(u_h)^* \varphi, Sz)_\delta = J(\varphi) \quad \forall \varphi \in V \times L^2. \tag{4.4}$$

The resulting weighted a posteriori estimate is

$$|J(e_u)| \leq \sum_{T \in \mathbb{T}_h} \left\{ \varrho_T^{(1)} \omega_T^{(1)} + \varrho_T^{(2)} \omega_T^{(2)} + \varrho_T^{(3)} \omega_T^{(3)} + \dots \right\}, \tag{4.5}$$

with the local residual terms

$$\varrho_T^{(1)} := \|F - Lu_h\|_T, \quad \varrho_T^{(2)} := \tfrac{1}{2}\nu\|[\partial v_h]\|_{\partial T}, \quad \varrho_T^{(3)} := \|\nabla v_h\|_T,$$

and weights

$$\omega_T^{(1)} := \|w - w_h\|_T + \delta_T\|v \cdot \nabla(w - w_h) + \nabla(q - q_h)\|_T,$$
$$\omega_T^{(2)} := \|w - w_h\|_{\partial T}, \quad \omega_T^{(3)} := \|q - q_h\|_T.$$

The dots "..." stand for additional terms measuring the errors in approximating the inflow and the curved cylinder boundary. For more details on this aspect, we refer to [7] and [4]. The bounds for the dual solution $\{w, q\}$ are obtained computationally by replacing the unknown solution u in the convection term by its approximation u_h and solving the resulting linearized problem on the same mesh. Suitable difference quotients are taken from the approximate dual solution in evaluating the weights $\omega_K^{(i)}$. The interpolation constant may again be determined analytically or be simply set to $c_{i,T}^{(j)} = 1$.

Table 1 shows the corresponding results for the pressure drop computed on (1) hierarchically refined meshes, starting from a coarse mesh with almost uniform mesh width, (2) hierarchically refined meshes starting from a coarse mesh which is hand-refined towards the cylinder contour, (3) adapted meshes using the global energy-error estimator (4.3) with the additional prescription that in each refinement cycle all cells along the cylinder are refined, and (4) adapted meshes using the weighted point-error estimator (4.5). These results demonstrate clearly the superiority of the weighted error estimator in computing local quantities. It produces an error of less than 1% already after 6 refinement cycles on a mesh with about 20000 unknowns while the other algorithms need at least 75000 unknowns to achieve the same accuracy. Finally, Figure 7 shows optimized meshes for the computation of pressure drop, drag, and lift, respectively.

Uniform Refinement, Grid1			Uniform Refinement, Grid2		
L	#unknowns	Δp	L	#unknowns	Δp
1	2268	0.109389	1	1296	0.106318
2	8664	0.110513	2	4896	0.112428
3	33840	0.113617	3	19008	0.115484
4	133728	0.115488	4	74880	0.116651
5	531648	0.116486	5	297216	0.117098
Adaptive Refinement, Grid1			Weighted Adaptive Refinement		
L	#unknowns	Δp	L	#unknowns	Δp
2	1362	0.105990	4	650	0.115967
4	5334	0.113978	6	1358	0.116732
6	21546	0.116915	9	2858	0.117441
8	86259	0.117379	11	5510	0.117514
10	330930	0.117530	12	8810	0.117527

Table 1: Results of the cylinder flow computations for various types of mesh refinements (reference value $\Delta p = 0.11752016......$)

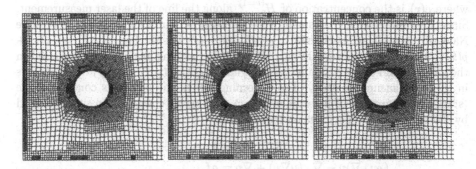

Figure 7: Optimized meshes for the computation of the pressure drop (left), the drag (middle), and the lift (right)

4.2 Chemically reactive flow

The results in this section are taken from Waguet [24] and Braack [9]. We consider a laminar flow reactor for determining the reaction velocity of elementary wall-desactivation reactions (slow chemistry) sketched in Figure 8. More complex combustion processes (fast chemistry) like the ozone recombination or an even more complex model of methane combustion (15 species and 84 reactions) have been treated by the same methods in [9], see also [5].

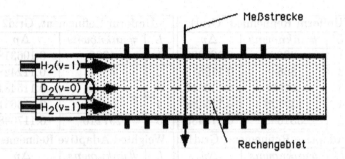

Figure 8: Configuration of flow reactor

The quantity to be computed is the CARS signal (Coherent Anti-Stokes Raman Spectroscopy)

$$J(c) = \kappa \int_{-R}^{R} \sigma(s) c(r-s)^2 \, ds,$$

where $c(r)$ is the concentration of $H_2^{(\nu=1)}$ along the line of the laser measurement. Since the inflow velocity is small, a low-Mach approximation of the compressible Navier-Stokes equations is used, i.e., the pressure is split like $p(x,t) = P_*(t) + p(x,t)$ into a thermodynamic part $P_*(t)$ which is constant in space and used in the gas law, and a much smaller hydrodynamic part $p(x,t) \ll P_*(t)$ which occurs in the momentum equation. The governing system of equations consists of the (stationary) equation of mass, momentum and energy conservation supplemented by the equations of species mass conservation:

$$\nabla \cdot (\varrho v) = 0,$$
$$(\varrho v \cdot \nabla) v - \nabla \cdot (\mu \nabla v) + \nabla p = \varrho f_e,$$
$$\varrho v \cdot \nabla T - c_p^{-1} \nabla \cdot (\lambda \nabla T) = c_p^{-1} f_t(T, w), \qquad (4.6)$$
$$\varrho v \cdot \nabla w_i - \nabla \cdot (\varrho D_i \nabla w_i) = f_i(T, w), \quad i = 1, ..., n,$$
$$\varrho = \frac{P_* \bar{M}}{RT}.$$

Due to exponential dependence on temperature (Arrhenius law) and polynomial dependence on w, the source terms $f_i(T, w)$ are highly nonlinear. In general, these zero-order terms lead to a coupling between all chemical species mass fractions. For robustness the resulting system of equations is to be solved by an implicit and fully coupled process which uses strongly adapted meshes.

The discretization of the flow system above uses continuous Q_1-finite elements for all unknowns and employs least-squares stabilization for the velocity-pressure coupling as well as for the transport terms. We do not state the corresponding discrete equations since they have the same structure as already seen in the preceding section for the incompressible Navier-Stokes equations. The derivation of the related (linearized) dual problem and the resulting a posteriori error estimates follows the same line of argument. For details, we refer to [9] and [24].

Table 2 contains results obtained by our approach for the computation of the mass fraction of $H_2^{(\nu=1)}$ and $H_2^{(\nu=0)}$. The comparison is against computations on heuristically refined tensor-product meshes. We observe improved accuracy on the systematically adapted meshes, particularly monotone convergence of the quantities is achieved.

Heuristic refinement				Adaptive refinement			
L	N	$H_2^{(\nu=0)}$	$H_2^{(\nu=1)}$	L	N	$H_2^{(\nu=0)}$	$H_2^{(\nu=1)}$
1	137	0.654322	0.005294	1	137	0.6556	0.005294
2	481	0.73735	0.00661	2	282	0.7382	0.006063
3	1793	0.796212	0.007096	3	619	0.7958	0.007132
4	6913	0.817192	0.007434	4	1368	0.8149	0.007323
5	7042	0.819748	0.007419	5	3077	0.8257	0.007457
6	7494	0.824006	0.007473	6	6800	0.8295	0.007534
7	8492	0.826994	0.007504	7	15100	0.8317	0.007564
8	10482	0.828583	0.0075207	8	33462	0.8328	0.007587
9	15993	0.828535	0.0075448	9			

Table 2: Results of simulation for the $H_2^{(\nu=1)} \rightarrow_{wall} H_2^{(\nu=0)}$ experiment on hand-adapted (left) and on automatic-adapted (right) meshes

4.3 Elasto-plasticity

The results in this section are taken from Suttmeier [22], and Rannacher, Suttmeier [19], [20]. The fundamental problem in the (static) deformation theory of linear-elastic perfect–plastic material (stationary *Hencky* model) in classical notation reads

$$\operatorname{div} \sigma = -f, \quad \varepsilon(u) = A : \sigma + \lambda \quad \text{in } \Omega,$$
$$\lambda : (\tau - \sigma) \leq 0 \quad \forall \tau \text{ with } \mathcal{F}(\tau) \leq 0, \tag{4.7}$$
$$u = 0 \text{ on } \Gamma_D, \quad \sigma \cdot n = g \text{ on } \Gamma_N,$$

where σ and u are the stress tensor and displacement vector, respectively, while λ denotes the plastic growth. This system describes the deformation of an elasto-plastic body occupying a bounded domain $\Omega \subset \mathbb{R}^d$ ($d = 2$ or 3) which is fixed along a part Γ_D of its boundary $\partial\Omega$, under the action of a body force with density f and a surface traction g along $\Gamma_N = \partial\Omega \setminus \Gamma_D$. The displacement u is supposed to be small in order to neglect geometric nonlinear effects, so that the strain tensor can be written as $\varepsilon(u) = \frac{1}{2}(\nabla u + \nabla u^T)$. The material tensor A is assumed to be

272

symmetric and positive definite. Here, we consider the case of the linear–elastic isotropic material law $\sigma = 2\mu\varepsilon^D(u) + \kappa\operatorname{div} uI$, with material dependent constants $\mu > 0$ and $\kappa > 0$. The perfect plastic behavior is expressed by the von Mises flow rule $\mathcal{F}(\sigma) := |\sigma^D| - \sigma_0 \leq 0$, with $\sigma_0 > 0$ and σ^D being the deviatoric part of the stress tensor σ.

We consider a common benchmark problem (c.f. [22]): A geometrically two-dimensional square disc with a hole is subjected to a constant boundary traction acting upon two opposite sides. We use the two-dimensional plain-strain approximation, i.e., the components of $\varepsilon(u)$ in z-direction are assumed to be zero, and assume perfectly plastic material behavior. In virtue of symmetry the consideration can be restricted to a quarter of the domain shown in Figure 9. The height and width of the quarter corresponding to lines $\overline{45}$ and $\overline{15}$ are 100, and the radius of the hole is 10. The material parameters are taken as $\kappa = 164206$, $\mu = 80193.8$ and $\sigma_0 = 450$ (values of aluminum), and the flow rule is $\mathcal{F}(\sigma) = |\sigma^D| - \sqrt{2/3}\sigma_0 \leq 0$. The boundary traction is constant $g = 450$. Among the quantities to be computed are the component σ_{22} of the stress tensor at point 2, and the horizontal deflection u_1 at points 5 and 2. The result on a very fine adapted mesh with about 200000 cells is taken as reference solution u_{ref}.

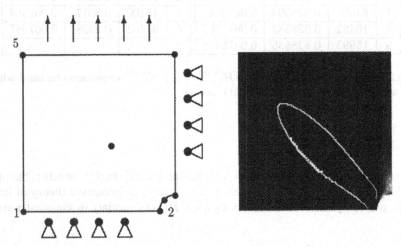

Figure 9: Geometry of the benchmark problem and plot of $|\sigma^D|$ (plastic region black, transition zone white) computed on a mesh with $N \approx 10000$ cells

The *primal* variational formulation of problem (4.3) seeks a displacement $u \in V := \{u \in H^1(\Omega, \mathbb{R}^d),\ u_{|\Gamma_D} = 0\}$, satisfying

$$A(u;\varphi) = (f,\varphi) + (g,\varphi)_{\Gamma_N} \quad \forall \varphi \in V, \qquad (4.8)$$

with the semi-linear form

$$A(u;\varphi) := (C(\varepsilon[u]),\varphi) = (\Pi(2\mu\varepsilon^D(u)), \varepsilon(\varphi)) + (\kappa\operatorname{div} u, \operatorname{div}\varphi),$$

and the projection

$$\Pi(2\mu\varepsilon^D(u)) := \begin{cases} 2\mu\varepsilon^D(u) & \text{, if } |2\mu\varepsilon^D(u)| \leq \sigma_0, \\ \dfrac{\sigma_0}{|\varepsilon^D(u)|}\varepsilon^D(u) & \text{, if } |2\mu\varepsilon^D(u)| > \sigma_0. \end{cases}$$

The finite element approximation of problem (4.8) reads

$$A(u_h; \varphi_h) = (f, \varphi_h) + (g, \varphi_h)_{\Gamma_N} \quad \forall \varphi_h \in V_h, \tag{4.9}$$

where V_h is a finite element space of piecewise bilinear shape functions as described above. Having computed the displacement u_h, we obtain a corresponding stress by $\sigma_h := C(\varepsilon[u_h])$. Details of the solution process can be found in [22].

Given an error functional $J(\cdot)$, we have the corresponding a posteriori error estimate

$$|J(e)| \leq \eta(u_h) := \sum_{T \in \mathcal{T}_h} h_T^4 \omega_T \varrho_T, \tag{4.10}$$

with the local residuals and weights

$$\varrho_T := h_T^{-1}\|f - \operatorname{div} C(\varepsilon(u_h))\|_T + \tfrac{1}{2}h_T^{-3/2}\|n \cdot [C(\varepsilon(u_h))]\|_{\partial T},$$

$$\omega_T := \max\left\{ h_T^{-3}\|z - z_h\|_T, h_T^{-5/2}\|z - z_h\|_{\partial T} \right\},$$

where $C(\epsilon) := \Pi(2\mu\varepsilon^D) + \kappa\,tr(\epsilon)$.

We compare this *weighted* error estimator against two of the traditional ways of estimation of the stress error $e_\sigma := \sigma - \sigma_h$:

1) The ZZ-approach (Zienkiewicz, Zhu [25]): The (heuristic) error indicator uses the idea of higher–order stress recovery by local averaging,

$$\|e_\sigma\| \approx \eta_{ZZ} := \left(\sum_{T \in \mathcal{T}_h} \|\mathcal{M}_h \sigma_h - \sigma_h\|_T^2 \right)^{1/2}, \tag{4.11}$$

where $\mathcal{M}_h \sigma_h$ is a local (super-convergent) approximation of σ.

2) An energy error estimator (Johnson, Hansbo [15]): Let Ω_h^e and Ω_h^p denote the union of elements where the discrete solution behaves elastic and plastic, respectively. The (heuristic) error estimator reads

$$\|e_\sigma\| \approx \eta_E := C_I \left(\sum_{T \in \mathcal{T}_h} \eta_T^2 \right)^{1/2}, \tag{4.12}$$

with the local error indicators defined by

$$\eta_T^2 := h_T^4 \varrho_T^2, \ T \in \Omega_h^e \qquad \eta_T^2 := h_T^2\|\mathcal{M}_h \sigma_h - \sigma_h\|_T \varrho_T, \ T \in \Omega_h^p.$$

274

Some of the results of the benchmark computations are summarized in Figure 10. They show again that the weighted a posteriori error bounds are rather sharp and yield economical meshes, particularly if high accuracy is required. For more details as well as for further results also for the time-dependent *Prandtl-Reuss* model in perfect plasticity, we refer to [19] and [20]. Finally, in Figure 11, we show the distribution of the weights ω_T for the computation of u_1 at point 5 and point 2. This illustrates the nontrivial effect of the irregularity along the elasto-plastic transition zone on the point-error evaluation.

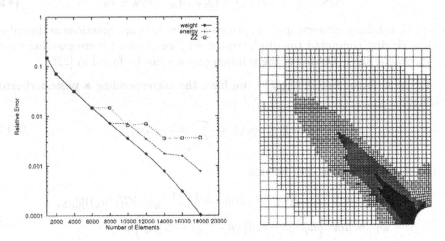

Figure 10: Relative error for computation of σ_{22} at point 2 using different estimators and "optimal" grid with about 10000 cells

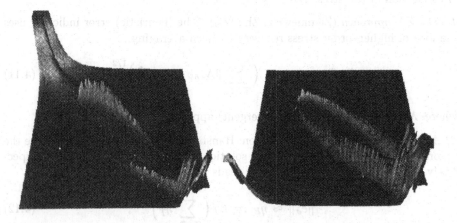

Figure 11: Distribution of weights ω_T on optimized grids with about 10000 cells in the computation of u_1 at point 5 (left) and at point 2 (right)

4.4   Radiative transfer

The results in this section are taken from Kanschat [16], [17], and Führer, Kanschat [12]. The emission of light of a certain wave length from a cosmic source is described by the *radiative transfer equation* (neglecting frequency coupling)

$$\theta \cdot \nabla_x u + (\kappa + \mu)u = \mu \int_{S_2} K(\theta, \theta')u \, d\theta' + B \quad \text{in } \Omega \times S_2, \qquad (4.13)$$

for the radiation intensity $u = u(x, \theta)$. Here, $x \in \Omega \subset \mathbb{R}^3$, a bounded domain, and $\theta \in S_2$, the unit-sphere in $\mathbb{R}^3$. The usual boundary condition is $u = 0$, on the "inflow" boundary $\Gamma_{\text{in},\theta} = \{x \in \partial\Omega, n \cdot \theta \leq 0\}$. The absorption and scattering coefficients κ, μ, the redistribution kernel $K(\cdot, \cdot)$, and the source term B (Planck function) are given. In interesting applications these functions exhibit strong variations in space requiring the use of locally refined meshes.

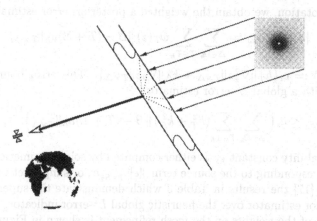

Figure 12: Observer configuration of radiation emission

We consider a prototypical example from astrophysics. A satellite-based observer measures the light (at a fixed wave length) emitted from a cosmic source hidden in a dust cloud. A sketch of this situation is shown in Figure 12. The measurement is correlated with results of a (two-dimensional) simulation which assumes certain properties of the coefficients in the underlying radiative transfer model (4.13). Because of the distance to the source, only the mean value of the intensity emitted in the observer direction θ_{obs} can be measured. Hence, the quantity to be computed is

$$J(u) = \int_{\{n \cdot \theta_{\text{obs}} \geq 0\}} u(x, \theta_{\text{obs}}) \, ds,$$

where $\{n \cdot \theta_{\text{obs}} \geq 0\}$ is the outflow boundary of the computational domain $\Omega \times S_1$ (here $\Omega \subset \mathbb{R}^2$ a square) containing the radiating object.

276

The finite element Galerkin formulation of (4.13) reads

$$((T + \Sigma)u_h, \varphi_h)_{\Omega \times S_2} = (B, \varphi_h)_{\Omega \times S_2} \quad \forall \varphi_h \in V_h, \tag{4.14}$$

where

$$Tu_h := \theta \cdot \nabla_x u_h, \quad \Sigma u_h := (\kappa + \mu)u_h - \mu \int_{S_2} K(\theta, \theta')u_h \, d\theta',$$

and $V_h \subset H^1(\Omega) \times L^2(\Omega)$ is a proper finite element subspace. The discretization uses standard (continuous) Q_1-finite elements in $x \in \Omega$, on meshes $\mathbb{T}_h = \{T\}$ with local width h_T, and (discontinuous) P_0-finite elements in $\theta \in S_2$, on meshes $\mathbb{D}_k = \{\Delta\}$ of uniform width k_Δ. The x-mesh is adaptively refined, while the θ-mesh is kept uniform (suggested by a priori error analysis). The refinement process is organized as described before. The associated dual problem reads

$$z \in V : \quad (z, (T + \Sigma)\varphi)_{\Omega \times S_2} = J(\varphi) \quad \forall \varphi \in V. \tag{4.15}$$

Using this notation, we obtain the weighted a posteriori error estimate

$$|J(e)| \leq \eta_\omega := \sum_{\Delta \in \mathbb{D}_k} \sum_{T \in \mathbb{T}_h} \omega_T(z) \, \|B - (T + \Sigma)u_h\|_{T \times \Delta}, \tag{4.16}$$

where $\omega_T(z) := c_i \{h_T^2 \|\nabla_x^2 z\|_{T \times \Delta} + k_\Delta \|\nabla_\theta z\|_{T \times \Delta}\}$. This error bound has to be compared with a global L^2-error estimator

$$\|e\|_{\Omega \times S_2} \leq c_s \Big(\sum_{\Delta \in \mathbb{D}_k} \sum_{T \in \mathbb{T}_h} (h_T^2 + k_\Delta^2) \, \|B - (T + \Sigma)u_h\|_{T \times \Delta}^2 \Big)^{1/2}, \tag{4.17}$$

where the stability constant c_s is either computed by solving numerically the dual problem corresponding to the source term $\|e\|_{\Omega \times S_2}^{-1} e$, or simply set to $c_s = 1$. We report from [17] the results in Table 3 which demonstrate the superiority of the weighted error estimator over the heuristic global L^2-error indicator. The effect of the presence of the weights on the mesh refinement is shown in Figure 13.

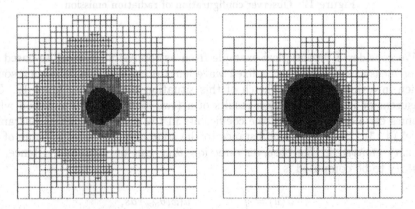

Figure 13: Optimized meshes for the radiative transfer problem generated by the weighted error estimator (left) and the (heuristic) L^2-error indicator (right)

	L^2-indicator		weighted estimator			
L	N_x	$J(u_h)$	N_x	$J(u_h)$	η_ω	$\eta_\omega/J(e)$
1	564	0.181	576	0.417	3.1695	23.77
2	1105	0.210	1146	0.429	1.0804	8.62
3	2169	0.311	2264	0.461	0.7398	7.11
4	4329	0.405	4506	0.508	0.2861	3.94
5	8582	0.460	9018	0.555	0.1375	3.33
6	17202	0.488	18857	0.584	0.0526	2.39
7	34562	0.537	39571	0.599	0.0211	1.76
8	68066	0.551	82494	0.608	0.0084	1.41
			∞	0.618		

Table 3: Results obtained for the radiative transfer problem by the (heuristic) L^2-error indicator and the weighted error estimator, the total number of unknowns being $N_{tot} = N_x \cdot 32$

References

[1] Ainsworth, M. and Oden, J.T. (1997) A posteriori error estimation in finite element analysis, *Comp. Methods Appl. Mech. Engrg.* 142, pp. 1-88.

[2] Backes, E. (1997) *Gewichtete a posteriori Fehleranalyse bei der adaptiven Finite-Elemente-Methode: Ein Vergleich zwischen Residuen- und Bank-Weiser-Schätzer*, Diploma Thesis, Institut für Angewandte Mathematik, Universität Heidelberg.

[3] Becker, R. (1995) *An Adaptive Finite Element Method for the Incompressible Navier-Stokes Equations on Time-Dependent Domains*, Thesis, Universität Heidelberg.

[4] Becker, R. (1998) Weighted error estimators for finite element approximations of the incompressible Navier-Stokes equations, Preprint 98-20, SFB 359, Universität Heidelberg.

[5] Becker, R., Braack, M. and R. Rannacher (1997) Adaptive finite elements for reactive flows, *Proc. Conf. Numer. Modelling in Continuum Mechanics*, Aug. 1997, Prague, and Proc. ENUMATH-97, Sept. 1997, Heidelberg, World Sci. Pub., to appear.

[6] Becker, R. and Rannacher, R. (1995) Weighted a posteriori error control in FE methods, ENUMATH-95, Paris, 1995, *Proc. ENUMATH-97*, World Sci. Pub., to appear.

[7] Becker, R. and Rannacher, R. (1996) A feed-back approach to error control in finite element methods: basic analysis and examples, *East-West J. Numer. Math.* 4, pp. 237-264.

[8] Becker, R. and Rannacher, R. (1997) A general concept of adaptivity in finite element methods with applications to problems in fluid and structural mechanics, *IMA-Report*, to appear.

[9] Braack, M. (1998) *An Adaptive Finite Element Method for Reactive Flow Problems*, Thesis, Universität Heidelberg.

278

[10] Brenner, S. and Scott, R.L. (1994) *The Mathematical Theory of Finite Element Methods*, Springer, Berlin-Heidelberg-New York.

[11] Eriksson, K., Estep, D., Hansbo, P. and Johnson, C. (1995) Introduction to adaptive methods for differential equations, *Acta Numerica* 1995 (A. Iserles, ed.), pp. 105-158, Cambridge University Press.

[12] Führer, C. and Kanschat, G. (1997) Error control in radiative transfer, *Computing* 58, pp. 317-334.

[13] Hartmann, R. (1998) *A posteriori Fehlerschätzung und adaptive Schrittweiten- und Ortsgittersteuerung bei Galerkin-Verfahren für die Wärmeleitungsgleichung*, Diploma Thesis, Institut f"ur Angewandte Mathematik, Universität Heidelberg.

[14] Johnson, C. (1987) *Numerical Solution of Partial Differential Equations by the Finite Element Method*, Cambridge University Press, Cambridge-Lund.

[15] Johnson, C. and Hansbo, P. (1992) Adaptive finite element methods in computational mechanics, *Computer Methods in Applied Mechanics and Engineering 101*, North-Holland, Amsterdam.

[16] Kanschat, G. (1996) *Parallel and Adaptive Galerkin Methods for Radiative Transfer Problems*, Thesis, Universität Heidelberg.

[17] Kanschat, G. (1997) Efficient and reliable solution of multi-dimensional radiative transfer problems, *Proc. Multiscale Phenomena and Their Simulation*, Bielefeld, Sept. 30 - Oct. 4, 1996, World Sci. Pub.

[18] Rannacher, R. (1998) A posteriori error estimation in least-squares stabilized finite element schemes, *Special Issue Advances in Stabilized Methods in Computational Mechanics*, Comp. Methods Appl. Mech. Engrg., to appear.

[19] Rannacher, R. and Suttmeier F.-T. (1997) A feed-back approach to error control in finite element methods: Application to linear elasticity, *Comp. Mech.* 19, pp. 434-446.

[20] Rannacher, R. and Suttmeier, F.-T. (1998) A posteriori error control in finite element methods via duality techniques: Application to perfect plasticity, *Comp. Mech.* 21, pp. 123-133.

[21] Schäfer, M. and Turek, S. (1996) The benchmark problem "flow around a cylinder", Report of the DFG Priority Research Program *Flow Simulation with High-Performance Computers* (E.H. Hirschel, ed.), Notes Comp. Fluid Mech., Vieweg.

[22] Suttmeier, F.-T. (1997) *Adaptive Finite Element Approximation of Problems in Elasto-Plasticity Theory*, Thesis, Universität Heidelberg.

[23] Verfürth, R. (1996) *A Review of A Posteriori Error Estimation and Adaptive Mesh-Refinement Techniques*, Wiley/Teubner, New York-Stuttgart.

[24] Waguet, C. (1998) *Adaptive Finite Element Computation of Chemical Flow Reactors*, Thesis, Universität Heidelberg, in preparation.

[25] Zienkiewicz, O.C. and Zhu, J.Z. (1987) A simple error estimator and adaptive procedure for practical engineering analysis, *Int. J. Numer. Methods Engrg.* 24, pp. 337-357.

VERIFIED SOLUTION OF LARGE LINEAR AND NONLINEAR SYSTEMS

SIEGFRIED M. RUMP

Inst. f. Informatik III Technical University Hamburg-Harburg, Eißendorfer Str. 38,

21071 Hamburg, Germany, rump@tu-harburg.de

Abstract. In this paper we describe some of the principles of methods for the verified solution of large systems of linear and nonlinear equations. Verified solution means to produce results with correct error bounds using floating point arithmetic, possibly with directed rounding. Some background is given why and for what kind of problems the methods work, and also why other approaches do not work.

1. Introduction

In numerical analysis it is usually distinguished between backward and forward error analysis. Given an approximation of the true solution of a given problem, the frequently used backward analysis asks: "How large a perturbation of the original input data is necessary such that the computed approximation is the exact solution of the perturbed problem?" For many numerical algorithms this perturbation is small. Consider, for example, the solution of a linear system $Ax = b$ with $A \in \mathbf{M}_n(\mathbf{R}), b \in \mathbf{R}^n$. If $\tilde{L}, \tilde{U} \in \mathbf{M}_n(\mathbf{F})$, where $\mathbf{F}$ denotes the set of floating point numbers, are the computed factors and $\tilde{x} \in \mathbf{F}^n$ is the approximate solution computed by Gaussian elimination with partial pivoting, then $\tilde{A}\tilde{x} = b$ where

$$|\tilde{A} - A| \le (3\,\varepsilon' + \varepsilon'^2)\,|\tilde{L}|\,|\tilde{U}| \quad \text{and} \quad \varepsilon' := n\varepsilon/(1 - n\varepsilon) \quad .$$

Here ε denotes the relative rounding error unit. This estimation ([23], [18]) is rigorous provided $n\varepsilon < 1$. If the factor in backward analysis is small, as for Gaussian elimination, the used algorithm is stable. More precisely, the algorithm does not introduce additional instability. This is not always true. A familiar example is the solution of least squares problems by normal

279

H. Bulgak and C. Zenger (eds.), Error Control and Adaptivity in Scientific Computing, 279–298.

equations instead of some orthogonalization technique, thus unnecessarily squaring the condition number of the problem.

On the other hand, backward analysis does not tell the condition of the problem. In case of Gaussian elimination, for example, the condition can be read from the growth factor. It is well known that for Gaussian elimination with partial pivoting the growth factor may grow exponentially with the dimension. On the other hand, until the work by Wright ([24], see also [7]) it was common belief that this occurs only in constructed, nonpractical examples. Since then the growth factor is monitored in LAPACK routines [3].

In forward error analysis the question is: "How far is the computed solution from the true solution of the problem?" including the question "Is the problem solvable?" At first sight this seems to be the more natural question to ask. This is not necessarily always the case. In many cases a good backward error is totally sufficient. If it is important to know the forward error of an approximation, for example, in problems where safety is important or controllability has to be assured, traditional estimations contain the condition number of the problem. For Gaussian elimination, for example, it is

$$\frac{||x - \tilde{x}||}{||x||} \leq \frac{\varepsilon \operatorname{cond}(A)}{1 - \varepsilon \operatorname{cond}(A)}$$

provided $Ax = b$, $(A + E)\tilde{x} = b$ and $||E|| \leq \varepsilon ||A||$, $\varepsilon \operatorname{cond}(A) = \varepsilon ||A|| \, ||A^{-1}|| < 1$ (cf. [10]). A finite condition number implies solvability of the problem. For many problems the reciprocal of the condition number is proportional to the distance to the nearest ill-posed problem. This is well known for normwise distances [6], and has recently proved to be also true for linear systems and componentwise distances [22]. However, for a valid estimate of the forward error we need an upper bound for the condition number, and this is frequently as difficult to compute as to solve the problem.

Traditional forward error analysis gives general bounds, independent of the actual data of the problem. Probably the oldest bounds of this kind are in [16]. Those bounds turned out to be very pessimistic, possibly one reason that today mostly backward bounds are used. In the following we will discuss possibilities to compute forward error bounds for a given problem taking into account the actual data. These bounds will cover all intermediate errors such as representation errors, rounding errors and all kinds of computational errors. The bounds are completely rigorous.

A major ingredient of those methods is the possibility to estimate the error of a sequence of operations and to estimate the range of a function. This is to be described in the next section.

2. Range of a function

One possibility to estimate the error of an individual operation is interval arithmetic. Consider the set of nonempty real intervals $\mathbf{IR} := \{[a_1, a_2] \mid a_1, a_2 \in \mathbf{R}$ and $a_1 \leq a_2\}$. Power set operations over real intervals are defined by

$$A, B \in \mathbf{IR} \quad \Rightarrow \quad A \circ B := \{a \circ b \mid a \in A, b \in B\}$$

where $\circ \in \{+, -, \cdot, /\}$. This definition can also serve as definition of interval operations because for $A, B \in \mathbf{IR}$ the result of the power set operations is always a real interval (provided the denominator does not contain zero in case of division). For $A = [a_1, a_2], B = [b_1, b_2]$, this definition is equivalent to

$$A \circ B = [\min(a_1 \circ b_1, a_1 \circ b_2, a_2 \circ b_1, a_2 \circ b_2), \max(a_1 \circ b_1, a_1 \circ b_2, a_2 \circ b_1, a_2 \circ b_2)]$$
$$(1)$$

For intervals with floating point endpoints, the same definition can be used provided operations for the left bound are executed with rounding downwards, and operations for the right bound are executed with rounding upwards. For details see a standard book on interval analysis, among them ([2],[15]). We want to stress that (1) is *not* used for computation of an interval result. Here much faster methods are available, see Section 6.

The definition can be extended to complex operations, to vector and matrix operations, and to real and complex standard functions. The actual computation of the result for complex standard functions is nontrivial (see [4], [12]).

All interval operations satisfy one fundamental property, the isotonicity. It is

$$\forall a \in A \quad \forall b \in B : \quad a \circ b \in A \circ B \qquad (2)$$

for all suitable A,B and all suitable interval operations $\circ$. For monadic operations isotonicity is formulated similar to (2).

Having isotonicity at hand, it seems natural to apply interval arithmetic in a so-called *naive way*: Replacing every operation by the corresponding interval operation produces a final result which includes the true solution of the problem. This follows trivially by successively applying the isotonicity property (2). However, other than in special cases, this produces a severe overestimation of the result. One reason is the so-called wrapping effect. Consider the iteration

$$z_0 = 1; \quad z_{\nu+1} := (a + ib) \cdot z_\nu \quad \text{where} \quad a = b = 1/\sqrt{2}. \qquad (3)$$

Evaluation of (3) in floating point introduces a small roundoff error in the computation of a and b. In the following iterations, the interval inclusion of z_ν is always turned by $45°$. The interval inclusion is always a rectangular parallel to the axes. That means, in every iteration step the true length of the axes is prolonged by a factor $\sqrt{2}$, which means more than a factor 10^{16} after 100 iterations. Another reason for overestimation is the fact that one interval is used several times. For example, using interval arithmetic it is

$$e^X - X = [0, e] \qquad \text{for } X = [0, 1] \quad,$$

instead of $[1, e-1]$ using power set operations. This is the dependency problem. If every interval quantity occurs exactly once in an expression, then the final result shows no overestimation. Otherwise, except special circumstances, the result is afflicted with overestimation. As a more practical example consider Gaussian elimination in interval arithmetic, frequently called IGA (Interval Gaussian Elimination). That is Gaussian elimination with partial pivoting is performed by replacing every operation by its corresponding interval operation.

For the following table we generated a random $n \times n$ matrix with entries uniformly distributed in $[-1, 1]$. We performed IGA for the original matrix, for the matrix preconditioned with an approximate inverse of its L-factor, and preconditioned with an approximate inverse of the entire matrix. In the second case, the preconditioned matrix is approximately the U-factor, in the third case approximately the identity matrix. We used INTLAB for the computation, see Section 6. The input is

```
for n=10:10:100; a=2*rand(n)-1;
    [l u]=luwpp(intval(a)); r1=rad(u(n,n));
    [l u]=lu(a);
    [l u]=luwpp(inv(l)*intval(a)); r2=rad(u(n,n));
    [l u]=luwpp(inv(a)*intval(a)); r3=rad(u(n,n));
    [n r1 r2 r3]
end
```

The output display only the radius of the U_{nn} element, where U is the U-factor produced by IGA, an interval matrix. The result is as follows.

10	1.6642e-013	2.8755e-014	1.7764e-015
20	2.7654e-010	3.9713e-012	3.8303e-015
30	5.8932e-008	2.5538e-011	1.0991e-014
40	8.2912e-005	4.318e-009	7.9381e-015
50	0.027519	1.9748e-007	1.0936e-014
60	NaN	1.8863e-005	1.1879e-014
70	NaN	0.00036907	1.4211e-014
80	NaN	0.026651	1.5155e-014
90	NaN	0.092109	1.7042e-014
100	NaN	NaN	2.0428e-014

Table 2.1. *Radius of U_{nn} for IGA*

Obviously, the table displays the exponential growth of interval radii produced by IGA for the first and second case. An output NaN means that computation broke down due to the fact that all pivoting elements contained zero. Only in the third case, where there is almost nothing to do, the algorithm seems to work. Note that the reason for the observed effect is *not* the condition number, it is the mere number of operations. The condition number in all cases is less than 10^3, where computation is performed in double precision (16 decimal places).

On the other hand, interval arithmetic offers the possibility to estimate the range of a function without further knowledge about Lipschitz constants or others. Consider, for example, the following function taken from a paper by Broyden [5]

$$y = \left(\begin{array}{c} 0.5 \sin(x_1 x_2) - x_2/4\pi - x_1/2 \\ (1 - 1/4\pi)(e^{2x_1} - e) + e \cdot x_2/\pi - 2ex_1 \end{array} \right) . \tag{4}$$

Then the INTLAB function

```
function y = f(x)
  Pi = typeof(intval('3.14159265358979_'), x);
  E = exp(typeof(intval(1)), x);
  y(1) = .5*sin(x(1)*x(2)) - x(2)/(4*Pi) - x(1)/2;
  y(2) = (1-1/(4*Pi))*(exp(2*x(1))-E)+E*x(2)/Pi-2*E*x(1);
```

calculates the function (4). The call

```
x = [ .5 ; 3.14 ]; y = f(x)
```

calculates the function in pure floating point, whereas the call

```
x = intval([ .5 ; 3.14 ]); y = f(x)
```

calculates the function with verified bounds. If x is specified to be an interval like $x = $ midrad ([.5; 3.14], $1e-5$), then the range of the function over the input interval is estimated. The function typeof(a,b) gives back the value of a with type adjusted to that of b.

Again, the estimation of the range will be afflicted with an overestimation. The objective of verification methods is to diminish the effect of overestimation. One of the possibilities to do that is to formulate the problem in such a way that all interval quantities are multiplied by a small factor. This is possible as demonstrated in the next section.

3. Verified solution of dense linear systems

Consider a (dense) system of linear equations

$$Ax = b \quad \text{for} \quad A \in M_n(\mathbf{R}), b \in \mathbf{R}^n . \tag{5}$$

It turns out that the use of fixed point theorems can take advantage of interval arithmetic and the possibility to estimate the range of a function. We reformulate (5) into the fixed point form

$$g(x) := x + R(b - Ax) \quad , \tag{6}$$

where $R \in \mathbf{M}_n(\mathbf{R})$ is a preconditioning matrix to be specified. Suppose for some interval vector $X \in \mathbf{IR}^n$, which is nonempty, closed, bounded and convex, it is

$$g(X) \subseteq X \quad . \tag{7}$$

Then Brouwer's Fixed Point Theorem implies existence of a fixed point $\hat{x} \in X$ of g, and (6) implies

$$R(b - A\hat{x}) = 0 \quad . \tag{8}$$

If the preconditioner is nonsingular, then a solution of the linear system (5) is enclosed in X. The proof of nonsingularity of the preconditioning matrix R together with that of A is given by the following lemma.

Lemma 3.1. *Let* $X, Z \in \mathbf{IR}^n$, $\mathcal{C} \in \mathbf{IM}_n(\mathbf{R})$ *and suppose*
$$Z + \mathcal{C} \cdot X \subseteq \text{int}(X).$$
Then $\rho(C) < 1$ *for all matrices* $C \in \mathcal{C}$.

For the proof of this and the following results in this section see [19]. The naive way to check (7) is to replace every operation by its corresponding interval operation. If $X - R(b - A \cdot X) \subseteq \text{int}(X)$, then $g(X) \subseteq \text{int}(X)$. However, this will never be satisfied because adding some quantity to an interval can never shrink its diameter. Instead, we use the so-called Krawczyk operator [13]:

$$\tilde{x} + R(b - A\tilde{x}) + (I - RA)(X - \tilde{x}) \subseteq \text{int}(x) \Rightarrow g(x) \subseteq \text{int}(x) \quad ,$$

where $\tilde{x} \in \mathbf{R}^n$. The proof follows by isotonicity:

$$\forall x \in X: \quad g(x) = x + R(b - Ax)$$
$$= \tilde{x} + R(b - A\tilde{x}) + (I - RA)(x - \tilde{x})$$
$$\subseteq \tilde{x} + R(b - A\tilde{x}) + (I - RA)(X - \tilde{x})$$
$$\subseteq \text{int}(X) \quad .$$

Together with Lemma 3.1 this implies nonsingularity of the matrices R and A, and therefore $A^{-1}b \in X$. For a practical application it turns out to be superior to enclose the error with respect to an approximate solution of the linear system. This gives better inclusions for less work. Moreover, it is easy to derive a corresponding theorem for linear systems with uncertain data.

Theorem 3.2 *Let* $[A] \in \mathbf{IM}_n(\mathbf{R}), [b] \in \mathbf{IR}^n, R \in \mathbf{M}_n(\mathbf{R}), \tilde{x} \in \mathbf{R}^n$ *and* $X \in \mathbf{IR}^n$ *be given. Suppose*

$$R \cdot ([b] - [A]\tilde{x}) + (I - R \cdot [A]) X \subseteq int(X) \quad . \tag{9}$$

Then the matrix R and all matrices $A \in [A]$ are nonsingular, and

$$\forall A \in [A] \quad \forall b \in [b] : \qquad A^{-1}b \in \tilde{x} + X \quad . \tag{10}$$

In a practical computation, the optimal preconditioner is the inverse of the midpoint matrix. Therefore, usually an approximate midpoint inverse is used. For the quality of the inclusion, consider the inclusion formula (9). The critical term is $(I - R[A])X$. However, this is the product of two small quantities: For a linear system with precise date, $I - RA$ is of the order $\varepsilon \cdot \|A\|$, and X is an inclusion of the *error* of $\tilde{x}$, which is also small if the problem is not too ill-conditioned.

It turns out in practice that the same limit $1/\varepsilon$ for the condition number applies to the verification algorithm. Therefore, one may ask what verification is useful for. The major difference to a pure floating point algorithm is that every result is verified to be correct. In pure floating point methods one might take a poor approximation to be a correct answer to a problem, see [24], [7]. This is not possible for verification methods.

Still the question remains: If a verification method delivers a poor answer, is this due to overestimations and dependencies or, is it due to the poor condition of the problem. This is solved by the following observation.

Theorem 3.3 *With the assumptions of Theorem 3.2, for all $i, 1 \leq i \leq n$ there exist $A_1, A_2 \in [A]$ and $b_1, b_2 \in [b]$ such that with*

$$Z := R \cdot ([b] - [A]\tilde{x}) \quad and \quad \Delta := (I - R[A]) X$$

it is

$$(A_1^{-1}b_1)_i \leq \inf(Z_i) + \sup(\Delta_i) \quad and \quad \sup(Z_i) + \inf(\Delta_i) \leq (A_2^{-1}b_2)_i \quad .$$

Note that Theorem 3.2 implies

$$\inf(Z_i) + \inf(\Delta_i) \leq (A^{-1}b)_i \leq \sup(Z_i) + \sup(\Delta_i)$$

for all $A \in [A], b \in [b]$ and for all $i, 1 \leq i \leq n$. Henceforth, the quality of the inclusion depends on the width of Δ, and due to the preceeding discussion this is the product of two small quantities and therefore small.

This is the reason why inclusions obtained by Theorems 3.2 and 3.3 are very sharp. We mention that an intrinsic assumption is that all uncertain data are independently varying within the given tolerances. If there are dependencies between input data, consult [11] and [19].

As an example consider the INTLAB input

```
n=500; A=2*rand(n)-1; b=A*ones(n,1); cond(A)
X = verifylss(A,b); X(1), X(n)
```

producing the output

```
8.753696537725864e+002
[    0.999999999989908,    1.000000000010085]
[    0.999999999990334,    1.000000000009646]
```

According to the condition number, the radius 10^{-11} is about what to expect in double precision. If we afflict the input data with tolerances like

```
X = verifylss(midrad(A,1e-10),b); X(1), X(n),
```

which means all matrix entries are afflicted with an absolute uncertainty of 10^{-10}, then the result is as follows:

```
[    0.999997448889150,    1.000002551110845]
[    0.999997561317640,    1.000002438682342]
```

The radii $2 \cdot 10^{-6}$ of the inclusion reflect the condition of the problem. Note that usually only approximate information about the condition number is available. Those approximations may be inaccurate (see the discussion in [10], Chapter 14 and the many papers cited over there).

4. Verified results for sparse linear systems

The methods described in the previous section are not suitable for sparse systems because an approximate inverse is used as preconditioner. For sparse systems we use an estimation of the smallest singular value of the matrix. In the following $|| \cdot ||$ denotes always the spectral norm; the results may be formulated for other norms as well. For $A\tilde{x} = b$ and a given approximation $\tilde{x}$ it is

$$||\tilde{x} - \hat{x}|| = ||A^{-1}(A\tilde{x} - b)|| \leq \sigma_n(A)^{-1} \cdot ||A\tilde{x} - b|| \quad , \tag{11}$$

where $\sigma_n(A)$ denotes the smallest singular value of A. For the moment we suppose A to be symmetric positive definite. In the course of estimations it will be *proved* that A has this property. Later, the concepts will be extended to general symmetric and unsymmetric matrices.

For the estimation we need a verified *lower* bound for the smallest singular value of A. Such a positive bound implies nonsingularity of A. One possibility is to compute an approximation s for $\sigma_n(A)$, and to verify that $A - sI$ is positive definite. This in turn is true if a Cholesky decomposition of $A - sI$ exists. One might try to prove this by performing an interval Cholesky decomposition. However, the same remarks as for IGA apply: due to overestimation and dependencies this works only for small dimension. Consider

$$A = 0.1 \cdot GG^T \quad \text{with} \quad G = \begin{pmatrix} 1 & & & & \\ 1 & 1 & & & 0 \\ 1 & 1 & 1 & & \\ & 1 & 1 & 1 & \\ 0 & & \ddots & \ddots & \ddots \end{pmatrix}.$$

We perform a Cholesky decomposition of A with interval operations and monitor the radius of the lower right element of the Cholesky factor. The result is as follows.

n	10	20	30	40
rad(G_nn)	7e-13	1e-8	2e-4	failed
cond(A)	200	600	1000	2000

Again, the failure of the approach is due to the number of operations, not due to the condition of the matrix. In contrast we use perturbation analysis of the symmetric eigenvalue problem. It is well known that for symmetric A and E and suitable numbering the following bounds are valid:

$$|\lambda_i(A + E) - \lambda_i(A)| \leq \|E\|_2 \quad \text{for } 1 \leq i \leq n \quad . \tag{12}$$

Here, λ_i denotes the i-th eigenvalue of a (symmetric) matrix. Suppose, s is an approximation to the smallest singular value of a symmetric matrix, and $HH^T \approx A - sI$ is an *approximate* Cholesky decomposition. If H has no zero on the diagonal, HH^T is positive definite, and by (12) it is for all i

$$|\lambda_i(A - sI)| \geq -\|\Delta\|, \quad \text{where } \Delta := A - sI - HH^T .$$

This implies

$$\sigma_n(A) \geq s - \|\Delta\| ,$$

a lower bound for the smallest singular value of A. The only computation to be performed in interval arithmetic in this verification process is the computation of Δ. Therefore, a verification may look as follows.

1. Compute an approximate Cholesky decomposition $A \approx GG^T$
2. Use G to compute an approximate solution $\tilde{x}$
3. Use G and inverse power iteration to compute an approximation t to the smallest singular value of A
4. Set $s := 0.9t$
5. Compute an approximate Cholesky decomposition $A - sI \approx HH^T$
6. Compute an upper bound δ of $||A - sI - HH^T||$
7. If $s - \delta > 0$, then A is positive definite and $||A^{-1}b - \tilde{x}|| \leq (s - \delta)^{-1} \cdot ||A\tilde{x} - b||$.

Algorithm 4.1. *Verified solution of s.p.d. linear systems*

The applicability of Algorithm 4.1 is limited by $\text{cond}(A) <\approx \varepsilon^{-1}$, the same bound as for any floating point algorithm. Algorithm 4.1 may also serve as a verified condition estimator. Sometimes it is even faster than an approximate condition estimator. Consider a linear system with $A := LL^T$, where L is an $n \times n$ random band matrix with band width 7. Therefore, A is of bandwidth 14. We calculate the right hand side such that the true solution of the linear system is $\hat{x}_i = (-1)^i/i$. In the following table we display the maximum error achieved by Algorithm 4.1, the condition number of A and the ratio

$$\rho := \frac{t(\text{LAPACK condition estimator})}{t(\text{verified inclusion by Algorithm 4.1})},$$

the ratio between the time for the LAPACK condition estimator divided by the total time to achieve the verified inclusion by Algorithm 4.1.

n	$||\hat{x} - \tilde{x}||_\infty / ||\hat{x}||_\infty$	cond(A)	ρ
1 000	2.0e-9	2.5e7	1
10 000	2.4e-6	3.5e10	40
100 000	3.0e-5	4.3e11	1100

Table 4.2. *Computing time for s.p.d. systems*

The main principle of this and many other verification methods is that interval computations are used only if absolutely necessary. Almost everything is done in floating point; interval arithmetic is only used to verify validity of certain assumptions of an inclusion theorem. To cite Wilkinson: "In general it is the best in algebraic computations to leave the use of interval arithmetic as late as possible so that it effectively becomes an a posteriori weapon."

For symmetric indefinite systems, the eigenvalues and singular values do not coincide. The problem can still be solved by perturbation theorems for symmetric matrices. Suppose, we have an approximation s to the smallest singular value of A. We perform an LDL^T-decompositions of $A - sI$ and $A + sI$:

$$L_1 D_1 L_1^T \approx A - sI \quad \text{and} \quad L_2 D_2 L_2^T \approx A + sI \ .$$

We use symmetric pivoting together with the method of Bunch-Kaufmann ([8], Chapter 4.4). Furthermore, we use a limited pivoting strategy to reduce fill-in [14]. By the perturbation result for symmetric eigenvalue problems, the inertia of $A - sI$ and $A + sI$ cannot differ by more than

$$\delta := \max \left(\|A - sI - L_1 D_1 L_1^T\|, \ \|A + sI - L_2 D_2 L_2^T\| \right)$$

from the inertia of D_1 and D_2. If the inertias of D_1 and D_2 are equal, then it is not difficult to see that $s - \delta$ is a lower bound for the smallest eigenvalue in absolute value of A, which is the smallest singular value of A. Therefore, the approach is similar to Algorithm 4.1 with the exception that the verification process is two-fold, the factorization of $A - sI$ and $A + sI$.

For indefinite systems we use the fact that the set of eigenvalues of

$$B := \begin{pmatrix} 0 & A^T \\ A & 0 \end{pmatrix}$$

is $\pm\sigma_i(A)$. Thus the condition number of B and A are the same. The verification process uses the above approach for symmetric linear systems. A band minimization can be performed in advance by taking advantage of the special structure of the matrix B.

Computational results for Gregory/Karney test cases 4.16 and 4.20 are as follows. The first matrix is positive definite, the second indefinite. The column "error" shows $\|\hat{x} - \tilde{x}\|_\infty / \|\hat{x}\|_\infty$.

	Example (4.16)		Example (4.20)	
n	cond(A)	error	cond(A)	error
1000	1.7e11	3.8e-14	6.2e2	1.0e-18
10000	1.6e15	5.4e-10	5.5e3	7.6e-17
20000	2.6e16	1.8e-8	1.1e4	2.6e-16
50000	1.0e18	failed	3.2e4	4.1e-15
100000			6.3e4	7.6e-14

Table 4.3 *Verified solution of banded systems*

5. Verified solution of nonlinear systems

The methods derived in the previous sections can be used for the inclusion of the solution of nonlinear systems of equations. The following results hold in a more general setting; for simplicity, we state the results with rather strong preassumptions. Let $f : \mathbf{R}^n \to \mathbf{R}^n$ be a continuously differentiable function, and let $\tilde{x} \in \mathbf{R}^n$ and $X \in \mathbf{IR}^n$ be given. We call $S(\tilde{x}, X) \subseteq \mathbf{M}_n(\mathbf{R})$ an *expansion* of f in X with respect to $\tilde{x}$ if

$$\forall\, x \in X \,\exists\, M \in S(\tilde{x}, X) : \quad f(x) = f(\tilde{x}) + M \cdot (x - \tilde{x}) \ .$$

Such an expansion can be computed by means of automatic differentiation ([17], [9]), automatic slopes ([15], [20]) or other techniques. Given such an expansion the following is true.

Theorem 5.1. *Let $f : \mathbf{R}^n \to \mathbf{R}^n$ be a continuously differentiable function, let $\tilde{x} \in \mathbf{R}^n$, $x \in \mathbf{IR}^n$, $R \in \mathbf{M}_n(\mathbf{R})$ and $S(\tilde{x}, x) \subseteq \mathbf{M}_n(\mathbf{R})$ be given. If*

$$-R \cdot f(\tilde{x}) + (I - R \cdot S(\tilde{x}, x)) \cdot X \subseteq \mathrm{int}(X) \ , \tag{13}$$

then there exists some $\hat{x} \in \tilde{x} + X$ with $f(\hat{x}) = 0$.

The proof can be found in [19]. Condition (13) can be verified on the computer. Again, the critical part $(I - R \cdot S(\tilde{x}, x)) \cdot X$ subject to overestimation is the product of two small quantities. In a practical application, the matrix R is usually choosen to be a good preconditioner, for example an approximate inverse of the midpoint of the expansion matrix $\mathrm{mid}(S(\tilde{x}, x))$.

For sparse nonlinear systems this is not possible because the inverse of the Jacobian is likely to be a full matrix. Nevertheless, we choose the preconditioner to be the *optimal* preconditioner, namely the *exact* inverse of $\mathrm{mid}(S(\tilde{x}, x))$. Define

$$R := \mathrm{mid}(S(\tilde{x}, x))^{-1} \ .$$

Then the basic rules of interval analysis ([2], [15]) yield

$$\begin{aligned} I - R \cdot S(\tilde{x}, x) \ &= I - R \cdot [\mathrm{mid}(S(\tilde{x}, x)) \pm \mathrm{rad}(S(\tilde{x}, x))] \\ &= [-|R| \cdot \mathrm{rad}(S(\tilde{x}, x)), \ +|R| \cdot \mathrm{rad}(S(\tilde{x}, x))] \ . \end{aligned}$$

A short calculation proves the following corollary of the preceeding theorem.

Corollary 5.2 *Let $f : \mathbf{R}^n \to \mathbf{R}^n$ be a continuously differentiable function, let $\tilde{x} \in \mathbf{R}^n$ and $0 < \rho \in \mathbf{R}$ be given and define $X := \{x \in \mathbf{R}^n : \|x\| \leq \rho\}$. Furthermore, let $S(\tilde{x}, x) \in \mathbf{IM}_n(\mathbf{R})$ be given with*

$$\forall\, x \in X \,\exists\, M \in S(\tilde{x}, x) : \quad f(x) = f(\tilde{x}) + M \cdot (x - \tilde{x}) \ .$$

If $0 < \tau \in \mathbf{R}$ satisfies $\tau < \sigma_n(\mathrm{mid}(S(\tilde{x}, x)))$ and

$$\|f(\tilde{x})\| + \|\mathrm{rad}(S(\tilde{x}, x))\| \cdot \rho \leq \tau \cdot \rho \ ,$$

then there exists $\hat{x} \in \mathbf{R}^n$ *with* $||\hat{x} - \tilde{x}|| \leq \rho$ *and* $f(\hat{x}) = 0$.

The advantage of the formulation of Corollary 5.2 is that it is suitable for sparse systems, and that it uses the optimal preconditioner without explicitly computing it. Note that if f is sparse, the expansion interval matrix $S(\tilde{x}, x)$ is also sparse.

As an example consider Emden's equation

$$-\Delta U = U^2 \qquad \text{where } U = 0 \text{ on } \partial\Omega \text{ for } \Omega = (0, l^{-1}) \times (0, l) \quad .$$

This equation is not too difficult to solve on the unit square; for values of l larger than 2 the discretized system becomes rapidly ill-conditioned. For an approximate solution $\tilde{u} := \lambda x_1^2 (1 - x_1)^2 x_2^2 (1 - x_2)^2$ with suitable parameter λ we perform some Newton steps and apply Corollary 5.2. The discretization uses central differences. The result for different values of l for the final error $e = ||\hat{u} - \tilde{u}|| / ||\tilde{u}||$ is as follows.

l	N	m1	m2	cond	e
1	32385	255	127	1.3e4	7.7e-13
2	32385	255	127	8.0e6	4.5e-10
2.5	32385	255	127	7.9e9	4.6e-7

Table 5.3 *Computational results for Emden's equation*

Note that verification of the result applies to the discretized system. The column N denote the total dimension of the nonlinear system, columns $m1$ and $m2$ denote number of grid points in the two variables, and *cond* the condition number of the nonlinear system.

For larger values of l the discretized system becomes truely ill-conditioned. Moreover, plenty of solutions appear and turn out to be ghost solutions produced by floating point errors. For example, consider $l = 3.1$, $m_1 = 127$, $m_2 = 31$ such that $N = 3937$. Denote the floating point iterates by $\tilde{u}^k$. Then the weighted residue shows the following behaviour.

k	$||f(\tilde{u}^k)|| / ||\tilde{u}^k||$
1	1.3e-3
2	9.8e-5
3	2.6e-7
4	3.1e-12
5	1.2e-4

Table 5.4 *Floating point iteration for Emden's equation*

What looks like an almost quadratic convergence finally turns out to be a fake. However, a stopping criterion may stop at iteration 4 and give back the computed approximation.

6. Implementation issues

Implementation of interval arithmetic is more involved than it looks at first sight. Consider, for example, the realization of (9). The computing intensive part in interval arithmetic is the multiplication $T = R \cdot [A]$, a real matrix times an interval matrix. A top-down implementation would be

```
T = zeros(n);
for i=1:n
  for j=1:n
    for k=1:n
      T(i,j) = T(i,j) + R(i,k)*A(k,j);
    end
  end
end
```

However, the most inner loop contains an interval multiplication and an interval addition. This has terrible consequences for the performance. First, the multiplication and addition each require two switches of the rounding mode, totally $4n^3$. Second, the compiler looses every possibility to optimize the inner loop, where this is exactly the strength of a floating point inner product. The latter would in fact use a multiply-and-add instruction, what is of course impossible in the above approach.

A much better way is the following. First, rewrite the interval matrix $[A]$ in midpoint-radius form, then multiply using BLAS routines. Suppose, A.inf and A.sup represent the matrices of lower and upper bounds of $[A]$, respectively. Then an algorithm looks as follows.

```
SetRoundUp
Amid = A.inf + 0.5*(A.sup-A.inf);
Arad = Amid - A.inf;
Trad = abs(R)*Arad;
Tsup = R*Amid + Trad;
SetRoundDown
Tinf = R*Amid - Trad;
SetRoundNear
```

Algorithm 6.1 *Real matrix times interval matrix*

The algorithm uses totally three matrix multiplications. However, these are calls to BLAS routines and are very fast on a variety of machines.

Apparently this seems to be the fastest way to perform interval matrix multiplication. Note that the compiled Tinf, Tsup satisfy for all $A \in [A]$

$$\text{Tinf} \leq R * A \leq \text{Tsup} \quad .$$

For the other operations similar considerations apply. Those operations are collected in a Matlab toolbox called INTLAB [21]. The main advantage of INTLAB is that it is entirely written in Matlab code, with the exception of exactly the 3 routines for switching the rounding mode. The above examples are calculated in INTLAB.

An example of INTLAB code is the following algorithm for solving dense systems of equation according to Theorem 3.2. The following is original INTLAB code.

```
function X = denselss(A,b)          % linear system solver
  R = inv( mid(A) ) ;               % for dense matrices
  xs = R * mid(b) ;
  Z = R * (b-intval(A)*xs) ;
  C = speye(size(A)) - R*intval(A);
  Y = Z;
  E = 0.1*rad(Y)*hull(-1,1) + midrad(0,10*realmin);
  k = 0; kmax = 15; ready = 0;
  while ~ready & k<kmax
    k = k+1;
    X = Y + E;
    Y = Z + C * X;
    ready = in0(Y,X);
  end
  if ready
    X = xs + Y;
  else
    disp('no inclusion achieved for \');
    X = NaN;
  end
```

Algorithm 6.2 *Solution of dense linear systems*

A timing in seconds on a 120 Mhz Pentium I Laptop for Algorithm 6.2 is as follows.

294

dimension	pure floating point	verified point	verified interval
100	0.09	0.53	0.70
200	0.56	3.35	4.23
500	8.2	50.9	67.6

Table 6.3 *Solution of real linear systems*

Looking into the code of Algorithms 6.1 and 6.2 reveals that the total number of operations for Algorithm 6.2 for interval linear systems is $4n^3$, n^3 for the matrix inversion and $3n^3$ for the point matrix times interval matrix multiplication $R \cdot A$. Gaussian elimination needs $n^3/3$ operations, but the measured computing times show only a factor of 7 to 8 instead of 12. The reason is that matrix multiplication does not take 3 times as long as Gaussian elimination. Consider

```
n=200; A=2*rand(n)-1; b=A*ones(n,1); k=10;
tic; for i=1:k, inv(A); end, toc/k
tic; for i=1:k, A*A; end, toc/k
tic; for i=1:k, A\b; end, toc/k
```

producing

1.24 seconds   for matrix inversion,
0.78 seconds   for matrix multiplication, and
0.56 seconds   for Gaussian elimination. Finally, we display the

code for the solution of a system of nonlinear equations according to Theorem 5.1.

```
function [ X , xs ] = verifynlss(f,xs)
%VERIFYNLSS   Verified solution of nonlinear system
%
%   [ X , xs ] = verifynlss(f,xs)
%
% f is name of function, to be called by f(x),
% xs is an approximation
% optional output    xs   improved approximation
%

% floating point Newton iteration
  n = length(xs);
  xsold = 2*xs;
  k = 0;
  while ( norm(xs-xsold)>1e-10*norm(xs) & k<10 ) | k<1
```

```
      k = k+1;                % at most 10,
      xsold = xs;             % at least 1 iteration performed
      x = initvar(xs);
      y = feval(f,x);
      xs = xs - y.dx\y.x;
    end

% interval iteration
  R = inv(y.dx);
  Z = - R * feval(f,intval(xs));
  X = Z;
  E = 0.1*rad(X)*hull(-1,1) + midrad(0,realmin);
  ready = 0; k = 0;
  while ~ready & k<10
    k = k+1;
    Y = hull( X + E , 0 );    % epsilon inflation
    Yold = Y;
    x = initvar(xs+Y);
    y = feval(f,x);           % f(x) and Jacobian by
    C = eye(n) - R * y.dx;    % automatic differentiation
    i=0;
    while ~ready & i<2        % improved interval iteration
      i = i+1;
      X = Z + C * Y;
      ready = in0(X,Y);
      Y = intersect(X,Yold);
    end
  end
  if ready
    X = xs+Y;                 % verified inclusion
  else
    X = NaN;                  % inclusion failed
  end
```

Algorithm 6.4. *Verified solution of nonlinear systems*

Again this is original INTLAB code. We want to stress that the function f need not to be changed to cover floating point arguments or interval arguments. The operator overloading generates automatically correct operations and code.

As an example consider a problem given by Abbott and Brent in [1], the discretization of

$$3 \ y'' \ y + y'^2 = 0 \quad \text{with} \quad y(0)=0; \ y(1)=20;$$

We solve the discretized system, not the continuous equation. In the paper, the initial (poor) approximation is a vector all entries of which are equal to 10, the true solution is 20*x^.75. The discretized problem is specified by the following INTLAB function.

```
function y = f(x)
  y = x;
  n = length(x); v=2:n-1;
  y(1) = 3*x(1)*(x(2)-2*x(1)) + x(2)*x(2)/4;
  y(v) = 3*x(v).*(x(v+1)-2*x(v)+x(v-1))...
          + (x(v+1)-x(v-1)).^2/4;
  y(n) = 3*x(n).*(20-2*x(n)+x(n-1)) + (20-x(n-1)).^2/4;
```

Note the vectorized formulation of the function. The timing on the 120 Mhz Pentium I Laptop by the following statement

```
tic; X = verifynlss('f',10*ones(n,1)); toc
```

is

5.4 seconds	for dimension n=50,
8.5 seconds	for dimension n=100, and
20.3 seconds	for dimension n=200.

The first and last components of the inclusion are

```
X(1:4)
intval ans =
    0.346256418326_
    0.6045521734322
    0.8305219234696
    1.0376691412984
X(197:200)
intval ans =
    19.7005694833674
    19.775568557350_
    19.8504729393822
    19.9252832242374
```

An "_" in the output means that subtracting 1 from and adding 1 to the last displayed figure before the underscore produces a correct interval for the result. An output without underscore is correct up the the last digit. The output is also written in Matlab and is rigorous and correct.

7. Conclusion

Self-validating methods are designed to produce rigorous error bounds for the solution of numerical problems. We want to stress again that those methods are *not* thought to replace existing numerical methods whatsoever. Traditional numerical methods usually produce correct results within a certain error margin. There are situations where numerical methods produce poor or sometimes erreneous results. However, those situations are rare. To cite Vel Kahan: "Numerical errors are rare, rare enough not to care about them all the time, but yet not rare enough to ignore them."

Self-validating methods should be used when another degree of safety is necessary and/or if there is doubt about reliability of some approximation. The scope of applicability of self-validating methods is still limited, although always increasing. The very large problems treated today in numerical analysis are clearly out of the scope of current self-validating algorithms. However, there is the possibility of treating data with uncertainties which may also be advantageous beside the correctness of all results.

References

1. Abbott, J.P. and Brent, R.P., (1975) Fast Local Convergence with Single and Multi-step Methods for Nonlinear Equations, *Austr. Math. Soc. 19 (Series B)*, pp. 173–199
2. Alefeld, G. and Herzberger,J.(1983) *Introduction to Interval Computations*, Academic Press, New York
3. Anderson, E., Bai, Z., Bischof, C., Demmel, J., Dongarra, J., Du Croz, J., Greenbaum, A., Hammarling, S., McKenney, A., Ostrouchov, S. and Sorensen D.C. (1995) *LAPACK User's Guide, Resease 2.0*, SIAM Publications, Philadelphia, Second edition
4. Braune, K.D. (1987) *Hochgenaue Standardfunktionen für reelle und komplexe Punkte und Intervalle in beliebigen Gleitpunktrastern*, Dissertation, Universität Karlsruhe
5. Broyden, C.G. (1969) A new method of solving nonlinear simultaneous equations, *Comput. J.*, **12**, pp. 94–99
6. Demmel, J.B. (1987) Condition Numbers and the Distance to the Nearest Ill-posed Problem, *Numer. Math.*, **51**, pp. 251–289
7. Foster, L.V. (1994) Gaussian elimination with partial pivoting can fail in practice, *Siam J. Matrix Anal. Appl.*,**14**, pp. 1354–1362
8. Golub, G.H. and Van Loan, C. (1989) *Matrix Computations*, John Hopkins University Press, Second edition
9. Griewank, A. (1989) On Automatic Differentiation, *Mathematical Programming 88*, Kluwer Academic Publishers, Boston
10. Higham, N.J. (1996) *Accuracy and Stability of Numerical Algorithms* SIAM Publications, Philadelphia
11. Jansson, C. (1991) Interval Linear Systems with Symmetric Matrices, Skew-Symmetric Matrices, and Dependencies in the Right Hand Side, *Computing*, **46**, pp. 265–274
12. Krämer, W. (1987) *Inverse Standardfunktionen für reelle und komplexe Intervallargumente mit a priori Fehlerabschätzung für beliebige Datenformate*, Dissertation, Universität Karlsruhe
13. Krawczyk, R. (1969) Newton-Algorithmen zur Bestimmung von Nullstellen mit Fehlerschranken, *Computing 4*, pp. 187–201
14. Leistikow, T. (1998) *Verifizierte Lsung groer Gleichungssysteme mit Bandstruktur*, Technical report, Inst. f. Informatik III, TU Hamburg-Harburg
15. Neumaier, A. (1990) *Interval Methods for Systems of Equations, Encyclopedia of Mathematics and its Applications*, Cambridge University Press
16. Neumann, J.v. and Goldstine, H.H. (1947) Numerical Inverting of Matrices of High

Order, *Bull. Amer. Math. Soc.*, **53**, pp. 1021–1099

17. Rall, L.B. (1981) Automatic Differentiation: Techniques and Applications, *Lecture Notes in Computer Science*, **120** Springer Verlag, Berlin-Heidelberg-New York

18. C. Reinsch (1979) Die Behandlung von Rundungsfehlern in der numerischen Analysis, *Jahrbuch Überblicke Mathematik 1979*, S. D. Chatterji and others, Bibliographisches Institut, Mannheim, Wien, Zürich, pp. 43–62

19. Rump, S.M. (1994) Verification Methods for Dense and Sparse Systems of Equations, *Topics in Validated Computations — Studies in Computational Mathematics*, Herzberger, J., Elsevier, Amsterdam, pp. 63–136

20. Rump, S.M. (1996) Expansion and Estimation of the Range of Nonlinear Functions, *Mathematics of Computation*, **65(216)**, pp. 1503–1512

21. Rump, S.M. (1998) INTLAB - Interval Laboratory, submitted for publication, http://www.ti3.tu-harburg.de/rump/intlab/index.html

22. Rump, S.M. (1999) Ill-conditioned Matrices are componentwise near to singularity, *SIAM Review, to appear*

23. Sautter, W. (1971) *Fehlerfortpflanzung und Rundungsfehler bei der verallgemeinerten Inversion von Matrizen*, Dissertation, TU München

24. Wright, S.J. (1993) A collection of problems for which Gaussian elimination with partial pivoting is unstable, *SIAM J. Sci. Comput.*, **14(1)**, pp. 231–238

THE ACCURACY OF NUMERICAL MODELS FOR CONTINUUM PROBLEMS

STANLY STEINBERG
Department of Mathematics and Statistics
University of New Mexico
Albuquerque NM 87131-1141 USA

Abstract.

I present here my rather personal view on what are the important points to consider in building simulation codes for continuum mechanics problems and, to some extent, electromagnetics. I focus lots of attention on geometry and grids as constructing this part of the model easily consumes much of the modeling effort. I also place a strong emphasis on understanding the conservation and material laws that underly the modeling, and because these laws are so different, that they must be discretized differently.

I also believe it is important to automate most code generation, that is, use a problem solving environment (PSE) to write efficient and error-free code.

1. Introduction

A large class of continuum models consist of a region in one, two, or three dimensional space that represents the object being modeled, a system of partial differential equations the describes the physics in the interior of the region, boundary conditions that describe the physics at the boundary of the region, and initial conditions that describe the initial state of the physical system. This is called an initial-boundary value problem and its solution represents a model for the underlying physical problem.

We want to understand, at least in general terms, what kinds of errors occur during the modeling. That is, how much error is there in modeling the physical situation with an initial boundary value problem and how much error is there in solving the mathematical model.

299

H. Bulgak and C. Zenger (eds.), Error Control and Adaptivity in Scientific Computing, 299–323.

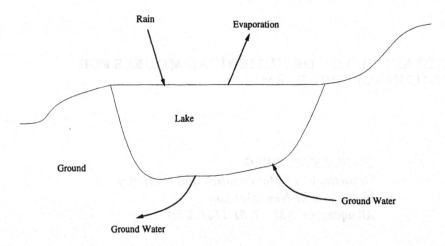

Figure 1.1. A Lake

These notes are directed towards students who have some modest or no experience in modeling, particularly numerical modeling. In the following we will discuss parts of the modeling problem and try to understand what types of errors occur in that part. Throughout we emphasize the importance of conservation and constitutive laws.

1.1. A SIMPLE EXAMPLE

Consider a lake and the conservation of water in the lake. Water can enter and leave the lake by various means:

- rain;
- evaporation;
- rivers in and out;
- flow thru the ground;
- someone puts a hose into the lake and pumps out water;
- ...

It seems reasonable to believe that the more accurately we account for the amount of water in the lake at some starting time and the sources and losses of water, the more accurately we determine the amount of water in the lake at any given future time. We believe that water cannot just appear or disappear, but can only enter or leave the lake for a reason. The level of accuracy could, in principle, be taken to the point of counting molecules of water. Surely this level of accuracy is impossible in such a natural setting, but is practical in the laboratory. We now want to look at what ideas go into modeling the amount of water in the lake: the continuum hypothesis, conservation laws, and initial and boundary conditions or source terms. By the way, if we know the rate at which the water leaves the lake and enters the ground, this is a boundary condition, whereas the rain falling on the lake is considered a source term at the boundary, and the extraction of water from the lake using a hose, is considered a sink, that is, negative source term.

We will always think of space as three dimensional, with one and two dimensional models representing three dimensional models where there are symmetries that eliminate the need for one or more variables to describe the region. For example extremely thin regions can be described using two variables or problems where the solution doesn't depend on of the variables can be described using the remaining two variables.

1.2. THE CONTINUUM HYPOTHESIS

The continuum hypothesis is discussed in many books on continuum mechanics (particularly fluid mechanics and solid mechanics books). From our point of view this hypothesis means that materials can be describe by integrable function that are scalar, vector, tensor, $\cdots$ valued and the amount of stuff described by the function f in a region R is given by

$$\int_R f \, dV, \tag{1.1}$$

where dV is the differential of volume. However, we will also consider surfaces and the amount of material passing through the surface if either the material or the surface or both are moving. So if a material is moving with a velocity $\vec{v}$ relative to a smooth surface S and $\vec{n}$ is a smooth unit normal to the surface, then the amount of material moving though the surface in the direction $\vec{n}$ is described by

$$\int_S f \vec{v} \cdot \vec{n} \, dS, \tag{1.2}$$

where dS is the differential of surface area. For this integral to make sense mathematically, the function f must satisfy additional modest smoothness requirements with continuity being sufficient. These smoothness requirements play a critical role in our modeling.

Now the above discussion about the lake indicates that the volume and surface integrals do not make sense when the volume or surface is smaller than a molecule, but with the apropriate probabilistic model for parts of a molecule, even smaller volumes and surfaces make mathematical sense. However, if the physical phenomena is on the scale of the size of molecules then the mathematical model must be changed.

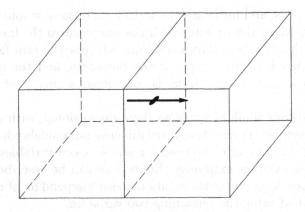

Figure 1.2. Computational Cells and Conservation

In any case, all models have their limits of applicability. I use the continuum hypothesis and and study problems in *continuum mechanics*. There are many restrictions on the problems that can be done without modifying the standard theory:

1. reasonable temperatures
2. reasonable pressures
3. moderate velocities
4. moderate forces
5. etc.

It is with in the context of the continuum hypothesis that the conservation and constitutive laws we will discuss make sense.

1.3. CONSERVATION LAWS

Let us return to the conservation of water in the lake. The main errors in this model are in determining the amount of water in the lake at some starting time, that is errors in the initial conditions, and in determining how much water enters and leaves the lake, at the boundaries of the lake, that is errors in the boundary conditions.

Now consider dividing the water in the lake into *conceptual* cells as one would do for a numerical simulation. The water in one cell can change only by what enters from or leaves to a neighboring cell or from some external source as illustrated in Figure 1.2. Now this more abstract conservation law is certainly accurate down to counting molecules, and with a reasonable interpretation, even parts of molecules. So the conservation laws in continuum mechanics, electomagnetics, ⋯ really are very accurate!

How can we capture this accuracy in a numerical simulation? Well there are a number of numerical methods based on modeling the conservation laws:

- FINITE VOLUME methods are based on approximating the conservation laws. This would include the finite-volume element methods.
- MIMETIC methods can exactly represent the conservation laws!

  - EXACTLY represent the conservation laws!
  - Yes really exactly, without numerical error!

Mimetic methods can be used to compute quantities such as the amount of stuff in a cell and the amount of stuff passing through a the surface of a cell. So it is obvious we can write conservation laws exactly! So there must be other problems with mimetic methods? Yes, they do require approximations, but in places more apropriate than the conservation laws. So this is really and advertisement for such methods.

1.4. CONSTITUTIVE LAWS

Constitutive laws describe how a material behaves if we push or pull on it, heat it, expose it to electromagnetic radiation and so on. For example Hook's law for a spring and Newton's law of cooling are constitutive laws.

We can illustrate the connection between the conservation laws and the constitutive laws using a simple one dimensional point particle of unit mass whose position at time t is given by $x = x(t)$ and the particle is being acted on by a force $f(x)$. (This is then not a continuum model, but a point mass model.) In this case, Newton's law for the motion of a particle under a force gives us a description of the motion of the particle:

$$\frac{d^2 x}{dt^2} = \ddot{x} = f(x). \tag{1.3}$$

In this situation, the energy is the sum of the kinetic energy and potential energies:

$$E(x) = \frac{1}{2}(\dot{x})^2 + P(x), \tag{1.4}$$

where

$$P'(x) = -f(x). \tag{1.5}$$

Now

$$\dot{E}(x) = \dot{x}(t)\ddot{x}(t) - f(x)\dot{x}(t) = \dot{x}(t)(\ddot{x}(t) - f(x)). \tag{1.6}$$

so if $x(t)$ is a solution of Newton's equation, then $\dot{E} = 0$ so $E(x(t))$ is constant and the energy is conserved.

Next, assume that $x(t)$ minimizes the energy, that is, for all $y(t)$ that satisfy $y(0) = 0$ and $y(T) = 0$,

$$\int_0^T E(x + \epsilon\, y)\, dt \qquad (1.7)$$

is a minimum when $\epsilon = 0$. Differentiating this with respect to ϵ and setting $\epsilon = 0$ gives

$$\int_0^T (\dot{x}\,\dot{y} + f(x)\, y)\; dt = 0. \qquad (1.8)$$

Integrating by parts gives

$$\int_0^T (-\ddot{x} + f(x))\; y\, dt = 0. \qquad (1.9)$$

and because this holds for all y, x must satisfy Newton's equation.

So Newton's law implies the conservation of energy and the minimization of the energy implies Newton's laws. Again we see that the conservation law is very important for describing the physical system.

For a spring under moderate displacement, we can take $f(x)$ to be Hook's law:

$$f(x) = -k\, x, \qquad (1.10)$$

where k is a constant. Hook's law is one of the most basic constitutive laws. How good is this law? Well not very good at all for complex springs or non-trivial motions! So we are in a situation where the conservation laws are very good, but the constitutive law has problems.

Now recall, from the elementary theory of differential equations, that we can understand the solutions of Newton's equation by plotting orbits in the phase plane, that is, we plot the curves $(x(t), \dot{x}(t))$ in a plane with the axes labeled by x and $\dot{x}$. When the force is given by hooks law, then the conservation of energy implies that the curves in the phase plane are ellipses, that is, the motion is periodic. However, if the potential energy $P(x)$ is merely concave up, then the curves in the phase plane will still be closed figures and the motion will be periodic. So I conclude that the exact nature of the constitutive laws is not critical for understanding the long time behavior of the system (the solutions are periodic), but the conservation law is critical. However, if we wish to predict the position or frequency of an oscillating weight on a spring with high precision, then we really need a high-precision Hook's law!

By the way, the same ideas apply if the energy is not conserved, but energy is added or dissipated or both. For example, we can add a resistance term to Newton's equation:

$$\frac{d^2 x}{dt^2} = \ddot{x} = f(x) + g(\dot{x}). \qquad (1.11)$$

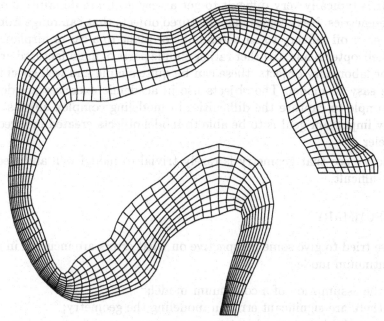

Figure 1.3. Grid for the Continental Shelf of the Gulf of Mexico

in which case the energy dissipates,

$$\dot{E}(x) = -\dot{x}(t)g(\dot{x}),\qquad(1.12)$$

if $g(x) > 0$. So the energy will dissipate independently of the detailed structure of g or f, and the orbits in the phase plane will decay into the origin. So still the energy plays a more important role than the details of the forces.

1.5. GEOMETRY

To model a system we must have a description of the geometry the system. There is a convenient hierarchy of geometries based on the type of system being modeled:

- Natural objects (e.g. lake)
- Manufactured objects (e.g. car engine)
- Laboratory objects (e.g. devices used in experiments)
- Academic objects (e.g. square, cube, etc.)

A sample of important natural objects that are frequently modeled is the earth's atmosphere, the oceans or parts of the ocean (see Figure 1.3 from [10]), a lake, ground water flow, the heart and other parts of organisms, etc. It is typically very difficult to get a very accurate description of natural geometries. Next comes manufactured objects that can range from a sky scraper or oil refinery to an automobile or plane to a car or airplane engin, to a helicopter blade, to other rather simple parts of manufactured artifacts. As for laboratory objects, these can be quite simple and well specified and quite easy to model. The objects use in academic exercises are definitely too simple to capture the difficulties in modeling complex objects. A currently important goal is to be able to model objects created in a geometric modeler.

So we see that geometry is rarely trivial to model well and frequently very difficult.

1.6. SUMMARY

I have tried to give some perspective on what errors are incurred in making a continuum model:

- the assumption of a continuum model;
- there are significant error in modeling the geometry;
- material laws are subject to many errors;
- initial conditions are difficult to determine accurately;
- source terms are also difficult to determine;
- determining the boundary conditions.

On the other hand, there are no significant errors in the conservation laws. The boundary conditions also have a conservation, constitutive, and source parts, with the conservation part being accurate, but the constitutive and source parts being difficult to determine accurately. This is my personal starting point when thinking about modeling a system.

So what do we mean by modeling something accurately when we are faced with so many errors, and we still haven't gotten to discretization errors and machine precision errors. And what about programming errors? How does all of this fit together.

We will decompose the accuracy question into two parts: the *verification* and *validation* of floating-point codes the implement discretizations of continuum models. Here I follow the terminology clearly explained in the book *Verification and Validation in Computational Engineering* written by Patrick J. Roache in 1998 [13], and which I recommend as a the book to start reading if you are interested in the problem of understanding when a floating-point code is good. We will not discusses the *validation* of code here. Validation essentially means to check the results of the simulations given by the code against experimental data, a problem even more difficult than *verification* of codes which is our main interest. One of the fundamental difficulties validation of codes is that experimental data can, and frequently does, contain more error than reasonable simulations.

Another important feature of modeling is just how much of a given system are we going to model, some complete system or some small part of a larger system? It is currently an important goal of business and government to model complete systems, say to model the operation of a factory, including distribution and sales and in the same modeling system model the complete design and manufacturing process. Such projects will take large teams, substantial computing power and lots of funds. It appears that many industries and government labs have a good start on such systems and that researchers at universities are now studying them. How can we measure the accuracy of such systems? Here we will restrict ourselves to the verification of codes for simulating continuum processes.

2. Numerical Modeling

Another apropriate title for this section would be *discrete modeling* to distinguish this material from the continuum or continuous modeling. If one is studying a situation where there are a small number of discrete items in the model, e.g. a small set of point masses, then one can deal with a fixed discrete model and both analyze the model and simulate the model numerically. In continuum models, there are so many molecules involved that it is forever hopeless to model the individual molecules. This doesn't preclude looking at small parts of a continuum and making a molecular model. Typical discrete models of the continuum involve dividing the continuum into a modest number of points or cells, as in finite-difference and volume methods, a modest number of elements as in the finite element method, or into a finite number of modes as in Fourier, spectral, and wavelet methods. By the way, we will measure the size of discretization by the diameter h of a cell, which is of the order of $1/n^k$ if the discretization contains n^k grid points in k dimensional space.

In the previous section, we have pointed out that there are many errors made in using initial boundary value problems to model continuum problems. However, there is a wide range in the amount of error made in estimating parameters in continuum models, and scientist and engineers are always improving how well they can measure these parameters. Really, what I am trying to say is that it is very difficult to predict how much accuracy will be required of a discrete model. So the first requirement on any *good* discrete model for the continuum is that it can be made arbitrarily accurate. Now current experience shows, that for modeling problems of modest complexity, that numerically simulating typical three-dimensional discrete models of reasonable accuracy easily swamps the largest computers. So there is an incessant demand for discrete models that can achieve high accuracy at more modest simulation cost.

The way that we will deal with this situation is to assume that we are given an accurate continuum model, a model that approximates some physical situation with a accuracy far greater than we can hope to achieve using simulations based on discrete models. So our task is then to create accurate discrete approximations of the continuum model and then simulate the model accurately and rapidly on a floating-point computer.

It turns out that there are some subtle points in measuring accuracy. To be more precise about accuracy, we will to distinguish between two types of norms for measuring errors. In the continuum, they are called *supremum* or *sup* norms and *integral norms*. For discrete data, these are *maximum* or *max* and *sum* norms. The first requirement is that the sum norms be normalized so that they converge to a nontrivial continuum norm (the max norm does this automatically). Because of this normalization of the sum norms, if we take a sequence of discrete vector converging to a continuum function, and then change one value of a discrete vector to 17 at one point, then both vectors converge, in sum norm, to the same function. For example, suppose that our vectors are approximating a one-dimensional function and have n components,

$$v = (v_1, v_2, \cdots), \qquad (2.13)$$

and that we use the sum norm that gives the mean square norm, normalized correctly,

$$\|v\| = \frac{1}{n} \sqrt{v_1^2 + v_2^2 + \cdots} \qquad (2.14)$$

and then we choose a vector of length n,

$$v_n = (1, 1, 1, \cdots), \qquad (2.15)$$

and then change first location in v to 17,

$$w_n = (17, 1, 1, \cdots). \qquad (2.16)$$

Then

$$\|v_n - w_n\| = \frac{16}{n}, \qquad (2.17)$$

and both sequences of vectors converge to the same function, namely $f \equiv 1$, at a *first-order* rate,

$$\|v_n - w_n\| = O(\frac{1}{n}), \qquad (2.18)$$

that is, the convergence is linear in $h = 1/n$. The failure of this and and all other common sum (or integral) norms to detect this arbitrary change in the vector says that if we use of integral norms then, at first order, we will miss such errors!

In higher dimensions, if h measures the maximum cell size then saying that a sequence of discrete vectors converges at order k,

$$\|v_n - v\| = O(h^k),\qquad(2.19)$$

leaves the possibility that the discrete solution is only converging at order $k - 1$ on some lower dimensional subset. This cannot happen with the sup or max norms. However, even the sup norm is dangerous at first order. For example, take a first order-order code and replace some random $i + 1$ by $i - 1$, which is all to easy to do by accident, and then code may still converge at first order, because $f(x + h) - f(x - h)$ is first order in h.

So checking that a code converges at a second order rate in the sup norm is a much more stringent than to checking the second-order convergence rate using an integral norm or the first-order convergence rate using the sup norm. We will adopt this as our minimal requirement for testing a code. Now we are not alone in this opinion. In [13] Roache gives a *Policy Statement on the Control of Numerical Accuracy* on page 410 that is used in an engineering journal to guide authors and supports our point of view. In fact, in the first place, my view was influenced by this statement.

We need to make another distinction: our floating point codes are implementations of algorithms, and algorithms have two main important properties: accuracy and stability. These properties can be analyzed theoretically or tested numerically using a code that implements the given algorithm. Both approaches have drawbacks. It is quite easy to make an error in the analysis, and the floating point implementation can also have errors. So we try to do both: make a theoretical analysis and then used guide and confirm that analysis using a floating-point implementation of the algorithm.

So now we can restate our above remarks more precisely. We require that our algorithms have second-order accuracy in max norm.

Stability also has some subtlies too! Let $f(t)$ be the solution of a time dependent initial boundary value problem and f^n, $0 \le n \le N$ be an approximation of $f(t)$ on the the interval $0 \le t \le T$. It is common to have problems where we have a important norm, typically and energy norm, where

$$\|f(t)\| = \|f(0)\|, \quad t \ge 0,\qquad(2.20)$$

that is all solutions of the continuum problem conserve the energy norm. However the standard numerical stability condition tells us that, for some discrete norm analogous to the energy norm, there are constants C and K such that

$$\|f^n\| \le Ce^{Kn}.\qquad(2.21)$$

So the numerical solution can grow exponentially in time while the continuum solution is bounded in time. The constant K typically becomes small as the discretization parameter h goes to zero, but to obtain a reasonable K can require an h much smaller than the accuracy requirement, and thus excessive computation is required for the simulation. In fact, fixing the bound on f^n and letting T go to infinity well require h goes to zero, increasing the computation cost of accurate long-time simulations. I think this is not good! It is better to have a numerical algorithm with a discrete analog of the energy norm that satisfies

$$\|f^n\| = \|f^0\|, \quad n \geq 0. \tag{2.22}$$

So we are looking at finite-volume an mimetic methods as being the methods of choice for long time simulations.

In fact, in many simulations the energy decreases or increases, such as when there is some energy dissipation to heat due to resistance (and heat is not accounted for in the model) or there is an energy source. In these cases, we want the energy in the discrete model to dissipate or increase at the same rate as in the continuum.

In any case, we now have a second requirement: the discretized model should have analogs of all of the conservation laws of the continuum systems. Moreover, if important functionals of the solution such as energy are increasing or decreasing at some rate in the continuum, then the discrete problem should have an analogous function that is decreasing or increasing at the same rate. This typically implies a strong form of stability for the discretized system. In fact, many engineering and science problems require understanding what happens to a system when it is starts in son ɔ state and evolves into some behavior for moderates times, for example, the airflow around a helicopter blade after it has started turning and has reached some rotation rate. If one wants to know how the system behaves as it starts up, then it seems any accurate numerical method will be useful, perhaps even unstable methods (see Roache [14]).

To make a point, I have known engineers to say, jokingly, "I don't care if the code is accurate, I just don't want it to blow up." I interpret this to mean that they need codes that, no matter what problem they are given, are stable and that modest rather than high accuracy is ok. It seems that we can actually build codes that are stable for wide classes of problems and provide good accuracy for modest problem and modest accuracy for really difficult problems.

So what is it that makes a problem difficult? I will make a small tour to get an overview on creating a floating-point code to model difficult problems.

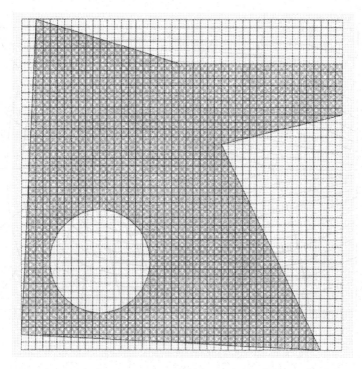

Figure 2.4. An Overlay Grid

2.1. GEOMETRY AND GRIDS

There are many resources for grid generation methods, but the best place to start is to look at the web site run by R. Schneiders [16]. I, as many others, consider discretizing the geometry the most challenging and time consuming part in creating a floating-point modeling codes. Here we will discuss planer grid generation. Three-dimensional volume and surface grid generation is significantly more difficult and an active research and development area.

We can begin with the simplest possible grids, *overlay* grids, an example of which is shown in Figure 2.4 (see also [3]). Such grids are very simple to generate, but require special programming for even simple boundary conditions, and implementing accurate boundary conditions is quite difficult. Should the region change shape during the simulation, then these problems are even more difficult. The special structure of the boundary conditions adversely impacts the vectorization of codes using such grids.

When working with a region that models some physical situation, the first question to ask is what is the connectivity of the region, that is, how many holes does the region have in it? Nice regions, of course, have no holes! So in the plane, a circle and a square have no holes, while an annulus

312

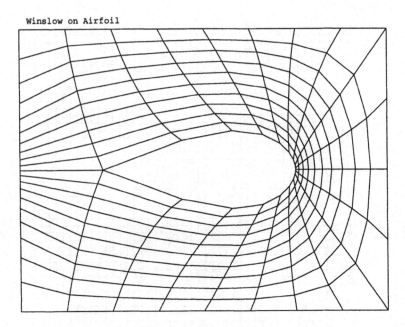

Figure 2.5. An Airfoil

does. In Figure 2.5, we display an airfoil with a C-type grid around it, that is, one set of grid lines look like the letter C written backwards. Another possible grid is an O-type grid where the grid lines wrap around the airfoil. In any case, one must decide what type of grid to use in such a situation. Is such a small space, I cannot give much guidance on which type of grid to choose, but we can immediately see some of the difficulties. For the C-type grid in figure airfoil, the line at the trailing edge is called a *cut* and the C lines come together and are essentially parallel at the cut; the remaining grid lines are nearly perpendicular to the cut. These nearly perpendicular lines meet at the trailing edge of the air foil but when crossing the cut, they are continuous but not differentiable. So the numerical scheme near the cut will suffer some loss of accuracy. On the other hand, the O-type grid lines will be severely "pinched" at the trailing edge, causing a loss of accuracy. For non-trivial regions, neither type of grid is trivial to generate and both types have an adverse impact on the discretization of the partial differential equations and the boundary conditions. One hole in the region makes things more difficult, many holes makes the situation much more difficult.

There are two major types of planar grids, triangular and logically rectangular. Triangular grids divide a region into a union of triangles, while logically-rectangular grids, which are very special quadrilateral grids, di-

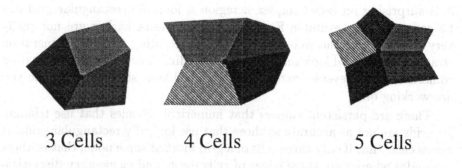

3 Cells　　　　　　　4 Cells　　　　　　　5 Cells

Figure 2.6.　Quadrilateral Grids

vide a region into quadrilaterals. General quadrilateral grids [6] are also of significant interest and should work well with finite-volume methods, but I don't know of any definitive study of this. In general quadrilateral grids, three or more quadrilaterals can come together at a node (point) as illustrated in Figure 2.6. In logically-rectangular grids, each node touches exactly four cells. More precisely, the points in a logically-rectangular grid can be written as $(x(i,j), y(i,j))$, $0 \leq i \leq I$, $0 \leq j \leq J$, for some I and J. I do believe that the book [10] by Knupp and myself has an excellent introduction to structured grid generation. The paper [5] by Castillo and Otto presents some elementary variational methods that show considerable promise. Note that logically-rectangular grids can be made into triangular grids by drawing a diagonal in the quadrilaterals, which should rightfully be called logically triangular.

A main point about the "logical" grids is that they can be described using simple data structures such as arrays (technically, no indirect addressing is needed). Other grids require special data structures, e.g. in a general triangular grid one needs to look up all triangles the have a node in common and all nodes on a given triangle. This requires special data structures which means that the codes using these grids are considerably more costly to run than those with a logical structure. However, there are several other considerations that play an important role in the choice of a grid.

For complex regions, after overlay grids, triangular grids are the easiest to generate, while logically-rectangular grids are the most difficult. So if a regions is complex, say some natural object like England or has been generated by a geometric modeling program, the grid of choice seems to be triangular. If the region is of some intermediate complexity, then triangular, quadrilateral, or logically-rectangular grids all can be used, while if the region is simple, then logically rectangular is the best choice. However, it is surprising on how complex a region a logically-rectangular grid can be generated, for example England or the Havana harbor are not really very difficult (see the web site [2]). Before deciding on a grid-generation strategy, you should look through the proceedings of several grid-generation conferences and several texts looking for problems similar to the one you are working on.

There are persistent rumors that numerical schemes that use triangular grids are not as accurate as those that use logically-rectangular grids. It seems clear that if only three cells of a grid meet at some node, that is, there are nodes where only three edges of cells meet, and elementary discretization methods are being used, then this will cause a reduction in accuracy. Of course this can be compensated for by using higher-order discretizations, but there are additional computational and programming costs for doing this. Anyway, grids with one or more "three-edge nodes" should be used with caution.

There are lots of "rumors" about grids and numerical methods; chasing them down a checking their validity would be useful for the numerical simulation community, and some would even make good research projects. One is that skewed grids, triangular grids where the triangles depart significantly from equiangle or quadrilateral grids where the cells depart significantly from rectangles, cause large errors in simulations. My experience doesn't support this conclusion! I have tested grids where the angle between two grids lines is a small as 2 or 3 degrees without significant problems and have successfully used grids with a cell aspect ratio of 10^6 (this is the ratio of the longest side of the cell to the shortest side). As it must be, the error does grow with decreasing angle θ between the grid lines, I believe proportional to θ^p with p around 3 or 4, but apparently with a small proportionality constant.

There are also rumors that it is important to have the grid lines perpendicular at the boundaries. My experience supports this, but the problem could be that I do not know how to discretize the boundary conditions accurately when the grids are not close to orthogonal at the boundary. My experience is supported by others though. It is not possible to use triangular grids and have the grid lines perpendicular at the boundary, but some subset of the lines can be made orthogonal, which should do the trick? If you are considering using non-trivial boundary conditions, e.g. mixed or Robin conditions, then be careful about this point.

2.1.1. *Adaptive Grids*

The grids we have been talking about are properly call *geometry adaptive* grids. It is also important to construct *solution adaptive* grids. We call most solution adaptive methods *solution feature adaptive* or just feature adaptive grids because they are adapted to some feature of the solution of a differential equation. This means, for example, that the grid cells are small where the gradient or curvature of the solution is large. The most fundamental notion is to choose a grid so that some measure of the error in the solution is equidistributed, that is, some measure of the error in each grid cell or at each grid point is constant. However, this can be quite difficult compared to feature adaption, so feature adaption is commonly chosen and seems to perform well in the sense that errors are reduced significantly.

There are two main types of solution adapted grids, *moving grids* and *restructured grids*. Moving grid adaption is most suited to structured grids, and the idea is to move the grids points while maintaining the grid's logical structure. I particularly like using variational grid generation techniques for feature or equidistribution-of-errors moving-grid adaption (see [10]). The restructuring methods methods add grid points and cells where the grid is too coarse and remove points and cells where the grid is too fine. When working with quadrilateral grids, the the most common restructuring method is called *quad-tree* because of the data structure used.

I believe that it is important to be able not only to control the size of cells but also their orientation. In particular, a good grid adaption method should be able to align the sides of grid cells tangential or normal to a vector field (see Knupp [12]). Also the production of convex cells is important (see Barrera and Tinoco [21]).

2.1.2. *Conclusion*

For grid generation, I take view that I will try to break a region into a small number of simple regions, then generate a grid on each of the region, then regenerate the grid on the full region so that the grids between the regions are compatible. One important aspect of grids is that they can be computed along with the solution of some other problem such in Lagrangian fluid dynamics codes and consequently one have little control over the grids. Also, in non-trivial problems, goods grids are hard to generate, so some compromises must be made, and the resulting grids will have problems. It is important to

BE PREPARED TO DEAL WITH BAD GRIDS

when you choose a discretization method.

2.2. DISCRETIZATION OF CONTINUUM PROBLEMS

When choosing discretization methods it is important to understand what time interval is of importance: do you want to understand how the system behaves for a short time after it is started, that is understand some transient behavior; for a long time as it settles into some non-transient behavior; or what the system will look like for large times, that is what is the steady state behavior? My experience is mostly with the long time but not steady state simulations.

When we are studying discretizations of continuum partial differential equations, we have an important theorem due to Lax, often called the Lax Equivalence Theorem [19], that implies the following. If we have a stable discretization that is consistent with the continuum problem, then the solution of the discrete problem converges to a solution of the continuum problem. This is usually the way the result is stated for finite-difference methods, but there are equivalent results for finite-element and other methods. In any case, the results for time dependent problems are all similar and say that if we pick a starting time, say 0, and a final time T, and then solve our discrete problem for $0 \leq t \leq T$, then the this solution converges to the solution of the continuum problem as we refine the spatial and time grid.

The first problem with the Lax result is that we must fix T and then refine the grid. It is a common practice in engineering simulations to fix the grid in the spatial region, choose a time step, and the run a simulation until "things settle down." For example, in simulating the flow around the tip of a helicopter blade, where the tip oscillates periodically with each rotation of the blade, we would run a simulation for about five rotations to understand the flow pattern about the blade tip. (More rotations would have been better, but too costly for the additional information.) So we really do not pick a T and then refine the discretization, we pick a discretization and run the simulations.

The second problem is that the notion of stability in the Lax result and many others allows the solution to grow exponentially. For many of our simulation we know or expect that the solution remains bounded, but in fact the typical simulation produces a solution that does grow exponentially. In short time simulations this can be an asset for we can monitor some functional of the solution such as the energy that is growing exponentially and use it to assess the accuracy of the solution; we can expect the error in the solution to be on the order of the growth in the functional. Support operators and mimetic methods use a stronger notion of stability and avoid solution growth by design.

The solution is steady-state problems is an art onto itself that we won't go into here; there are many books on this topic. Suffice it to say for now that for all discretization methods, accuracy and stability play a central role as in time dependent problems, but here stability means something a bit different: the solutions of the approximate problems cannot "grow" as the discretization spatial discretization is refined. This concept is closely connected to the notion of ellipticity in continuum problems: the discrete problems should be elliptic and the ellipticity is "uniform" in the discretization.

Many, probably all, texts on numerical partial differential equations emphasize accuracy first and stability second. However, engineers on the whole prefer just the opposite: stable codes of minimal accuracy will typically provide useful information, while unstable codes, codes that blow up, are really useless, unless of course, the physical problem blows up. Anyway, for such reasons my coworkers and I have tried to emphasize stability first and then accuracy second, and the type of stability is different. If we know some important functional of a continuum problem, such as total mass or energy, is bounded, decays exponentially, or even grows exponentially, then our methods are designed to retain this property in the discretization. Finite volume methods do this for some functionals, support operators and mimetic methods, the methods that interest us, extend the finite volume ideas to all "important" functionals of the solution. Having a positive-definite functional that has controlled growth implies the stability needed for the the Lax result. For non-trivial problems, the accuracy of support-operator or mimetic methods are difficult to prove analytically, but are easy to determine numerically. We are very interested in improving the existing tools for analyzing accuracy, including accuracy for rough grids and solutions that are not smooth.

2.2.1. An Important Example

TABLE 2.1. Notation for the Generalized Laplacian

V	some region
∂V	boundary of V
$\nabla \cdot$	divergence operator
∇	gradient operator
$\mathcal{K}$	symmetric tensor
u	function to be solved for
f	given forcing function
$\vec{n}$	unit outward normal to ∂V
α	function given on ∂V
β	function given on ∂V
ψ	function given on ∂V

It is important to understand that the motion of materials (solids, fluids and gases) or the motion of electro-magnetic fields can be described in terms of the divergence, gradient, and curl. These differential operators are invariant, that is, have a coordinate-free descriptions, which is part of why they are so fundamental. After some algebraic manipulations, most continuum problems will have an associated equation that involves a generalized Laplacian, which we will describe below. Support operator methods are designed to produce discretizations of the generalized Laplacian that have the same important properties as the continuum Laplacian, where the properties of the discrete problem are uniform in the spatial discretization size.

The generalized Laplacian is given by $\nabla \cdot \mathcal{K} \nabla$ where the notation is described in Table 2.1. The standard boundary-value problem associated with this operator is to find a function u satisfying

$$-\nabla \cdot \mathcal{K} \nabla u = f \quad \text{in } V, \tag{2.23}$$

$$\beta \langle \mathcal{K} \nabla u, \vec{n} \rangle + \alpha u = \psi \quad \text{in } \partial V, \tag{2.24}$$

where (2.23) is the differential equation and (2.24) is the boundary condition. If we add to the left-hand side of the differential equation (2.23) one of the terms

$$\frac{\partial u}{\partial t}, \quad \frac{\partial^2 u}{\partial t^2} \tag{2.25}$$

where t is the time variable and add one or two initial conditions, then we get an initial-boundary value problem for the *diffusion* or *wave* equation. The analysis of these time-dependent equations is rather easy once the above steady-state problem is understood, so we look at the steady state equation more carefully.

The flux form of the steady-state equation is quite useful and is given by

$$\nabla \cdot \vec{w} = f \quad \text{in } V, \tag{2.26}$$

$$\vec{w} = -\mathcal{K} \nabla u \quad \text{in } V, \tag{2.27}$$

$$-\beta \langle \vec{w}, \vec{n} \rangle + \alpha u = \psi \quad \text{in } \partial V, \tag{2.28}$$

Equation (2.26) is the conservation law, while Equation (2.27) is the material law which gives the flux, that is, the amount stuff going through surfaces. The flux is the crux of the matter! The flux is the product of the material properties *tensor* and the gradient.

First, we consider the steady-state problem trivial if both the tensor $\mathcal{K}$ is the identity and either $\alpha \equiv 0$ (Neumann boundary conditions) or $\beta \equiv 0$ (Dirichlet boundary conditions). There are many intermediate cases of varying difficulty, for example smoothly varying $\mathcal{K}$. My interest has been in problems where $\mathcal{K}$ is a symmetric positive definite tensor (matrix) where the tensor is a discontinuous function of the spatial variables, the boundary condition is neither Dirichlet nor Neumann, and the grid is of poor quality. Many discretization methods perform well on the trivial problem; few methods perform well on this more general problem. Support-operator and mimetic methods do well on the general problem.

We discretize such problems using the mixed formulation (2.26) and (2.27), where we use a finite-volume or conservation approach to the conservation law (2.26). For the flux equation (2.27), we can separately estimate the gradient and the *conductivity* tensor $\mathcal{K}$, or estimate them together. If we use support-operators ideas to discretize them both together, we obtain a useful generalization of the harmonic-averaging idea so effectively used in one dimensional problems [15].

If x is the vector of spatial variables, t is the time variable, and u is the solution variable, and

$$\mathcal{K} = \mathcal{K}(x, t, u) \tag{2.29}$$

then there are many important applications where $\mathcal{K}$ is not a *continuous* function of of the variables x, t, and u. There are applications where $\mathcal{K}$ has a finite number of jump discontinuities on smooth curves, periodic jumps, and random variation:

- soil and rocks for groundwater modeling;
- layered material;
- uniform material with random contaminants;
- manufactured devices, e.g. klystron tube with vacuum and steel.

There are also many applications when $\mathcal{K}$ is not a scalar.

Mimetic methods will allow us to estimate the gradient in many different ways:

- finite differences (interpolation);
- finite elements (finite element-volume methods);
- spectral (we will try this);
- wavelet (seem hard to mesh mimetic and finite-volume);
- particle methods.

There are also many ways to estimate $\mathcal{K}$ and in fact this is a major industry:

- Averaging: linear, harmonic, geometric;
- homogenization;
- sub-scale modeling (e.g. turbulence);
- interpolation;
- ....

My feeling is that the right way to do this is to simulate small blocks of the material in parallel and then use the results of these simulation in a algorithm for large-scale simulation.

2.2.2. *Conclusions*

We have tried to give some overview of a methodology for discretizing differential equations based on separating the differential equation into conservation laws and material laws. It is natural to discretize the conservation laws using finite-volume ideas and there is substantial experience to show that this is a good idea. The situation for the material laws is far less clear.

As soon as we cannot assume the material laws are smooth, none of the classical methods of analyzing a numerical algorithm applies, so how can we tell if a code is good? First, the typically algorithm applies when the material laws are smooth, so the algorithm should provide solutions that are at least second-order accurate in the maximum norm. Next the algorithm should remain stable when the material laws are discontinuous, and the accuracy should degrade "gently" as the smoothness of the material laws degrade or the grid degrades.

3. Creating a Numerical Simulation Code

There are two main types of time-stepping algorithms for time-dependent problems: explicit and implicit. If there are diffusive effects in the model, then they typically diffuse higher frequencies components of the solution more rapidly than lower frequencies, suggesting that the time-stepping is *stiff*. As the spatial grid is resolved, higher frequencies can be represented on the grid, so the time-stepping algorithm becomes stiffer. This implies that for explicit algorithms, stability constraints on the time step dominate accuracy constraints, so that implicit methods must be used. If one is studying pure wave equations or Maxwell's equation, then explicit methods are tempting, but otherwise, implicit methods are required to overcome the stability constraint and have the step size only determined by accuracy constraints.

In implicit methods, typically a nonlinear system of equations must be solved for each time step. The first thing to note is that this system is not the same as the steady state system as it typically has the form: solve

$$f + \epsilon A(f) = g \qquad (3.30)$$

for f given A, ϵ and g. Here ϵ is a scalar parameter that goes to zero as the time step goes to zero and A is a positive operator. The analogous steady state problem has the form

$$A(f) = g. \qquad (3.31)$$

In the time stepping problem, the solution from the previous time step must be a good guess for the solution at the current time step or the time step is too large. Also it is typically a simple matter to extrapolate an even better guess for the solution at the current time step. So in the steady-state problem, one usually has a poor initial guess for the solution, while in the time stepping problems one must have a good starting value for any iterative solution method. For simple problems, any iterative solver can find the solution in a few iterations, so it is quite difficult to understand which solver is likely to be the most efficient, but small efficiencies can have a big impact if there are a large number of time steps.

In any case, an important point is that the solution of the system of equations is usually by far the most expensive part of solving a time-stepping problem, so the greatest impact on efficiency is to choose a discretization that produces a systems of equations that is easiest to solve and then choosing a good solver. Fortunately there is considerable information available about efficient solvers and a good place to start looking for nonlinear solvers is the book [9] by Kelly, and for linear solvers, the book [4] written by the "gang," gives an excellent overview of the commonly used methods with the exception of multigrid methods. Multigrid [11] methods are rumored to be the best for discretizations of partial differential equations, and this actually seems to be true for scalar linear steady-state problems. But for nonlinear problems, problem with vector solutions, and time stepping problem this is far from clear.

Along with several colleagues and student I have been studying what are called implicit-explicit time discretization methods. This is an operator splitting method where part of the time discretization is done implicitly while the remaining part is done implicitly. This is rather commonly done in engineering codes, but has only been studied seriously by Ascher and colleagues [1]. This class of algorithms allows one to take advantage of the best properties of both implicit and explicit methods and to keep the system of equations that needs to be solved simple. I believe this class of methods shows real promise, and the engineering literature I am familiar with supports this conclusion.

Now, at least on simple serial computing machines, no other optimization has nearly the importance of finding a simple system of equations to solve and choosing a good solver. Moreover the other optimizations I am familiar with should be able to be done by a good optimizing compiler. Unfortunately, it is not a simple matter to understand which optimizations can be done by the rather large number of available compilers. But surely this stuff can be automated?

3.1. *Automatic Code Generation*

The is substantial agreement that it is best to record algorithms at as high a level of abstraction as is possible and then compile that description. Historically, this has been done through subroutine libraries such as those at the GAMS [8] site. More recently the C++ Class Libraries at NIST [20] have been used to recored algorithms in a way that is independent of the data structures used for storing vectors and matrices. In these libraries, algorithms such as conjugate gradient can be recorded in a form that is much like the pseudo-code given in textbooks.

A much better goal is to write problems in applied mathematics notation and then compile such descriptions. In fact, this is now possible for restricted classes of problems. Several systems have been built for the numerical solution of boundary value problems, two of which are:

– SciNapse [18]
– Ctadel [7]

Other projects are given on my web page. SciNapse is a *problem solving environment* for the numerical solution of initial-boundary value problems for partial differential equations and is capable of compiling standard applied math descriptions of time-dependent problems that involve several dependent variables in three spatial dimensions and using implicit methods. Ctadel emphasizes generating parallel code for weather forecasting. A very interesting point about Ctadel is that it can generate code that is more efficient that that written by experienced programmers. We will be able to make much faster progress on algorithms when these PSEs become generally available.

3.2. *Conclusions*

Choosing a good discretization, that is, one that produces the simplest systems of nonlinear or linear equations to solve is the most important efficiency consideration for producing efficient code. Experience shows that all other programming considerations can be left to problem solving environments (including the choice of solver). The use of problem solving environments frees modelers to focus on building accurate models, leaving the creation of accurate and efficient code to the PSE.

References

1. Ascher, U., Ruuth, S. and Wetton, B., Implicit-Explicit Methods for Time-Dependent PDE's, *SIAM Journal of Numerical Analysis*, **32** (1995), pp. 797-823.

   http://www.cs.ubc.ca/nest/scv/group-info/professors/ascher/research.html

2. Barrera, P. and Tinoco, G., UnaMalla (A Grid).

   http://www.mathmoo.unam.mx/unamalla

3. Berger, M., Cartesian Grid Methods for Complex Geometry.

   http://cs.nyu.edu/cs/faculty/berger/geom.html

4. Barrett, R., Berry, M., Chan, T.F., Demmel, J., Donato, J., Dongarra, J., Eijkhout, V., Pozo, R., Romine, C. and van der Vorst, H., Templates for the Solution of Linear Systems: Building Blocks for Iterative Methods, SIAM, 1993.

   http://http://www.siam.org/catalog/mcc01/barrett.htm

5. Castillo, J.E. and Otto, J.S., A Generalized Length Strategy for Direct Optimization in Planar Grid Generation, *Mathematics And Computers In Simulation*, **44** (1997), pp. 441-456.

6. Cubit Mesh Generation Toolkit.

    http://sass2537.endo.sandia.gov/SEACAS/CUBIT/Cubit.html

7. The Ctadel Problem Solving Environment.

    http://www.wi.leidenuniv.nl/CS/HPC/ctadel.html

8. GAMS: Guide to Available Mathematical Software.

    http://gams.nist.gov/

9. Kelley, C.T., Iterative Methods for Linear and Nonlinear Equations SIAM, 1995.

    http://www.siam.org/books/kelley/kelley.htm

10. Knupp, P.M. and Steinberg, S., The Fundamentals of Grid Generation, CRC Press, Boca Raton, 1993.

    http://sass2537.endo.sandia.gov/~pknupp/   http://math.unm.edu/~stanly

11. MgNet Multigrid Archive.

    file://na.cs.yale.edu/pub/mgnet/www/mgnet.html/

12. Knupp, P.M., Mesh Generation Using Vector Fields, *Journal of Computational Physics*, **119** (1995), pp. 142-148.

    http://sass2537.endo.sandia.gov/~pknupp/

13. Roache, P.J., Verification and Validation in Computational Science and Engineering, Hermosa Publishers, Albuquerque, 1998.

    http://www.hermosa-pub.com/hermosa/

14. Roache, P.J., Elliptic Marching Methods and Domain Decomposition. CRC Press, 1995.

    http://www.crcpress.com/

15. Hyman, J., Shashkov, M. and Steinberg, S., The Numerical Solution of Diffusion Problems in Strongly Heterogeneous Non-Isotropic Materials, *Journal of Computational Physics*, **132** (1997), pp. 130-148.

    http://cnls.lanl.gov/~shashkov/   http://math.unm.edu/~stanly/

16. Schneiders, R., Mesh Generation & Grid Generation on the Web.

    http://www-users.informatik.rwth-aachen.de/~roberts/meshgeneration.html

17. Shashkov, M., Conservative Finite-Difference Methods on General Grids, CRC Press, Boca Raton, 1995.

    http://www.crcpress.com/

18. SciComp, Inc., the developers of SciNapse.

    http://www.scicomp.com

19. Strikwerda, J.C., Finite Difference Schemes and Partial Differential Equations, Chapman and Hall, 1989.

20. SparseLib++.

    http://math.nist.gov/sparselib++/

21. Tinoco-Ruiz, J.G. and Barrera-Snchez, P., Smooth and Convex Grid Generation Over General Plane Regions, *Mathematics and Computers In Simulation*, **46** (1998), pp. 87-102.

    http://amnesia.fismat.umich.mx/~jtinoco/

22. Warsi, Z.U.A., Fluid Dynamics : Theoretical and Computational Approaches, CRC, 1992.

DOMAIN DECOMPOSITION METHODS FOR ELLIPTIC PARTIAL DIFFERENTIAL EQUATIONS

OLOF B. WIDLUND
Courant Institute of Mathematical Sciences
251 Mercer Street
New York, NY 10012, USA [†]

1. Introduction

These lecture notes concerns the iterative solution, by domain decomposition methods, of the often huge linear system of algebraic equations which arise when elliptic problems are discretized by finite elements. These algorithms are preconditioned conjugate gradient methods, or more generally, preconditioned Krylov space methods, where the preconditioner is constructed from smaller instances of a given discrete elliptic problem, typically defined by restricting the domain of definition to many small subregions of the given region on which the elliptic problem is defined. These algorithms are designed with parallel computing in mind.

The 1996 book by Smith, Bjørstad, and Gropp [40] provides a good introduction to the field. The book contains an outline of the basic theory, and also discusses many computer science and pragmatic aspects of domain decomposition methods. We also note that Smith and Gropp are members of a research group at Argonne National Laboratory that is developing the PETSc library; PETSc stands for Portable, Extensible Toolkit for Scientific Computation; cf.[1]. One of the main features of that system is that it supports parallel computing. [1] Other valuable surveys of important results on domain decomposition methods are given in Farhat and Roux [19] and Le Tallec [26]; see also the forthcoming book by Quarteroni and Valli [37].

In these lectures, we will focus on the mathematical foundation of domain decomposition methods concentrating our efforts on a study of low order finite element approximations of Poisson's equation. The ambition

---

[†] widlund@cs.nyu.edu, URL: http://cs.nyu.edu/cs/faculty/widlund/index.html
[1] URL for PETSc documentation: http://www.mcs.anl.gov/petsc/petsc.html

H. Bulgak and C. Zenger (eds.), Error Control and Adaptivity in Scientific Computing, 325–354.

is to provide the basics of the theory, giving several proofs, in order to demonstrate what mathematical tools are required and what can be done. In fact, only relatively few tools from mathematical analysis are required; a suitable background is provided by [9] or any other good mathematical book on finite elements. Familiarity with the basics of finite elements and Krylov space methods, in particular the conjugate gradient method, will be assumed. Some knowledge of Sobolev spaces, and related tools of mathematical analysis is also helpful; however, only a few basic results on the $H^s(\Omega)$ spaces are necessary for the understanding of the proofs. Naturally, we will draw on our own experience in developing the theory and, therefore, we will refer chiefly to our own papers and those of some close affiliates.

We note that some recent work on domain decomposition methods has focused on the extension of results from the lower order case to spectral elements; cf. [14, 13, 33, 35, 36, 34]. Other recent work on domain decomposition methods for higher order methods include a 1997 thesis by Bică [2], on the p-version finite element method and work by Guo and Cao [21, 22, 23, 24] on $h - p$ methods. The theory has also been extended to elliptic systems, [36, 34], and the $H(div)$ and $H(curl)$ spaces; see [41, 42, 44].

The main topics of the lecture notes are:

• Abstract Schwarz theory; cf., e.g. [18, 6, 31].

• Examples of Schwarz algorithms, including several multilevel methods; cf. [17, 15, 46, 48]. Only short outlines are given of how these results can be obtained in the framework of the abstract theory.

• A detailed examination of the two-level overlapping Schwarz method; see [17].

• A description and analysis of the balancing method, a *Neumann-Neumann* algorithm; see [29, 16, 18, 26]. This provides an example of the iterative substructuring methods, which form the second important family of domain decomposition methods.

2. Schwarz methods

The earliest domain decomposition method known to this author is the alternating method of Hermann Amandus Schwarz, [39], discovered in 1869. Schwarz used the algorithm to establish the existence of harmonic functions with prescribed boundary values on regions with non-smooth boundaries. The regions were constructed recursively by forming unions of pairs of regions starting with regions for which existence could be established by some more elementary means. At the core of this work is a proof that this iterative method converges in the maximum norm at a geometric rate.

For more than two subregions, we can in fact define a step of the algorithm by recursion: i) Solve on the first subregion; ii) Solve on the union of all other subregions, approximately, by recursively invoking the same algorithm.

As pointed out by Pierre-Louis Lions, in [27], the convergence of this algorithm can be established by two different methods: By a maximum principle and by using Hilbert spaces. The Hilbert space method is the most successful since much of the work relies on the classical calculus of variation and finite elements.

The classical Schwarz method can be described as follows: Consider Poisson's equation on a bounded Lipschitz region Ω with zero Dirichlet data given on $\partial\Omega$, the boundary of Ω. There are two fractional steps corresponding to two overlapping subregions, Ω_1' and Ω_2' of the original region $\Omega = \Omega_1' \bigcup \Omega_2'$. Given an initial guess u^0, the iterate u^{n+1} is determined from the previous iterate u^n in two sequential fractional steps in which the approximate solution on the two subregions is updated:

$$
\begin{aligned}
-\Delta u^{n+1/2} &= f && in\ \Omega_1', \\
u^{n+1/2} &= u^n && on\ \partial\Omega_1',
\end{aligned}
$$

and

$$
\begin{aligned}
-\Delta u^{n+1} &= f && in\ \Omega_2', \\
u^{n+1} &= u^{n+1/2} && on\ \partial\Omega_2'\ .
\end{aligned}
$$

Thus the Dirichlet data for these problems is obtained from the original data given on $\partial\Omega \cap \partial\Omega_i'$, and the values from the previous fractional step on the remaining part of $\partial\Omega_i$. We note that this algorithm also can be viewed as a mapping from $\partial\Omega_1' \setminus \partial\Omega$ to itself.

The algorithm is now rewritten in variational form. For the original problem we naturally use the space $H_0^1(\Omega)$ and the bilinear form

$$
a(u,v) = \int_\Omega \nabla u \cdot \nabla v\, dx
$$

obtained by using Green's formula. This formula defines the $H^1(\Omega)-$semi-norm. We also recall that $H_0^1(\Omega)$ is the subspace of elements of $H^1(\Omega)$ with vanishing trace on $\partial\Omega$, the boundary of Ω. We also introduce a finite element triangulation τ_h and the space of continuous, piecewise linear finite elements on τ_h and denote it by $V^h(\Omega)$.

We can now write Schwarz's method in terms of two orthogonal projections P_i, $i = 1, 2$, onto two subspaces $V_i^h = V^h \cap H_0^1(\Omega_i')$. Here, and in what follows, we regard $H_0^1(\Omega_i)$ as a subspace of $H^1(\Omega)$ obtained by extending any element of $H_0^1(\Omega_i)$ by zero in the complement of Ω_i. The projections are defined by

$$
a(P_i v_h, \phi_h) = a(v_h, \phi_h) \quad \forall \phi_h \in V_i^h\ .
$$

It is easy to show that the error propagation operator of this *multiplicative* Schwarz method is

$$(I - P_2)(I - P_1) = I - (P_1 + P_2 - P_2 P_1).$$

Therefore, the algorithm can be viewed as a simple iterative method for solving

$$(P_1 + P_2 - P_2 P_1) u_h = g_h.$$

With an appropriate right-hand side g_h, u_h will be the solution of the original finite element problem.

Once we have made this observation, it is natural to ask if we could use a different polynomial of the projections; this idea will soon be explored.

We note that the error propagation operator is not symmetric. However, if we repeat the first fractional step, we obtain $(I - P_1)(I - P_2)(I - P_1)$, and if we then subtract it from the identity, we obtain the symmetric operator

$$P_1 + (I - P_1) P_2 (I - P_1). \tag{1}$$

Since $I - P_1$ is a projection essentially no extra work is involved. This is seen by considering the elements of the Krylov space based on the operator given in (1) and observing that $(I - P_1)^2 = (I - P_1)$. However, this device no longer works if we have three or more fractional steps.

For practical reasons, it is indeed necessary to extend our discussion to the case of three or more, possibly very many subspaces. Let V a finite dimensional space and let $a(u, v)$ be a bilinear form inherited from a self adjoint elliptic problem. We consider:
Find $u \in V$ such that

$$a(u, v) = f(v) \qquad \forall v \in V. \tag{2}$$

Let

$$V = V_0 + V_1 + \cdots + V_N$$

be a decomposition of the space V. This decomposition is not necessarily a direct sum of subspaces; in many cases, the representation of an element of V in terms of components of the V_i is not unique. The first space V_0 is often a special, coarse subspace; sometimes it is left out.

An important variant of the Schwarz methods is the *additive Schwarz method*. We introduce it and, at the same time, some additional bilinear forms which will make it possible to accommodate, in the theory, the use of inexact solvers for the problems on the subspaces; cf., e.g. [40, 18]. Let $\tilde{a}_i(u, v)$, $i = 0, \cdots, N$, be symmetric, positive definite bilinear forms on $V_i \times V_i$. Introduce operators $T_i: V \to V_i$, by

$$\tilde{a}_i(T_i u, v) = a(u, v) \qquad \forall v \in V_i, \tag{3}$$

and let
$$T_{as} = T_0 + T_1 + \cdots + T_N.$$

We note that it is always possible to choose $\tilde{a}_i(u,v) = a(u,v)$. Then, $T_i = P_i$, the orthogonal projection with respect to the inner product $a(\cdot,\cdot)$ already introduced.

We can now replace the elliptic finite element problem by a transformed, preconditioned problem with the same unique solution as (2):

$$T_{as}u = g, \quad g = \sum_{i=0}^{N} g_i, \quad g_i = T_i u. \tag{4}$$

The right hand side g is obtained as $g = \sum_i g_i$ by solving

$$\tilde{a}_i(g_i, v_i) = a(u, v_i) = f(v_i) \quad \forall v_i \in V_i.$$

We can now use the conjugate gradient method, without any further preconditioning, using $a(\cdot,\cdot)$ as the inner product. The extreme eigenvalues of T_{as} provide a standard estimate of the rate of convergence of the conjugate gradient method: The iteration error, measured in the $a(\cdot,\cdot)-$ norm, of the conjugate gradient method will decrease by at least the factor

$$2(\frac{\sqrt{\kappa}-1}{\sqrt{\kappa}+1})^k \quad \text{where} \quad \kappa = \frac{\lambda_{max}(T_{as})}{\lambda_{min}(T_{as})}$$

in k steps; cf., e.g. [20].

3. Block Jacobi and the matrix form of the operators

It is useful to first consider a very simple example, namely the block Jacobi method, and try to understand how well it works and how it can be improved. Let us first turn the finite element variational problem into a linear system of algebraic equations, $Kx = b$. Here K is the stiffness matrix and b the load vector. The stiffness matrix is positive definite, symmetric, $K^T = K > 0$, which are properties inherited by any conforming finite element method, from the bilinear form $a(\cdot,\cdot)$.

We consider the block-Jacobi/conjugate gradient method: The stiffness matrix K is preconditioned by a matrix K_J, which is the direct sum of diagonal blocks of K. Each block corresponds to a set of degrees of freedom, which define a subspace V_i. The space V is a direct sum of the subspaces $V_i, i = 1, \cdots, N$. The choice of the subspaces is a key issue and so is the choice of basis of V, in particular in the spectral element case.

We associate an orthogonal projection P_i, defined exactly as in the previous section, to each subspace V_i, or we use an approximate projection defined by formula (3). It is of interest to consider matrix representations of the operators P_i and T_i. Thus, let V_i correspond to a set of variables at adjacent nodes and be associated with the subregion Ω_i', which is the union of the supports of the corresponding nodal basis functions; all degrees of freedom outside and on the boundary of Ω_i' vanish. After a suitable permutation of the variables, P_i can be represented by

$$y = P_i x = \begin{pmatrix} (K^{(i)})^{-1} & 0 \\ 0 & 0 \end{pmatrix} Kx,$$

where $K^{(i)}$ is the stiffness matrix of the Dirichlet problem on Ω_i'. We easily see that $\sum P_i$ corresponds directly to $K_J^{-1}K$.

An approximate projection T_i can be obtained by replacing $K^{(i)}$ by a suitable symmetric, positive definite preconditioner $\tilde{K}^{(i)}$ of the Dirichlet problem on Ω_i'.

The relevant spectrum is that of

$$T_{as} = \sum_{i=1}^{N} T_i.$$

A simple computation shows that the eigenvalues of $K^{-1}\tilde{K}_J$ are the stationary values of the generalized Rayleigh quotient

$$\frac{\sum_{i=1}^{N} \tilde{a}_i(u_i, u_i)}{a(u, u)}, \quad u = \sum_{i=1}^{N} u_i, \quad u_i \in V_i.$$

The upper bound is often the more challenging. The lower bound can be expressed in terms of *upper bounds* on $a(u, u)/\tilde{a}_i(u, u), \forall u \in V_i$. Considering the case when the region is subdivided into subregions of diameter H, and of good aspect ratio, we have to conclude for the case of exact solvers, and after some detailed work, that the condition number $\kappa(K_J^{-1}K)$, disappointingly, is on the order or $1/Hh$. We will see that the method can be greatly improved by increasing the overlap between the subregions and, in particular, by adding a coarse space; see further Subsection 5.1 and Section 6.

The method just outlined also works for any other *direct sum* decomposition of the space; we can introduce a basis of the space V by merging the bases of $V_1, \cdots, V_N$.

This class of preconditioners can be extended to the case when the finite element space no longer is a direct sum of subspaces corresponding to the blocks of a block Jacobi splitting. We can, e.g. add a global subspace, V_0, which can be a finite element space of degree one on a coarse triangulation of Ω, or, more generally, a subspace with one or just a few degrees of freedom for each coarse mesh element. We will examine this idea later. There are also many other possibilities.

The basic formula for the general case, which after dividing by $a(u, u)$ replaces the Rayleigh quotient given above, is

$$a(T_{as}^{-1}u, u) = inf_{u=\sum u_i} \sum \tilde{a}_i(u_i, u_i). \tag{5}$$

This formula is relatively easy to prove after noticing that the minimizing elements are given by $u_i = T_i T_{as}^{-1} u$. This formula appears first to have been given in [47].

Thus, $T_{as} = \sum T_i$ is again the operator relevant for the iterative method considered. There is freedom of choice in the representation of at least some u if the subspaces V_i do not form a direct sum. We note that it follows from formula (5) that the entire spectrum of T_{as} moves to the right when individual subspaces are enlarged, e.g. by increasing the overlap between subregions. This is often quite advantageous. We can, for example start with a direct sum decomposition of the finite element space and then enrich some or all of the subspaces, increasing their dimensions. We also note that if two subspaces are merged, then the right hand side of (5) does not increase, and in fact often decreases relative to $a(u, u)$. We can also strengthen a preconditioner by adding additional subspaces, e.g. a coarse global space because, again, there are more choices in the decomposition.

4. Abstract Schwarz theory

We will now develop a quite general theory that has proven very useful in the development and analysis of many methods. There are two main results in this theory. They are both expressed in terms of the same few parameters and the theoretical study of many domain decomposition methods can therefore simply be reduced to deriving good bounds for these parameters.

4.1. ADDITIVE SCHWARZ METHODS

Theorem 1 *Let there exist*
    (i) a constant C_0 such that for all $u \in V$ there exists a decomposition $u = \sum_{i=0}^{N} u_i$, $u_i \in V_i$, *such that*

$$\sum_{i=0}^{N} \tilde{a}_i(u_i, u_i) \le C_0^2 a(u, u);$$

(ii) a constant ω such that for $i = 0, \ldots, N$,

$$a(u, u) \le \omega \tilde{a}_i(u, u) \quad \forall u \in Range(T_i);$$

(iii) constants ϵ_{ij}, $i, j = 1, \ldots, N$, such that

$$a(u_i, u_j) \leq \epsilon_{ij} a(u_i, u_i)^{1/2} \, a(u_j, u_j)^{1/2} \quad \forall u_i \in V_i \ \forall u_j \in V_j.$$

Then, T_{as} is invertible and

$$C_0^{-2} a(u, u) \leq a(T_{as}u, u) \leq (\rho(\mathcal{E}) + 1)\, \omega\, a(u, u) \quad \forall u \in V. \qquad (6)$$

Here $\rho(\mathcal{E})$ is the spectral radius of the matrix $\mathcal{E} = \{\epsilon_{ij}\}_{i,j=1}^N$.

Note that i and $j \geq 1$, in assumption (iii). A bound on $\rho(\mathcal{E})$ can often be obtained quite easily for many domain decomposition methods. For multi-level and multigrid methods, it often involves *strengthened Cauchy-Schwarz inequalities*, of the form $\epsilon_{ij} \leq q^{-|i-j|}$ with $q < 1$.

Proof: Upper bound: We first note that it is easy to prove, using the definition of the T_i given in (3) and assumption (ii), that $\|T_i\|_a \leq \omega$. Indeed,

$$
\begin{aligned}
a(T_i u, T_i u) &\leq \omega \tilde{a}_i(T_i u, T_i u) = \omega\, a(u, T_i u) \\
&\leq \omega a(u, u)^{1/2} \, a(T_i u, T_i u)^{1/2}.
\end{aligned}
$$

Thus, $a(T_i u, T_i u) \leq \omega^2 a(u, u)$, which implies $\|T_i\|_a \leq \omega$.

Using assumptions (iii) and (ii) and the definition of the T_i, we obtain

$$a\left(\sum_{i=1}^N T_i u, \sum_{i=1}^N T_i u\right) = \sum_{i,j=1}^N a(T_i u, T_j u)$$

$$\leq \sum_{i,j=1}^N \epsilon_{ij} a(T_i u, T_i u)^{1/2} a(T_j u, T_j u)^{1/2}$$

$$\leq \sum_{i,j=1}^N \epsilon_{ij} (\|T_i\|_a \|T_j\|_a)^{1/2} a(T_i u, u)^{1/2} a(T_j u, u)^{1/2} \leq \rho(\mathcal{E}) \omega \sum_{i=1}^N a(T_i u, u).$$

We find,

$$a\left(\sum_{i=1}^N T_i u, \sum_{i=1}^N T_i u\right) \leq$$

$$\rho(\mathcal{E}) \omega \sum_{i=1}^N a(T_i u, u) \leq \rho(\mathcal{E}) \omega a(u, u)^{1/2} a\left(\sum_{i=1}^N T_i u, \sum_{i=1}^N T_i u\right)^{1/2}.$$

Hence, $a(\sum_{i=1}^N T_i u, \sum_{i=1}^N T_i u) \leq \rho(\mathcal{E})^2 \omega^2 a(u, u)$, and

$$a\left(\sum_{i=1}^N T_i u, u\right) \leq \rho(\mathcal{E}) \omega a(u, u). \qquad (7)$$

We finally note that

$$a(T_0u, u) \leq \omega a(u, u),$$

and complete the proof by adding the two last inequalities.

The upper bound of T_{as} can be replaced by $(\rho(\hat{\mathcal{E}}) + \|T_0\|_a)\, a(u, u)$, where the elements of $\hat{\mathcal{E}}$ are given by $\|T_i\|_a^{1/2}\epsilon_{ij}\|T_j\|_a^{1/2}$. We note that the rate of convergence of an additive Schwarz method is affected by the scaling of bilinear forms $\tilde{a}_i(\cdot, \cdot)$, relative to each other, i.e. by the scaling of the T_i; both the upper and lower bounds of T_{as} are affected. It is easy to show that if the norm of the T_i all are made equal, then the bound for the condition number of additive algorithm at least as good as before.

Lower bound: By the definition of T_i and assumption (i), we have

$$\begin{aligned}
a(u, u) &= \sum_{i=0}^{N} a(u, u_i) = \sum_{i=0}^{N} \tilde{a}_i(T_i u, u_i) \\
&\leq \left(\sum_{i=0}^{N} \tilde{a}_i(T_i u, T_i u) \right)^{1/2} \left(\sum_{i=0}^{N} \tilde{a}_i(u_i, u_i) \right)^{1/2} \\
&\leq \left(\sum_{i=0}^{N} a(u, T_i u) \right)^{1/2} C_0 (a(u, u))^{1/2}.
\end{aligned}$$

Hence, T_{as} is invertible and $a(u, u) \leq C_0^2 a(T_{as} u, u)$.

4.2. MULTIPLICATIVE SCHWARZ METHODS

There is a corresponding theory for *multiplicative* Schwarz methods. The principal contributors to the analysis of this case are Bramble, Pasciak, Wang, and Xu [6]; see also Cai and Widlund [11, 12] for variants of the theory for nonsymmetric and indefinite problems. In the multiplicative case, we need to provide an upper bound for the spectral radius, or norm, of the error propagation operator

$$E_N = (I - T_N) \cdots (I - T_0). \qquad (8)$$

We often wish to use many subspaces. To avoid a high degree of the polynomials, which define the Schwarz method and which can make the algorithm quite sequential, we use a simple graph theory tool called *coloring*. Subspaces that only intersect at the origin, such as those corresponding to disjoint subregions, can be grouped together into classes of subspaces each of which can be regarded as one subspace. This is done by coloring a graph with a vertex for each subspace and an edge for each pair of subspaces that have a nontrivial intersection by a minimal or good coloring. For each color, the subspaces are then merged.

Examining the factors of the product (8), we note that $\|I - T_i\|_a > 1$ if $\|T_i\|_a > 2$. Therefore an assumption of $\omega < 2$ is most natural. If ω is larger, we can scale the bilinear forms $\tilde{a}_i(\cdot, \cdot)$ to decrease the $\|T_i\|_a$ appropriately.

We also note that a multiplicative method can be very slow if some of the $\|T_i\|_a$ are very small; this means that there are only very small corrections in the corresponding subspace. As previously noted, the parameter C_0 changes if the T_i, and the $\tilde{a}_i(\cdot,\cdot)$, are rescaled.

Theorem 2 will, for technical reasons, be given in terms of $\hat{\omega} = \max(1,\omega)$ rather than ω. This bound is of interest only when the parameter $\omega < 2$ and is bounded away from zero from below. The result is expressed in terms of $\hat{\omega}$ and the two other parameters of Theorem 1 and is a variant of a result due to Bramble, Pasciak, Wang, and Xu.

Theorem 2 *The error propagation operator of the multiplicative Schwarz algorithm satisfies, with $\hat{\omega} = \max(1,\omega)$,*

$$\|E_N\|_a \leq \sqrt{1 - \frac{(2-\hat{\omega})}{(1 + 2\hat{\omega}^2 \rho(\mathcal{E})^2)C_0^2}} \ .$$

Proof: Our task is to estimate the norm of the error propagation operator E_N of the multiplicative Schwarz method. We begin by observing that with

$$E_j = (I - T_j) \cdots (I - T_0), \quad E_{-1} = I,$$

$$\text{and} \quad R_j = 2T_j - T_j^2 = (2 - T_j)T_j,$$

we have

$$E_j^T E_j - E_{j+1}^T E_{j+1} = E_j^T R_{j+1} E_j.$$

Here and in what follows, the transpose T is with respect to the bilinear form $a(\cdot,\cdot)$. This leads to the identity

$$I - E_N^T E_N = \sum_{j=0}^{N} E_{j-1}^T R_j E_{j-1}. \tag{9}$$

A satisfactory upper bound for $\|E_N\|_a$ is obtained by showing that the operator on the right hand side of (9) is sufficiently positive definite. We note that, for $\omega = \max \|T_j\|_a < 2$, the operators R_j are positive semidefinite, and that

$$R_j \geq (2 - \omega)T_j \geq (2 - \hat{\omega})T_j.$$

Therefore,

$$I - E_N^T E_N \geq (2 - \hat{\omega}) \sum_{j=0}^{N} E_{j-1}^T T_j E_{j-1}. \tag{10}$$

A direct consequence of the definition of the operator E_j is that

$$I = E_{j-1} + \sum_{k=0}^{j-1} T_k E_{k-1} = E_{j-1} + T_0 + \sum_{k=1}^{j-1} T_k E_{k-1}. \tag{11}$$

For $j > 0$,

$$a(T_j u, u) = a(T_j u, E_{j-1} u) + a(T_j u, T_0 u) + \sum_{k=1}^{j-1} a(T_j u, T_k E_{k-1} u).$$

This expression can, by using Schwarz's inequality, the upper bound on $\|T_i\|_a$, and the definition of the ε_{ij}, be bounded from above by

$$a(T_j u, u)^{1/2} (a(T_j E_{j-1} u, E_{j-1} u)^{1/2} + a(T_j T_0 u, T_0 u)^{1/2}$$

$$+ \omega \sum_{k=1}^{j-1} \varepsilon_{jk} a(T_k E_{k-1} u, E_{k-1} u)^{1/2}).$$

We note that $\varepsilon_{jj} = 1$ and that we can combine the first and third terms.
Denote by c a vector with the components

$$c_k = a(T_k E_{k-1} u, E_{k-1} u)^{1/2}, \quad k = 1, \cdots, N.$$

Cancelling a common factor and using elementary arguments, we find that

$$a(T_j u, u) \le 2\hat{\omega}^2 (\mathcal{E}c)_j^2 + 2a(T_j T_0 u, T_0 u).$$

Summing from $j = 1$ to N, using (7), and adding the term $a(T_0 u, u)$ to both sides, we obtain

$$a(T_{as} u, u) \le 2\hat{\omega}^2 \rho(\mathcal{E})^2 |c|_{\ell^2}^2 + (1 + 2\hat{\omega}^2 \rho(\mathcal{E})) a(T_0 u, u),$$

and finally,

$$a(T_{as} u, u) \le (1 + 2\hat{\omega}^2 \rho(\mathcal{E})^2) \sum_{j=0}^{N} a(E_{j-1}^T T_j E_{j-1} u, u).$$

The proof can now be completed by using (8) and the lower bound of Theorem 1.

4.3. ALTERNATIVE SCHWARZ METHODS

Other variants of the basic algorithms can be analyzed similarly. Each corresponds to some polynomial, without a constant term, in the operators T_i and a factorization of that polynomial into linear factors, or sums of products of linear factors, each of which corresponds to a fractional step of the Schwarz algorithm. The resulting operator equations can also be solved using different acceleration methods.

In practice, one of the most powerful algorithms is obtained by solving the equation

$$T_{ms}u = (I - E_N)u = \hat{g},$$

where $T_{ms} = I - (I - T_N) \ldots (I - T_0)$, by the GMRES method [38], or some other conjugate gradient type method for nonsymmetric problems. The standard conjugate gradient method can also be used with a symmetrized version of the multiplicative method. Then, the polynomial is given by

$$T_{sms} = I - (I - T_{ms})^T (I - T_{ms}) = T_{ms} + T_{ms}^T - T_{ms}^T T_{ms}.$$

We can simplify this algorithm by removing a factor $(I - T_N)$; if the exact projection P_N onto V_N is used, the algorithm remains exactly the same. In the general case, we can still obtain a somewhat weaker bound on the rate of convergence of the resulting algorithm by interpreting it as a multiplicative Schwarz method using the subspaces $V_0, V_1, \cdots, V_{N-1}, V_N, V_{N-1}, \cdots, V_1, V_0$.

We can also replace T_{sms} by the polynomial

$$T_{ms} + T_{ms}^T.$$

This new operator, which is also symmetric, involves only about half as many fractional steps per iteration as T_{sms} if different processors can be assigned to the two parts of the operator. It is easy to see that this operator has a larger smallest eigenvalue than T_{sms}.

Another variant has been introduced by Mandel [28, 29]. Let us consider a multiplicative two subspace method with

$$V = V_0 + V.$$

We solve the coarse problems, corresponding to V_0, exactly and use an additive Schwarz solver based on $V = V_1 + \cdots + V_N$ as an inexact solver for the second space. The polynomial is now

$$P_0 + T_1 + \cdots + T_N - P_0(T_1 + \cdots + T_N). \tag{12}$$

As pointed out in our discussion of the multiplicative methods with only two fractional steps, symmetry is not lost in this case; we can work with $I - (I - P_0)(I - T_1 + \cdots + T_N)(I - P_0)$. A particular example of this Schwarz method will be analyzed in some detail in Section 7.

Cai [10] advocates the use of a method derived from the polynomial

$$\gamma T_0 + I - (I - T_N) \cdots (I - T_1);$$

$\gamma > 0$ is a balancing parameter. It is then possible to take advantage of the generally more rapid convergence of a multiplicative method, while solving the special coarse problem at the same time as the local problems. One or several processors can work on the coarse problem while the rest of the processors are assigned to the local problems. We note that in a standard multiplicative algorithm, there is a potential bottleneck with many processors idly waiting for the solution of the coarse problem.

5. Examples of Schwarz methods

In this section, we will briefly describe several domain decomposition and multi-level methods in terms of the abstract theory just developed.

5.1. A TWO-LEVEL ADDITIVE SCHWARZ METHOD WITH OVERLAP

Let us consider Poisson's equation with zero Dirichlet conditions on a bounded polygonal or polyhedral region Ω in two or three dimensions. We introduce two triangulations, a coarse and a fine, which might be a refinement of the former. The resulting finite element space $V^h \subset H_0^1(\Omega)$ is the space of continuous, piecewise linear finite element functions on the fine triangulation. There is also a covering of the region by overlapping subregions Ω_i', each of which is the union of finite elements. By δ we denote the minimum of the width of the region common to neighboring subregions and we note that δ/H is then a measure of the relative overlap between adjacent subregions. We also assume the elements are shape regular, i.e. none of them are very flat, but we need not assume quasi-uniformity, i.e. all the elements of the triangulations need not be of comparable size.

The spaces chosen for this Schwarz method are

$$V_0 = V^H \quad \text{based on the coarse triangulation,}$$

and,

$$V_i = V^h \cap H_0^1(\Omega_i'), \quad i > 0.$$

Theorem 3 *The condition number of the additive Schwarz method satisfies*

$$\kappa(T_{as}) \leq C(1 + H/\delta).$$

The constant C is independent of the parameters H, h, and δ.

We will prove this result in Section 6. We also note that Brenner [8] has shown recently that this bound is best possible.

5.2. ITERATIVE SUBSTRUCTURING METHODS

This is the second important family of domain decomposition methods, which are also known as *Schur complement* methods. The region Ω is partioned into nonoverlapping substructures $\Omega_j, j = 1, \cdots, M$. In a first step,

338

the variables interior to all the Ω_j are eliminated using a direct method; these are local computations. The reduced linear system, the Schur complement system, is then solved by a Schwarz method. The remaining variables are associated with nodes on the interface Γ, the union of the interfaces that separate the subregions. The vector of nodal values associated with the closure of Ω_j will be denoted by $x^{(j)}$.

The stiffness matrix K can be built by subassembly from local matrices

$$x^T K x = \sum_{j=1}^{N} x^{(j)^T} K^{(j)} x^{(j)}.$$

We now order the interior variables of Ω_j first followed by those on $\partial\Omega_j$, which are the components of two subvectors $x_I^{(j)}$ and $x_\Gamma^{(j)}$, respectively,

$$K^{(j)} = \begin{bmatrix} K_{II}^{(j)} & K_{I\Gamma}^{(j)} \\ K_{I\Gamma}^{(j)^T} & K_{\Gamma\Gamma}^{(j)} \end{bmatrix}.$$

We eliminate the interior unknowns by solving a local Dirichlet problem for each substructure, and obtain local Schur complements

$$S^{(j)} = K_{\Gamma\Gamma}^{(j)} - K_{I\Gamma}^{(j)^T} K_{II}^{(j)^{-1}} K_{I\Gamma}^{(j)}.$$

The global Schur complement is also built by subassembly from local contributions

$$x_\Gamma^T S x_\Gamma = \sum_{j=1}^{N} x_\Gamma^{(j)^T} S^{(j)} x_\Gamma^{(j)}.$$

The reduced system $Sx = \tilde{b}$ can be solved by a Krylov space method, such as the conjugate gradient method with a preconditioner $\hat{S}$. Typically, these preconditioners are built from some coarse, global solver, and many local solvers which might correspond to individual edges or faces of substructures. In fact, the local solvers for a number of the iterative substructuring methods can be obtained by a block Jacobi splitting of the Schur complement S. A basic condition number bound of the form $C(1 + \log(H/h))^2$ holds for many of these methods.

The iterative substructuring methods can be viewed as Schwarz methods with overlap, cf. [16], where we work on the interface Γ instead of Ω. In a natural sense, the overlap is now only $\delta = h$. The appropriate space is no longer $H^1(\Omega)$ but $H^{1/2}(\Gamma)$, which is the space of traces of $H^1(\Omega)$ on the interface. A detailed discussion and comparison between this case and that of the previous subsection is given in [17].

We will return to iterative substructuring methods in Section 7. For an overview of many such methods; see [15]. For a quite recent survey article, see [45]. We note that both of these papers only discuss scalar elliptic problems.

5.3. YSERENTANT'S HIERARCHICAL BASIS METHOD

We will again consider the same elliptic problem in two dimensions, and introduce a nested multi-level triangulation. We begin by introducing a coarse mesh T_0 and then refine it recursively, ℓ times, by cutting all the triangles into four by joining the midpoints of the edges. The intermediary and final triangulations are denoted by $T_i, 1 \leq i \leq \ell$.

The subspaces chosen for this additive Schwarz methods are:

$$V_0 = V^H \text{ based on } T_0,$$

and

$$V_i = \{u \in V^{h_i}| \ u = 0 \text{ at all the nodes of } T_{i-1}\}.$$

These spaces form a direct sum decomposition of $V^h = V^{h_\ell}$. In the space V_i, we use the standard nodal basis functions defined on T_i, for the relevant, new nodal points, and with the exception of V_0, we replace the bilinear form $a(\cdot, \cdot)$ by a diagonal operator, i.e. we use point Jacobi. We can accomplish the same thing by splitting these subspaces further into one dimensional subspaces. The following theorem was first proven in [46].

Theorem 4 *The condition number of the additive Schwarz method satisfies*

$$\kappa(T_{as}) \leq C(1 + \ell)^2.$$

The constant C is independent of the parameters H and ℓ.

The decomposition used in the proof is unique and simply,

$$u_h = I_0 u_h + (I_1 - I_0)u_h + \dots + (I_\ell - I_{\ell-1})u_h, \tag{13}$$

where the operator I_k is the nodal interpolation operator onto the space V_k.

In view of the abstract theory, the main mathematical issue is whether the norm of any of these terms can be much larger than that of u_h, for some choice of u_h. We also need to estimate $\rho(\varepsilon)$ and ω.

We rely on a Sobolev-type inequality; see [3, 5, 43].

Lemma 1 *Let α be any convex combination of $u_h(x)$, $x \in \bar{\Omega}_i$, a bounded subregion in the plane of diameter H. Then, for all $u_h \in V^h$,*

$$\|u_h - \alpha\|^2_{L^\infty(\Omega_i)} \leq \text{const.} \ (1 + log(H/h))a_{\Omega_i}(u_h, u_h) \,.$$

The logarithmic factor cannot be removed.

The terms of the decomposition in (13) are estimated using this lemma. We find that the square of the energy norm of the individual operators in the formula can be bounded by $C(\ell - i)$ from which a bound proportional to ℓ^2 for the C_0^2 of the general theory easily follows. A uniform bound of $\rho(\mathcal{E})$ is obtained by relatively elementary, and local, arguments. A bound for ω is obtained quite easily.

5.4. THE V-CYCLE MULTI-GRID AND THE BPX METHODS

The multigrid V-cycle [4] and the BPX method [7] form a pair of multiplicative and additive methods that use the same subspaces and bilinear forms. We can now consider an elliptic problem in either two or three dimensions and will use the same type of multi-level triangulation as for Yserentant's method.

The spaces chosen for the Schwarz methods are

$$V_0 = V^H \text{ based on the coarsest triangulation,}$$

and

$$V_i = V^{h_i}.$$

We are now far from the direct sum case. We use the nodal basis on $\mathcal{T}_i$, for the subspace V^{h_i} and, with the exception of V_0, we replace the bilinear form by a diagonal operator, i.e. we use point Jacobi for the additive BPX method. We can also think of this method as resulting from a splitting of the different $V_i, i > 0$, into one dimensional subspaces for which we use exact solvers. Similarly, we obtain the standard $V-$cycle multigrid method from the basic multiplicative Schwarz method provided that we order the subspaces in a particular way. It is interesting to note, that the theory based on the Schwarz approach shows that the rate of convergence of any multiplicative method based on these subspaces has a rate of convergence that is bounded uniformly for any ordering of the subspaces and that this bound is independent of the coarse mesh size and ℓ, the number of levels of the triangulation; see further Theorem 5.

A proof that $\rho(\mathcal{E})$ is bounded independently of ℓ is given in Zhang [47, 48]. His proof is similar to, but involves more detail than, Yserentant's for the hierarchical basis method; it is based on a coloring argument and strengthened Cauchy-Schwarz inequalities.

The bound for ω presents no difficulties.

A uniform bound for C_0 can be obtained by using standard finite element tools, including the Aubin-Nietsche argument, if Ω is convex. We can then simply use the decomposition

$$u_h = P_0 u_h + (P_1 - P_0)u_h + ... + (P_\ell - P_{\ell-1})u_h,$$

where P_i is the H_0^1-projection onto V^{h_i}. These terms are $a(\cdot, \cdot)$-orthogonal from which it follows that they can be estimated individually. A decomposition borrowed from the proof of the two-level Schwarz method, see Section 6, can then be used to split the spaces V_i into a direct sum of one-dimensional spaces in a stable way. As shown by Zhang [48], the general case can be reduced to the case of a convex region by doing some interesting, hard work.

Theorem 5 *The BPX method, the additive method based on the spaces and bilinear forms just introduced, satisfies*

$$\kappa(T_{as}) \leq C.$$

The constant C is independent of the parameters H and ℓ.

The V-cycle multigrid method, based on the same subspaces, converges at a rate which is uniformly bounded in terms of the same parameters.

We note that the optimal result for the multi-grid case was first given by Oswald [30].

6. Proof for the two-level overlapping Schwarz method

In this section, we will outline the proof of Theorem 3. We will follow Dryja and Widlund [17] relatively closely. We note that the study that led to that paper was motivated by a desire to refine the analysis after that numerical experiments had shown that the method converges quite satisfactorily even with an overlap that is far from generous.

We recall that we have two triangulations, a coarse, and a fine, and we now assume that the latter is a refinement of the former. Neither of them has to be quasi-uniform but we assume that they are shape regular. The finite element space V^h is the space of continuous, piecewise linear finite element functions on the fine triangulation. The region is also covered by overlapping subregions Ω_i', each of which is assumed to be a union of elements. The overlap is measured by δ which stands for the minimum width of the regions that are common to more than one subregion. We assume that the Ω_i' have diameters which are not smaller than by a constant factor than those of the coarse mesh elements which they intersect. Our proof will be based exclusively on arguments that only involve one subregion and its next neighbors, which by assumption will all be of comparable size. Therefore, what matters is only the smallest ratio of the local minimum width of the overlap and the diameter of the local coarse mesh elements.

We recall that one of the spaces chosen for this Schwarz method is

$$V_0 = V^H \text{ based on the coarse triangulation .}$$

(If the coarse triangles were not unions of fine triangles, we would have to use $I^h : C \to V^h$, the standard interpolation operator, to map V_0 into a subspace of V_h.) The local spaces are

$$V_i = V^h \cap H_0^1(\Omega_i'), \quad i > 0.$$

The ideas behind the proof of Theorem 3 are equally valid for two and three dimensions.

We first prove a constant upper bound for the spectrum of $P = P_0 + P_1 + \cdots + P_N$; this is done without using the general theory. We note that $P_i, i > 0$, is also an orthogonal projection of $V^h(\Omega_i')$ onto V_i. Therefore,

$$a(P_i u_h, u_h) \leq a_{\Omega_i'}(u_h, u_h), \quad i > 0.$$

By construction, there is an upper bound, N_c, on the number of subregions to which any $x \in \Omega$ can belong; therefore

$$\sum_{i=1}^{N} a_{\Omega_i'}(u_h, u_h) \leq N_c\, a(u_h, u_h).$$

In addition, we use the fact that the norm of P_0 is equal to one and obtain

$$\lambda_{max}(P) \leq (N_c + 1).$$

The lower bound is obtained by estimating the parameter C_0 of the general theory. Given u_h, we first construct a suitable $u_{h,0} \in V_0$. It would be tempting to use the interpolant, i.e. $u_{h,0} = I^H u_h$. However as we have learned from Subsection 5.3 and Lemma 1, this could not lead to a uniform bound even in two dimensions. In three dimensions there would be a factor H/h. We see that by considering the energy of the coarse space interpolant of a standard nodal basis function of V^h; with the nodes selected the same, a nodal basis function in V^h is mapped onto a nodal basis function in V^H of considerably much larger energy.

Instead, we will first take averages of u_h over neighborhoods of the coarse nodal points and then interpolate the result. Consider, only to simplify our notations, the two dimensional case and a coarse triangle T with vertices $a, b,$ and c. It is easy to see that the energy contributed by a linear function with values $\bar{u}_h(a)$, $\bar{u}_h(b)$, and $\bar{u}_h(c)$, at these points, can be bounded by

$$(\bar{u}_h(a) - \bar{u}_h(b))^2 + (\bar{u}_h(b) - \bar{u}_h(c))^2 + (\bar{u}_h(c) - \bar{u}_h(a))^2.$$

Let B_x be a ball of area H^2 centered at the point x, to be used in the averaging, and assume that the points a and b are both on the x_1-axis. Then,

$$(\bar{u}_h(a) - \bar{u}_h(b))^2 = (H^{-2} \int_{B_a} u_h(x_1, x_2)dx - H^{-2} \int_{B_b} u_h(x_1, x_2)dx)^2$$

$$= (H^{-2} \int_{B_a} \int_0^{|b-a|} \partial u_h(x_1 + s, x_2)/\partial x_1 ds dx)^2.$$

Using Schwarz's inequality, it is now a quite simple matter to show that $(\bar{u}_h(a) - \bar{u}_h(b))^2$ can be estimated from above by the contribution to the energy of u_h from a neighborhood $\hat{T}$ of the triangle T, which also has a diameter on the order of H. Adding the contributions from all the coarse triangles and taking into account that there is a finite covering, we find

$$a(u_{h,0}, u_{h,0}) \leq Ca(u_h, u_h),$$

where the constant C depends only on the angles of the coarse triangles.

We can also show that

$$\|u_{h,0}\|^2_{L_2(T)} \leq C \|u_h\|^2_{L_2(\hat{T})}.$$

Then, trivially,

$$\|u_h - u_{h,0}\|^2_{L_2(T)} \leq CH^2(a_{\hat{T}}(u_h, u_h) + H^{-2}\|u_h\|^2_{L_2(\hat{T})}). \tag{14}$$

We wish to obtain

$$\|u_h - u_{h,0}\|^2_{L_2(T)} \leq CH^2 a_{\hat{T}}(u_h, u_h), \tag{15}$$

i.e. remove the L_2 term. This can be done by using Poincaré's inequality:

Lemma 2 Let $\hat{\Omega}$ be a region of diameter H. Then,

$$\sup_c \frac{a_{\hat{\Omega}}(v, v)}{\|v - c\|^2_{L_2(\hat{\Omega})}} \geq CH^{-2}. \tag{16}$$

We note that the maximizing c is the average of v over the region $\hat{\Omega}$. For an interior subregion, where we are away from the boundary where the Dirichlet condition is imposed, we can shift u_h by subtracting any constant; by our construction $u_{h,0}$ will be shifted by the same constant. The left hand side and the first term on the right hand side of (14) will not change while the last term becomes $H^{-2}\|u_h - c\|^2_{L_2(\hat{\Omega})}$. We are then ready to use Lemma 2. The case of subregions next to $\partial\Omega$ can be treated similarly using Friedrichs' inequality. We can therefore drop the last term in (14) at the expense of only increasing the constant in the inequality.

We note that Lemma 2 can be interpreted as saying that the second eigenvalue of the Neumann problem for the Laplace operator is bounded from below by a constant for a region of unit diameter. All that remains is a simple dilation argument. Friedrichs' inequality similarly addresses the case when we have a zero Dirichlet boundary condition on at least part of the boundary of $\hat{\Omega}$.

Next let $w_h = u_h - u_{h,0}$ and let $u_{h,i} = I^h(\theta_i w_h)$, $i = 1, \cdots, N$. Here I_h is the interpolation operator onto the space V^h, and the $\theta_i(x)$ define a partition of unity with respect to the Ω_i', i.e. $\sum_i \theta_i(x) \equiv 1$ with θ_i supported in Ω_i'. These functions are chosen as nonnegative elements of V^h. It is easy to see that

$$u_h = u_{h,0} + \sum u_{h,i}.$$

Let $\Gamma_{\delta,i} \subset \Omega_i'$ be the set of points which also belong to another Ω_j'. In the interior part of Ω_i', which does not belong to $\Gamma_{\delta,i}$, $\theta_i \equiv 1$. This function must decrease to 0 over a distance on the order of δ. It is in fact easy to construct a partition of unity with $0 \le \theta_i \le 1$ and such that

$$|\nabla \theta_i| \le \frac{C}{\delta} .$$

In order to use the abstract Schwarz theory, we first estimate $a(u_i, u_i)$ in terms of $a(w_h, w_h)$. We note that, trivially,

$$a_{\Omega_i' \backslash \Gamma_{\delta,i}}(u_i, u_i) = a_{\Omega_i' \backslash \Gamma_{\delta,i}}(w_h, w_h).$$

Let K be an element in $\Gamma_{\delta,i}$. Then, using the definition of u_i,

$$a_K(u_i, u_i) \le 2a_K(\bar{\theta}_i w_h, \bar{\theta}_i w_h) + 2a_K(I^h((\theta_i - \bar{\theta}_i)w_h), I^h((\theta_i - \bar{\theta}_i)w_h)),$$

where $\bar{\theta}_i$ is the average of θ_i over the element K. Using the fact that the diameter of K is on the order of h and the bound on $\nabla \theta_i$, we obtain, after adding over all the relevant elements, and using Lemma 4, given below, that

$$a_{\Gamma_{\delta,i}}(u_i, u_i) \le 2a_{\Omega_i'}(w_h, w_h) + \frac{C}{\delta^2}\|w_h\|^2_{L^2(\Gamma_{\delta,i})}.$$

To complete the proof, we need to estimate $\|w_h\|^2_{L^2(\Gamma_{\delta,i})}$. We note that each $x \in \Omega$ is covered only a finite number of times by the subregions. We apply Lemma 3, given below, to the function w_h, sum over i and use inequality (15) to complete the estimate of the parameter C_0^2 of the abstract Schwarz theory.

Lemma 3 *Let u be an arbitrary element of $H^1(\Omega_i')$. Then,*

$$\|u\|^2_{L^2(\Gamma_{\delta,i})} \le C\,\delta^2((1 + H/\delta)|u|^2_{H^1(\Omega_i')} + 1/(H\delta)\|u\|^2_{L^2(\Omega_i')}) .$$

Proof: We consider a square region $\Omega_i' = (0, H) \times (0, H)$ only; an extension of the discussion to a more complicated geometry and three dimensions is straightforward. Since,

$$u(x,0) = u(x,y) - \int_0^y \frac{\partial u(x,\tau)}{\partial y} d\tau , \tag{17}$$

we find, by elementary arguments, that

$$H \int_0^H |u(x,0)|^2 dx \leq 2 \int_0^H \int_0^H |u(x,y)|^2 dxdy + H^2 \int_0^H \int_0^H |\frac{\partial u}{\partial y}|^2 dxdy \, .$$

Therefore,

$$H \int_0^H |u(x,0)|^2 dx \leq 2\|u\|_{L^2(\Omega_i')}^2 + H^2 |u|_{H^1(\Omega_i')}^2 \, .$$

We now rearrange the terms of (17) and consider an integral over a narrow subregion, of width δ, next to one side and obtain,

$$\int_0^H \int_0^\delta |u(x,y)|^2 dxdy \leq \delta^2 |u|_{H^1(\Omega_i')}^2 + 2\delta \int_0^H |u(x,0)|^2 dx \, .$$

By combining this and the previous inequality, we obtain

$$\int_0^H \int_0^\delta |u(x,y)|^2 dxdy \leq \delta^2 |u|_{H^1(\Omega_i')}^2 + 2\delta(\frac{2}{H}\|u\|_{L^2(\Omega_i')}^2 + H|u|_{H^1(\Omega_i')}^2) \, ,$$

as required.

The modifications necessary for the case of an arbitrary, shape regular substructure and the extension of the proof to the case of three dimensions are routine.

As already indicated, we also need an additional result:

Lemma 4 *Let u_h be a continuous, piecewise quadratic function defined on the finite element triangulation and let $I^h u_h \in V^h$ be its piecewise linear interpolant on the same mesh. Then there exists a constant C, independent of h and H, such that*

$$|I^h u_h|_{H^1(\Omega_i')} \leq C|u_h|_{H^1(\Omega_i')} \, .$$

The same type of bound also holds for the L_2 norm.

Proof: It is elementary to show that,

$$|I^h u_h|_{H^1(\Omega_i')}^2 \leq 2(|I^h u_h - u_h|_{H^1(\Omega_i')}^2 + |u_h|_{H^1(\Omega_i')}^2).$$

Consider the contribution to the first term on the right hand side from an individual element K. We obtain

$$|I^h u_h - u_h|_{H^1(K)}^2 \leq Ch^2 |u_h|_{H^2(K)}^2 \leq C|u_h|_{H^1(K)}^2$$

by using a standard error bound and an elementary inverse inequality for quadratic polynomials. The bound in L_2 follows from the linear independence of the standard finite element basis for the space of quadratic polynomials.

7. A Neumann-Neumann algorithm

As an example of an important family of iterative substructuring method, we will consider the *balancing method* of Mandel and Brezina [28, 29] in some detail. This method belongs to the Neumann-Neumann family and also provides interesting insight into one of the hybrid Schwarz methods already discussed. We note that the Neumann-Neumann methods have been developed successfully for elliptic problems with arbitrary jumps in the coefficients across the boundaries of the subregions; in addition to [28, 29] see also [18, 26]. Here, to simplify our discussion and notations, we will consider Poisson's equation only. Iterative substructuring methods are often different in two and three dimensions, and we will therefore focus on the more challenging case of three dimensions. We consider only a Dirichlet problem discretized by piecewise linear finite elements.

The domain Ω is decomposed into nonoverlapping subregions Ω_i, also called substructures. For simplicity, we assume that they are shape regular tetrahedra, and that their diameters are on the order of H. The entire domain Ω is partitioned into shape regular tetrahedral elements of minimal diameter h, and we assume that the interface $\Gamma = \cup \partial \Omega_i \setminus \partial \Omega$ does not cut through any elements. The set of nodal points on $\partial \Omega_i$ is denoted by $\partial \Omega_{i,h}$, and the set of those on Γ by Γ_h, etc.

As pointed out before, the interior unknowns are eliminated when iterative substructuring methods are used, and the remaining linear equations form the *Schur complement* system. The interior spaces, defined as $V^h \cap H_0^1(\Omega_i)$, are $a(\cdot, \cdot)$—orthogonal to the space of *piecewise discrete harmonic* functions. A discrete harmonic finite element function is fully defined by its values on Γ_h. It is easy to show that the restriction of the quadratic function $x^T K x$ to the corresponding subspace of vectors is $x_\Gamma^T S x_\Gamma$. We also note that for boundary values w given at the nodes of $\partial \Omega_{i,h}$, the discrete harmonic extension $\mathcal{H}_i w$ to the interior of Ω_i minimizes the energy in the sense of $a_{\Omega_i}(\cdot, \cdot)$.

An important role is played by the *counting functions* ν_i which are associated with the individual $\partial \Omega_{i,h}$. They are defined by

$$\nu_i(x) = \text{ number of sets } \partial \Omega_{j,h} \quad \text{to which } x \in \partial \Omega_{i,h} \text{ belongs,}$$
$$\nu_i(x) = 0, \quad x \in \Gamma_h \setminus \partial \Omega_{i,h},$$
$$\nu_i(x) = 1, \quad x \in \partial \Omega_{i,h} \cap \partial \Omega_h.$$

Thus at the nodes of a *face* $\mathcal{F}^{ij}$, an open set common to two adjacent substructures Ω_i and Ω_j, $\nu_i(x) = 2$, while $\nu_i(x) > 2$ for any nodal point on the *wire basket* $\mathcal{W}_i$, formed by the union of the edges and vertices of the substructure Ω_i.

The pseudo inverses $\nu_i^\dagger$ are defined by

$$\nu_i^\dagger(x) = \nu_i(x)^{-1}, \ x \in \partial\Omega_{i,h}, \ \nu_i^\dagger(x) = 0, \ x \in (\Gamma_h \cup \partial\Omega_h) \setminus \partial\Omega_{i,h},$$

and they define a partition of unity, i.e.

$$\sum_i \nu_i^\dagger(x) \equiv 1, \ x \in \Gamma \cup \partial\Omega. \tag{18}$$

The *coarse space* V_0 is spanned by the $\nu_i^\dagger$ that correspond to the interior substructures, i.e. those with boundaries which do not intersect $\partial\Omega$. For the other substructures, we modify these functions by replacing the values of $\nu_i^\dagger$ by zero at the nodal points on $\partial\Omega$, the boundary of the original region Ω. These functions are then extended into the interior of the substructures as piece-wise discrete harmonic functions.

When we now turn to the *local spaces* V_i, we will only consider those associated with the interior substructures in any detail. The elements of V_i, associated with such an Ω_i, are piece-wise discrete harmonic functions defined by arbitrary values at the nodes of $\partial\Omega_{i,h}$ and which vanish at all points of $\Gamma_h \setminus \partial\Omega_{i,h}$. The bilinear form $\tilde{a}_i(u, v)$ for the subspace V_i is defined by

$$\tilde{a}_i(u, v) = a_{\Omega_i}(\mathcal{H}_i(\nu_i u), \mathcal{H}_i(\nu_i v)).$$

This form defines a projection-like operator T_i as in (3). We note that for an interior substructure, the operator $T_i : V^h \to V_i$, defined by

$$\tilde{a}_i(T_i u, v) = a(u, v), \ \forall v \in V_i, \tag{19}$$

is well defined only for $u \in V^h$ such that $a(u, v) = 0$ for all v such that $\mathcal{H}_i(\nu_i v)$ is constant on Ω_i. This condition is satisfied if $a(u, \nu_i^\dagger) = 0$; we note that this test function is a basis function for V_0. The right hand side of (19) is then said to be *balanced*. We make the solution $T_i u$ of (19) unique by imposing the constraint

$$\int_{\Omega_i} \mathcal{H}_i(\nu_i T_i u) dx = 0, \tag{20}$$

which just means that we select the solution orthogonal to the null space of the Neumann operator.

The bilinear form of the left hand side of (19) is defined in terms of the $H^1(\Omega_i)$–semi-norm and a diagonal scaling of the nodal values on $\partial\Omega_i$. This scaling has the great advantage that the decomposition,

$$u = \sum_{i=1}^{N} u_i, \ \text{with} \ u_i(x) = \mathcal{H}(\nu_i^\dagger u)(x),$$

which is valid for any piece-wise discrete harmonic function $u(x)$, satisfies,

$$\sum_{i=1}^{N} \tilde{a}_i(u_i, u_i) = a(u, u). \tag{21}$$

This is seen easily by using (18) and a simple computation.

We will now use P_0 and the T_i and the special hybrid Schwarz operator with the error propagation operator

$$(I - \sum_{i=1}^{N} T_i)(I - P_0),$$

or after an additional coarse solve,

$$(I - P_0)(I - \sum_{i=1}^{N} T_i)(I - P_0). \tag{22}$$

As pointed out before, this is a symmetric operator with which we can work essentially without any extra cost, since when forming powers of the operator (22), we can use the fact that $I - P_0$ is a projection.

We also note that the condition on the right hand side of (19) is satisfied for all elements in the range of $I - P_0$. The Schwarz operator is therefore well defined. We also note that given a right hand side, we could use any solution of (19) in our computations since any two such solutions will differ only by an element in the null space of $I - P_0$.

Subtracting the operator (22) from I, we obtain the operator

$$T_{hyb} = P_0 + (I - P_0)(\sum_{i=1}^{N} T_i)(I - P_0). \tag{23}$$

Recalling that just one good decomposition is needed, we can now use (21) and (5) and find that $\sum_{i=1}^{N} T_i \geq I$.

We note that $w = \sum_{i=1}^{N} T_i u$ is obtained by first solving a Neumann problem for each subregion Ω_i. If a standard unscaled local Neumann problem is used, the right hand side has to be scaled using the values of $\nu_i^\dagger$. The resulting local solutions w_i generally do not match across the interface but by construction they define elements in the V_i. This step of the algorithm is completed by computing w, a piece-wise harmonic extension of traces of the w_i averaged across the interface, by the formula

$$w = \sum_{i=1}^{N} \mathcal{H}(\nu_i^\dagger w_i).$$

We note that this requires the solution of a Dirichlet problem for each substructure. The study of this mapping could be a center piece of the analysis.

Instead we proceed somewhat differently to find upper and lower bounds on the spectrum of T_{hyb}. We split the space of piece-wise discrete harmonic functions into the range of P_0 and that of $I-P_0$; they are $a(\cdot,\cdot)$−orthogonal. The component in the range of P_0 can be handled trivially. On the range of $I-P_0$, we use that $\sum_{i=1}^{N} T_i \geq I$; the smallest eigenvalue of T_{hyb} therefore equals 1.

There remains to provide an upper bound on the eigenvalues of T_{hyb}. Since only a bounded number of subspaces V_i can contribute nontrivially to the value of $\sum v_i(x), v_i \in V_i$, at any given $x \in \Omega$, we see that what remains is to provide a bound on the $a(\cdot,\cdot)$−norm of the individual operators T_i. This is done by estimating ω in the inequality

$$a(u,u) \leq \omega \tilde{a}_i(u,u) \quad \forall u \in Range(T_i); \tag{24}$$

cf. Theorem 1. In this case $\omega \leq C(1 + \log(H/h))^2$.

Theorem 6 *The hybrid Schwarz method defined by the operator (23), and the spaces and bilinear forms of this section, satisfies*

$$I \leq T_{hyb} \leq C(1 + \log(H/h))^2 I,$$

where C is independent of the mesh sizes h and H, as well as the number of substructures.

Proof: There remains to show that the parameter ω in (24) is on the order of $C(1 + \log(H/h))^2$.

We will only consider an interior substructure Ω_i in some detail and we first note that an element $u \in V_i$ generally differs from zero in any substructure which is a neighbor of Ω_i. We therefore have to bound the energy of u, contributed by these substructures and by Ω_i itself, in terms of $\tilde{a}_i(u,u)$, which is the energy of $\mathcal{H}_i(\nu_i u)$ contributed by Ω_i. The following four auxiliary results, all proved in [15], are important.

Lemma 5 *Let $\theta_{F_{ij}}$ be the finite element function that is equal to 1 at the nodal points on the face $\mathcal{F}^{ij}$, which is common to two subregions Ω_i and Ω_j, vanishes on $(\partial\Omega_{i,h} \cup \partial\Omega_{j,h}) \setminus \mathcal{F}_h^{ij}$, and is discrete harmonic in Ω_i and Ω_j. Then,*

$$a_{\Omega_i}(\theta_{\mathcal{F}^{ij}}, \theta_{\mathcal{F}^{ij}}) \leq C(1 + \log(H/h))H,$$

and

$$\|\theta_{\mathcal{F}^{ij}}\|_{L^2(\Omega_i)}^2 \leq CH^3.$$

The same bounds also hold for the other subregion Ω_j.

The proof of Lemma 5 involves the explicit construction of a partion of unity constructed from functions $\vartheta_{\mathcal{F}^{ij}}$, with the same boundary conditions as the $\theta_{\mathcal{F}^{ij}}$, and which satisfies the first bound. This set of functions are well defined in the interior of the substructure where they form a partition of unity. The discrete harmonic function $\theta_{\mathcal{F}^{ij}}$ will have a smaller energy than $\vartheta_{\mathcal{F}^{ij}}$. Further details cannot be provided here; see, e.g. [40, Chapter 5.3.2].

We will also use the same set of functions in the following lemma:

Lemma 6 Let $\vartheta_{\mathcal{F}^{ij}}(x)$ be the function just discussed above, and let I^h denote the interpolation operator onto the finite element space V^h. Then,

$$\sum_j I^h(\vartheta_{\mathcal{F}^{ij}} u)(x) = u(x) \quad \forall x \in \overline{\Omega}_i \setminus \mathcal{W}_i$$

and

$$a_{\Omega_j}(I^h(\vartheta_{\mathcal{F}^{ij}} u), I^h(\vartheta_{\mathcal{F}^{ij}} u)) \le C(1 + \log(H/h))^2 (a_{\Omega_i}(u, u) + \frac{1}{H^2}\|u\|^2_{L_2(\Omega_i)}).$$

We will also need an additional Sobolev-type inequality for finite element functions. It is used to estimate the contributions to our bound from the values on the wire basket. It is closely related to Lemma 1.

Lemma 7 Consider the restriction of $u \in V^h$ to Ω_i with the wire basket $\mathcal{W}_i$. Then,

$$\|u\|^2_{L^2(\mathcal{W}_i)} \le C(1 + \log(H/h))(a_{\Omega_i}(u, u) + \frac{1}{H^2}\|u\|^2_{L_2(\Omega_i)}).$$

We also need the following result.

Lemma 8 Let $u \in V^h$ vanish at the nodal points on the faces of Ω_j and let it be discrete harmonic in Ω_j. Then,

$$a_{\Omega_j}(u, u) \le C\|u\|^2_{L^2(\mathcal{W}_j)}.$$

This result follows by estimating the energy norm of the zero extension of the boundary values and by noting that the harmonic extension has a smaller energy.

The bound on ω can now be obtained by considering the contributions of the energy $a(u, u)$, for $u \in Range(T_i)$, from the substructure Ω_i and its neighbors, one by one. On $\partial\Omega_i$, u and the function $\nu_i u$ differ by different constant factors on the faces, edges, and vertices that form the wire basket $\mathcal{W}_i$. Lemmas 7 and 8, and the triangle inequality provide a bound of the contribution of the energy from Ω_i of the form

$$C(1 + \log(H/h))^2 (\tilde{a}_i(u, u) + \frac{1}{H^2}\|\mathcal{H}_i(\nu_i u)\|^2_{L_2(\Omega_i)}).$$

The same type of bounds are obtained, by employing Lemmas 6, 7, and 8, for the neighboring subregions that share a face, an edge, or just a vertex with Ω_i. We note that on these neighboring substructures any element of V_i will vanish on the part of the boundary that does not intersect $\partial\Omega_i$; this is important since the right hand side of the inequality (24) depends only on the values on $\partial\Omega_i$.

Finally, we use Lemma 2 and condition (20) and can remove the L_2- term at the expense of only increasing the constant in front of the remaining term in the estimate. This completes our outline of the proof of the bound on the norm of T_i and the proof of Theorem 2.

It is possible to prove equally strong results for elliptic problems with large variations in the coefficients from one substructure to its neighbors; such general results are not available for the methods based on overlapping subregions. This work requires a change of the bilinear forms $\tilde{a}_i(\cdot,\cdot)$; the counting functions are replaced by weighted counting functions, with weights that depend on the coefficients. We also note that there has been considerable success with the balancing method for very hard practical problems including elliptic systems. The theory has been extended to *spectral element approximations* of elliptic equations by Pavarino [32].

We also note that a new, close connection has recently been discovered between the Neumann-Neumann and FETI methods and that this will be the topic of a forthcoming paper; see [25]. For an introduction to the FETI algorithms, see [19].

352

References

1. Balay, S., W. Gropp, L. C. McInnes, and B. F. Smith: 1998, 'PETSc, the Portable, Extensible Toolkit for Scientific Computation'. Argonne National Laboratory, 2.0.22 April 29, 1998 edition.

2. Bică, I.: 1997, 'Iterative Substructuring Algorithms for the p-version Finite Element Method for Elliptic Problems'. Ph.D. thesis, Courant Institute of Mathematical Sciences. Tech. Rep. 743, Department of Computer Science, Courant Institute.

3. Bramble, J. H.: 1966, 'A second order finite difference analogue of the first biharmonic boundary value problem'. *Numer. Math.* **9**, 236–249.

4. Bramble, J. H.: 1993, *Multigrid Methods.* Burnt Mill, Harlow, Essex CM20 2JE, England: Longman Scientific & Technical. Pitman Research Notes in Mathematics Series #294.

5. Bramble, J. H., J. E. Pasciak, and A. H. Schatz: 1986, 'The construction of preconditioners for elliptic problems by substructuring, I'. *Math. Comp.* **47**(175), 103–134.

6. Bramble, J. H., J. E. Pasciak, J. Wang, and J. Xu: 1991, 'Convergence Estimates for Product Iterative Methods with Applications to Domain Decomposition'. *Math. Comp.* **57**(195), 1–21.

7. Bramble, J. H., J. E. Pasciak, and J. Xu: 1990, 'Parallel Multilevel Preconditioners'. *Math. Comp.* **55**, 1–22.

8. Brenner, S. C.: 1998, 'Lower Bounds of Two-level Additive Schwarz Preconditinoners with Small Overlap'. Technical Report 1998:01, The University of South Carolina, Department of Mathematics.

9. Brenner, S. C. and L. R. Scott: 1994, *The Mathematical Theory of Finite Element Methods.* Berlin: Springer-Verlag.

10. Cai, X.-C.: 1993, 'An Optimal Two-level Overlapping Domain Decomposition Method for Elliptic Problems in Two and Three Dimensions'. *SIAM J. Sci. Comp.* **14**, 239–247.

11. Cai, X.-C. and O. Widlund: 1992, 'Domain Decomposition Algorithms for Indefinite Elliptic Problems'. *SIAM J. Sci. Statist. Comput.* **13**(1), 243–258.

12. Cai, X.-C. and O. Widlund: 1993, 'Multiplicative Schwarz Algorithms for Some Nonsymmetric and Indefinite Problems'. *SIAM J. Numer. Anal.* **30**(4), 936–952.

13. Casarin, M. A.: 1996, 'Schwarz Preconditioners for Spectral and Mortar Finite Element Methods with Applications to Incompressible Fluids'. Ph.D. thesis, Courant Institute of Mathematical Sciences. Tech. Rep. 717, Department of Computer Science, Courant Institute.

14. Casarin, M. A.: 1997, 'Quasi-Optimal Schwarz Methods for the Conforming Spectral Element Discretization'. *SIAM J. Numer. Anal.* **34**(6), 2482–2502.

15. Dryja, M., B. F. Smith, and O. B. Widlund: 1994, 'Schwarz Analysis of Iterative Substructuring Algorithms for Elliptic Problems in Three Dimensions'. *SIAM J. Numer. Anal.* **31**(6), 1662–1694.

16. Dryja, M. and O. B. Widlund: 1990, 'Towards a Unified Theory of Domain Decomposition Algorithms for Elliptic Problems'. In: T. Chan, R. Glowinski, J. Périaux, and O. Widlund (eds.): *Third International Symposium on Domain Decomposition Methods for Partial Differential Equations.* pp. 3–21.

17. Dryja, M. and O. B. Widlund: 1994, 'Domain Decomposition Algorithms with Small Overlap'. *SIAM J. Sci. Comput.* **15**(3), 604–620.

18. Dryja, M. and O. B. Widlund: 1995, 'Schwarz Methods of Neumann-Neumann Type for Three-Dimensional Elliptic Finite Element Problems'. *Comm. Pure Appl. Math.* **48**(2), 121–155.

19. Farhat, C. and F.-X. Roux: 1994, 'Implicit parallel processing in structural mechanics'. In: J. T. Oden (ed.): *Computational Mechanics Advances*, Vol. 2 (1). North-Holland, pp. 1–124.

20. Golub, G. H. and C. F. V. Loan: 1996, *Matrix Computations.* Johns Hopkins Univ. Press. Third Edition.

21. Guo, B. and W. Cao: 1996, 'A preconditioner for the $h - p$ version of the finite element method in two dimensions'. *Numer. Math.* **75**(1), 59–77.
22. Guo, B. and W. Cao: 1997a, 'Additive Schwarz Methods for the $h - p$ Version of the Finite Element Method in Two Dimensions'. *SIAM J. Sci. Comput.* **18**(5), 1267–1288.
23. Guo, B. and W. Cao: 1997b, 'An Iterative and Parallel Solver Based on Domain Decomposition of the $h - p$ Version of the Finite Element Method'. *J. Comp. Appl. Math.* **83**, 71–85.
24. Guo, B. and W. Cao: 1998, 'An Additive Schwarz Method for the $h - p$ Version of the Finite Element Method in Three Dimensions'. *SIAM J. Numer. Anal.* **35**(2), 632–654.
25. Klawonn, A. and O. B. Widlund: 1999, 'FETI and Neumann-Neumann Iterative Substructuring Methods: A Comparison and New Results'. Technical report, Computer Science Department, Courant Institute of Mathematical Sciences. To appear.
26. Le Tallec, P.: 1994, 'Domain decomposition methods in computational mechanics'. In: J. T. Oden (ed.): *Computational Mechanics Advances*, Vol. 1 (2). North-Holland, pp. 121–220.
27. Lions, P.-L.: 1988, 'On the Schwarz alternating method. I.'. In: R. Glowinski, G. H. Golub, G. A. Meurant, and J. Périaux (eds.): *First International Symposium on Domain Decomposition Methods for Partial Differential Equations*. Philadelphia, PA, pp. 1–42.
28. Mandel, J.: 1993, 'Balancing Domain Decomposition'. *Comm. Numer. Meth. Engrg.* **9**, 233–241.
29. Mandel, J. and M. Brezina: 1996, 'Balancing Domain Decomposition for Problems with Large Jumps in Coefficients'. *Math. Comp.* **65**, 1387–1401.
30. Oswald, P.: 1992, 'On Discrete Norm Estimates Related to Multilevel Preconditioners in the Finite Element Method'. In: K. Ivanov and B. Sendov (eds.): *Proc. Int. Conf. Constructive Theory of Functions, Varna 91.* pp. 203–241.
31. Oswald, P.: 1994, *Multilevel Finite Element Approximation, Theory and Applications*, Teubner Skripten zur Numerik. Stuttgart: B.G. Teubner.
32. Pavarino, L. F.: 1997, 'Neumann-Neumann algorithms for spectral elements in three dimensions'. *RAIRO Mathematical Modelling and Numerical Analysis* **31**, 471–493.
33. Pavarino, L. F. and O. B. Widlund: 1996, 'A Polylogarithmic Bound for an Iterative Substructuring Method for Spectral Elements in Three Dimensions'. *SIAM J. Numer. Anal.* **33**(4), 1303–1335.
34. Pavarino, L. F. and O. B. Widlund: 1997a, 'Iterative Substructuring Methods for Spectral Element Discretizations of Elliptic Systems. II: Mixed Methods for Linear Elasticity and Stokes Flow'. Technical Report 755, Dept. of Computer Science, Courant Institute of Mathematical Sciences.
35. Pavarino, L. F. and O. B. Widlund: 1997b, 'Iterative Substructuring Methods for Spectral Elements: Problems in Three Dimensions Based on Numerical Quadrature'. *Computers Math. Applic.* **33**(1/2), 193–209.
36. Pavarino, L. F. and O. B. Widlund: 1999, 'Iterative Substructuring Methods for Spectral Element Discretizations of Elliptic Systems. I: Compressible Linear Elasticity'. *SIAM J. Numer. Anal.* To appear.
37. Quarteroni, A. and A. Valli: 1999, *Domain Decomposition Methods for Partial Differential Equations*. Oxford Science Publications.
38. Saad, Y. and M. H. Schultz: 1986, 'GMRES: A generalized minimal residual algorithm for solving nonsymmetric linear systems'. *SIAM J. Sci. Stat. Comp.* **7**, 856–869.
39. Schwarz, H. A.: 1890, *Gesammelte Mathematische Abhandlungen*, Vol. 2, pp. 133–143. Berlin: Springer. First published in Vierteljahrsschrift der Naturforschenden Gesellschaft in Zürich, volume 15, 1870, pp. 272–286.
40. Smith, B. F., P. E. Bjørstad, and W. D. Gropp: 1996, *Domain Decomposition: Parallel Multilevel Methods for Elliptic Partial Differential Equations*. Cambridge

University Press.

41. Toselli, A.: 1997, 'Overlapping Schwarz Methdos for Maxwell's Equations in Three Dimensions'. Technical Report 736, Department of Computer Science, Courant Institute.

42. Toselli, A., O. B. Widlund, and B. I. Wohlmuth: 1998, 'An Iterative Substructuring Method for Maxwell's Equations in Two Dimensions'. Technical Report 768, Department of Computer Science, Courant Institute.

43. Widlund, O. B.: 1988, 'Iterative Substructuring Methods: Algorithms and Theory for Elliptic Problems in the Plane'. In: R. Glowinski, G. H. Golub, G. A. Meurant, and J. Périaux (eds.): *First International Symposium on Domain Decomposition Methods for Partial Differential Equations*. Philadelphia, PA.

44. Wohlmuth, B. I., A. Toselli, and O. B. Widlund: 1998, 'Iterative Substructuring Method for Raviart-Thomas Vector Fields in Three Dimensions'. Technical Report 775, Department of Computer Science, Courant Institute.

45. Xu, J. and J. Zou: 1998, 'Some Nonoverlapping Domain Decomposition Methods'. *SIAM Review* **40**, 857–914.

46. Yserentant, H.: 1986, 'On the Multi-Level Splitting of Finite Element Spaces'. *Numer. Math.* **49**, 379–412.

47. Zhang, X.: 1991, 'Studies in Domain Decomposition: Multilevel Methods and the Biharmonic Dirichlet Problem'. Ph.D. thesis, Courant Institute, New York University.

48. Zhang, X.: 1992, 'Multilevel Schwarz Methods'. *Numer. Math.* **63**(4), 521–539.